Evolutionary Economics and Social Complexity Science

Volume 12

The Japanese Association for Evolutionary Economics (JAFEE) always has adhered to its original aim of taking an explicit "integrated" approach. This path has been followed steadfastly since the Association's establishment in 1997 and, as well, since the inauguration of our international journal in 2004. We have deployed an agenda encompassing a contemporary array of subjects including but not limited to: foundations of institutional and evolutionary economics, criticism of mainstream views in the social sciences, knowledge and learning in socio-economic life, development and innovation of technologies, transformation of industrial organizations and economic systems, experimental studies in economics, agent-based modeling of socio-economic systems, evolution of the governance structure of firms and other organizations, comparison of dynamically changing institutions of the world, and policy proposals in the transformational process of economic life. In short, our starting point is an "integrative science" of evolutionary and institutional views. Furthermore, we always endeavor to stay abreast of newly established methods such as agent-based modeling, socio/econo-physics, and network analysis as part of our integrative links.

More fundamentally, "evolution" in social science is interpreted as an essential key word, i.e., an integrative and/or communicative link to understand and re-domain various preceding dichotomies in the sciences: ontological or epistemological, subjective or objective, homogeneous or heterogeneous, natural or artificial, selfish or altruistic, individualistic or collective, rational or irrational, axiomatic or psychological-based, causal nexus or cyclic networked, optimal or adaptive, micro- or macroscopic, deterministic or stochastic, historical or theoretical, mathematical or computational, experimental or empirical, agent-based or socio/econo-physical, institutional or evolutionary, regional or global, and so on. The conventional meanings adhering to various traditional dichotomies may be more or less obsolete, to be replaced with more current ones vis-à-vis contemporary academic trends. Thus we are strongly encouraged to integrate some of the conventional dichotomies.

These attempts are not limited to the field of economic sciences, including management sciences, but also include social science in general. In that way, understanding the social profiles of complex science may then be within our reach. In the meantime, contemporary society appears to be evolving into a newly emerging phase, chiefly characterized by an information and communication technology (ICT) mode of production and a service network system replacing the earlier established factory system with a new one that is suited to actual observations. In the face of these changes we are urgently compelled to explore a set of new properties for a new socio/economic system by implementing new ideas. We thus are keen to look for "integrated principles" common to the above-mentioned dichotomies throughout our serial compilation of publications. We are also encouraged to create a new, broader spectrum for establishing a specific method positively integrated in our own original way.

More information about this series at http://www.springer.com/series/11930

Takahiro Fujimoto • Fumihiko Ikuine
Editors

Industrial Competitiveness and Design Evolution

Springer

Editors
Takahiro Fujimoto
Graduate School of Economics
The University of Tokyo
Bunkyo-Ku, Tokyo, Japan

Fumihiko Ikuine
Faculty of Engineering, Information and Systems
University of Tsukuba
Tsukuba, Ibaraki, Japan

ISSN 2198-4204 ISSN 2198-4212 (electronic)
Evolutionary Economics and Social Complexity Science
ISBN 978-4-431-55144-7 ISBN 978-4-431-55145-4 (eBook)
https://doi.org/10.1007/978-4-431-55145-4

Library of Congress Control Number: 2018954734

This Springer imprint is published by the registered company Springer Japan KK, part of Springer Nature.
The registered company address is: Shiroyama Trust Tower, 4-3-1 Toranomon, Minato-ku, Tokyo 105-6005, Japan

Preface

Evolutionary Economics and Social Complexity Science **Series**
Evolution of Industries and Firms – Capability Building and Demand Creation

Purpose of the Present Book

The purpose of this edited book is to explore the evolution of industries and firms both theoretically and empirically. Theoretically, we will combine the theory of manufacturing capability in management studies, the concept of product-process architecture, classical (Ricardian and neo-Ricardian) trade theories, and a few other theoretical concepts in business and economics, integrating them into an evolutionary economics framework. As the common element of industries and firms is the so-called manufacturing site (*genba* in Japanese), i.e., the place where value-added flows, we will adopt the site (genba) as the basic unit of our analysis.

In this regard, this book includes chapters that deal with the following theoretical topics: evolutionary perspectives of capability architecture fit for industrial comparative advantage, design-based view of manufacturing, evolution of manufacturing capabilities, Ricardian comparative advantage with changing labor and material input coefficients, comparative design costs and selection of design locations, evolution of architectures within a product category, resource-based view of the firm and its growth, evolution of product architectures across industries (particularly in the open architecture environment), cumulative demand creation by means of network externality among complementary goods in the digital industries with open architecture platforms, evolution of product designs in the industrial lifecycle, evolution or expansion of product variety for effective demand creation, international division of labor that matches the distribution of knowledge, and simultaneous pursuit by firms of productivity increase and effective demand creation (or stable employment and markup ratio) at all levels, i.e., at the site-firm-industry-economy level.

Empirical and Historical Background: Globalization and Digitalization

Empirically, this book focuses primarily on industrial phenomena that occurred in the late twentieth century and early twenty-first century, including the Cold War period until the 1980s and the post–Cold War period after the 1990s. Most of our empirical research concentrates on the latter period (between the 1990s and the 2010s), during which global competition and digitalization emerged as main trends that deserve in-depth academic investigation.

Indeed, it is a remarkable coincidence that the post–Cold War global competition and the Internet-driven digitalization started to happen almost simultaneously in the 1990s. The global cost competition among the firms and sites in higher-wage advanced countries (e.g., the USA, Western Europe, and Japan) and those in lower-wage emerging countries (e.g., China, India, and Eastern Europe) had a major impact on the supply side of industries worldwide, so the firms and sites in high-wage nations were forced to accelerate their efforts for capability building in order to survive. The Internet-driven digitalization, on the other hand, brought about cumulative demand creation among complementary goods and services connected by industry-standard interfaces, rapid formation of open architecture platforms, and fundamental changes not only in our life and society but also in the rules of the game of industrial competition in this domain.

These two major aspects, capability building for global competition and demand creation in platform-driven digital industries, have been analyzed using different approaches and disciplines in the existing literature. For instance, the concept of lean production in the field of industrial engineering and technology-operations management has often been applied to the former, while the economic theories of network externalities among complementary goods, or the concepts of platform formation and leadership in strategic management, have been the core logic for investigating the latter.

This separation in existing academic research seems to be caused partly by the separation between the physical and digital layers of the industries themselves. That is, the third industrial revolution, driven by electronic technologies, progressed rapidly after the mid-twentieth century, but technological advancements in the digital layers (e.g., IBM mainframe computers, personal computers, Internet, and mobile phones) and in the physical layers (e.g., programmable automation, NC machine tools and robots, and factory automation) evolved rather separately. Although NC machines, factory automation, and automobiles in the physical layers were also controlled, at least partially, by digital technologies, the controlling system itself remained rather closed within a factory or a vehicle.

In the 2010s, however, the physical and digital layers became more tightly connected, as more and more physical artifacts, such as machines, devices, and vehicles, were equipped with sensors and transmitters that digitally linked these physical objects to Ethernet, Internet, and other standard networks in the digital layer. Using an analogy, we may say that this particular period is characterized by a trend toward integrating the *high sky* (i.e., the digital or cyber layer) with the *ground* (i.e., the physical layer), possibly via the *low sky* (i.e., interface or cyber-physical layer). Popular concepts of the 2010s, like connected factories, Internet of Things (IoT), Industry 4.0, connected cars, and automatic driving, are all related to this trend of digital-physical integration.

As the physical and digital layers are connected ever more tightly in the real world, researchers may need to develop analytical frameworks that can explain the total system of physical-digital industries in a reasonably consistent manner. Given the rapid changes in both manufacturing capabilities and product-platform architectures during the period in question, we think that an evolutionary economic approach that incorporates both organizational capabilities in manufacturing and architectures of products/components/processes/platforms may be a possible candidate for explaining the performance of today's industries and firms.

In sum, the present book proposes an evolutionary economic framework that includes capability and architecture performance as its main components.

Preliminary Analysis of Globalization, Trade, and Digitalization: A Brief History

Before tackling the main discussion, let us briefly introduce what we aim to analyze in the present book. Global competition, intra-industrial trade, and digital platform formation are the three main topics that we wish to shed light on.

The early part of the twenty-first century is *the era of post–Cold War global competition*, when tradable goods and services tend to be imported and exported worldwide, since international trade barriers, transportation costs, and information transmission costs have decreased significantly. The 200-year-old theories of trade, since the era of David Ricardo, tell us that the international division of labor, driven by the principle of comparative advantage, tends to progress when freer trade systems prevail.

It ought to be noted that today's trade phenomena have increasingly been affected by the *design* characteristics of tradable goods. This book tries to introduce design and capability, and their evolution, into the existing theories of trade to explain the reality of international trade in the early twenty-first century.

In the era of global competition, *comparative advantage* still remains one of the key principles to analyze trade and industrial structures. Today's international transactions are also characterized by *intra-industrial trade at the microscopic level*, in which, for instance, sheet steel for the inner door panels of an automobile is exported from country A to country B whereas that for the outer panels is exported from B to A. It may be difficult for existing trade theories to explain such phenomena with sufficient accuracy. To tackle this issue, we start from the simple fact that firms' selection of locations for designing products precedes that of locations for manufacturing them, and a new product's initial production location is usually the same as its design location. In other words, design matters in explaining today's trade phenomena. Thus, here we try to analyze product design and its evolution within the context of the comparative advantage theory.

Besides, we now recognize more sharply that, in the middle of the post–Cold War global competition, *relative productivities and wages* change significantly over time and across trading countries. In other words, Ricardo's input coefficients must in fact be treated as *variables*, which factories and firms try to improve through their capability-building efforts. This book argues that the concept of Ricardo's comparative advantage must be reinterpreted in a more dynamic way, with changing labor and material input coefficients driven by international capability-building competition among factories.

In the open architecture world of digital industries, on the other hand, a very different type of industrial evolution is occurring, i.e., competition among platforms. A small number of leading companies with core technologies (e.g., Intel, Microsoft, Apple, Google, and Qualcomm) have adopted *closed-inside-open-outside architectural strategies* and gained winners-take-it-all profits and revenues by making the most of network externality effects among complementing goods. They provide open standard interfaces and design information to other firms, so that the latter can develop or produce complementary goods or services with open-modular architectures, even if these firms have limited technological competences.

The lowering of technological barriers, caused by the platform-forming strategies of the leading firms, has made it possible for lower-wage firms to enter these open areas in the digital industries and, even without strong technological capabilities, produce simple, open-modular digital products. As a result, for the manufacturing firms and sites in high-wage countries, such as Japan, this open area has become a sort of *desert*, an extremely unattractive sector in which only low-wage factories in emerging countries are likely to survive during the period of post–Cold War global competition by relying on very scarce profits.

In this sense, the extremely adverse competitive environment for firms manufacturing digital products in high-wage nations, like Japan, has been generated by the combination of globalization and digitization, which simultaneously emerged in the 1990s. As the research results in this book suggest, many Japanese manufacturing firms and sites of physical goods have fought for survival by combining capability building for drastic productivity improvements, demand creation by increasing the variety of their products and businesses, and architectural strategies to effectively connect their products to the domain of digital platforms. Many have disappeared but just as many have survived, and, as of the 2010s, Japan is one of the few large, advanced nations that still possess a fairly large manufacturing sector, amounting to about 20% of its GDP.

To sum up, this book will try to explore the concepts of design, architecture, organizational capability, productivity, as well as their interactions and evolutions. We expect the dynamic fit between manufacturing sites' capabilities and product-process architectures to affect the locations of the design/production facilities and the comparative advantages of the products in question.

Structure of the Book

Part I: Overview and Framework In Part I, we present the theoretical framework to investigate the evolution of industries and companies based on the concept of site (manufacturing site, or *genba* in Japanese).

Chapter "A Design-Information-Flow View of Industries, Firms, and Sites (Fujimoto)": In this introductory chapter, we present the purpose, key concepts, and analytical framework of this book, which discusses the evolution of industries and firms, both theoretically and empirically, based on field studies of sites. Our framework, called design-information-flow view, can help understand the concept of site (manufacturing site, or genba in Japanese) as a place where value-added flows. We also adopt the notions of organizational capability and product-process architecture to analyze the competitiveness and evolution of sites. We illustrate empirical research about organizational capabilities and product architecture and then explain site competitiveness through the dynamic fit of both.

Chapter "The Nature of International Competition Among Firms (Shiozawa and Fujimoto)": Capability building for productivity improvements is critical for manufacturing firms and sites in high-wage countries that face intense global cost competition vis-à-vis their rivals in low-wage emerging countries. We can regard this as capability-building competition for higher physical productivities with international wage gaps as handicaps, which can be seen as a dynamic reinterpretation of the Ricardian model of international values and comparative advantage. This chapter shows that international values (a set of wage rates and prices) can be determined in the general case of an *N*-commodity, *M*-country economy, where input goods are freely traded across countries. There is no need to point out that the trade of input goods cannot be explained by means of traditional theories, which is a major shortcoming in the age of global supply chains. The new theory provides a framework suited to exploring the situation in which global supply chains play a vital role in the world economy.

This chapter also argues that Ricardo's theory of values and specializations can be mathematically reinterpreted as a microscopic model of comparative product costs at the manufacturing site level, in which comparing international wage gaps and physical productivity gaps is essential. Thus, the reinterpreted dynamic model of the Ricardian trade theory may be effectively used to explain capability-building competition by firms amid intense global cost competition.

Chapter "Product Variety for Effective Demand Creation (Shiozawa)": For firms pursuing survival, stability, and growth, capability building at their manufacturing sites is often complementary to demand creation in the market. Therefore, here, we introduce the theory of demand creation. The economic model illustrated in this chapter shows that a firm's additional product variety creates additional demand and that there may exist a specific optimal product variety for a firm seeking long-term profit maximization during the entire lifecycle of the products in question. An economic model with *expected coverage function* is proposed to shed light on these circumstances.

Chapter "Capability Building and Demand Creation in *Genba-Oriented Firms* (Fujimoto)": Following on from chapter "Product Variety for Effective Demand Creation", we look at the concept of demand creation in conjunction with capability building. We introduce a *PXNW model* and show that genba-oriented firms' manufacturing capability building and demand creation are mutually complementary for achieving target profit rates and stable employment at the same time. We then present some brief case studies. Finally, we argue that genba-oriented firms, which simultaneously build capabilities and create demand, tend to pursue *grassroots innovations*, motivated by the will to stabilize their employment in response to intense competitive environments.

Chapter "Evolution of Business Ecosystems (Tatsumoto)": This chapter presents newer approaches to explain the evolution of ecosystems in the digital industries with open-modular (open) architecture, platforms, and products. To understand the rapid demand creation in these digital industries with open architecture platforms and industry-standard interfaces linking a variety of products, we need to incorporate such economic concepts as complementary goods, network externality, and modularization into our evolutionary framework.

Part II: Capability Building in Global Competition In Part II of this book, we focus on empirical research about the multilayered capability building of manufacturing firms, sites, and work groups, mostly within the context of intense global competition between high-wage and low-wage countries.

Chapter "Evolution of Organizational Capabilities in Manufacturing – The Case of the Toyota Motor Corporation (Fujimoto)": In this chapter, we use the evolutionary theory to consider the process of capability building of sites. The subject of our empirical research is the Toyota Production System (TPS), which comprises three layers of organizational capabilities, i.e., routinized manufacturing capability, routinized learning capability, and evolutionary learning capability. In addition, we emphasize the fact that these capabilities have been generated through multipath system emergence in a consistent manner.

Chapter "The Nature of Surviving Japanese Factories in the Global Competition: An Empirical Analysis of Electrical and Electronics Factories (Fukuzawa, Inamizu, Shintaku, Yokozawa, and Suzuki)": Here, we present an empirical study of the organizational capabilities and competitive performance of electrical and electronics factories. Interviews with 8 successful Japanese electrical and electronics companies and a survey of 97 business units were conducted. Our findings show that, during the period under investigation, Japanese electrical and electronics companies retained competitive advantages at the shop-floor level, except for what concerns unit costs. We also identify some common aspects of such companies in relation to both demand creation and multifunctionality.

Chapter "The Effectiveness of Group Leaders in the Lean Production System: Time Study and Agent-Based Model of Leaders' Behavior (Inamizu and Fukuzawa)": This chapter sheds light on one of the main sources of organizational capability within a manufacturing site – group leaders of shop-floor work organizations. It adopts a mixed approach that combines simulation modeling and field studies conducted in a Japanese automobile assembly plant. The simulation results and empirical results suggest that the group leaders' behavior (e.g., help actions) may improve the performance of the manufacturing sites in terms of task flows, but the opposite may occur under certain condition of task uncertainty and problem frequency.

Chapter "The Diversity and Reality of Kaizen in Toyota (Iwao)": This chapter clarifies how continuous capability building, or *kaizen*, works in the real settings of better-performing factories. The empirical study carried out here observes continuous improvement projects in an automobile factory in Japan for an extended period of time. Based on these longitudinal observations, it describes seven case studies and presents new findings about the nature of kaizen. Furthermore, it discusses the important role of shop-floor engineers, who coordinate between shop floors and engineering departments on the basis of the *staff-in-line* structure of organizations.

Chapter "Balancing Standardization and Integration in ICT Systems: An Archi tectural Perspective (Park and Fujimoto)": This chapter discusses what kind of information systems supports the capability building of sites. Firstly, we look back on the integrated manufacturing IT system (IMIS) that Japanese companies have been building for a long time. Next, we describe the appearance of the so-called GSIS (Global Standard IT System) which is, from example, the ICT system used by Samsung and LG. Finally, we propose the adoption of a GIMIS (Global Integrated Manufacturing IT System) as a new ICT system integrating IMIS and GSIS.

Part III: Architecture and Competitiveness In Part III, we empirically explore capability building and demand creation in the open architecture environment.

Chapter "Creating New Demand: The Impact of Competition, the Formation of Submarkets and Platforms (Ikuine)": This chapter examines how firms effectively create demand in digital industries with open architecture platforms. We present the case of the home video game (console game) industry in Japan, where a completely new market formed in the early 1980s. Based on a case study, we conclude that the existence of platforms contributes to demand creation by bringing together different elements of a fragmented market. Firms can create effective demand through the power of platforms, which combine considerably different submarkets.

Chapter "Decline in Demand Creation: The Development Productivity Dilemma and Its Consequences (Ikuine)": Here, we look once again at the game software business in Japan and consider the consequences of demand creation by companies. The industry lifecycle model and the A-U model show that the power to spur demand diminishes over time. This is why we concentrate on the mechanisms that reduce the ability to drive demand in a certain market. The main mechanism is the development productivity dilemma, which emerges during the firms' product development activities. Furthermore, we discuss where the *lost* demand might be redirected when a certain industry enters its declining phase.

Chapter "Investigating the Creation and Diffusion of Knowledge for Demand Creation: The Case of the Telecommunications Industry (Wei, Yasumoto, and Shiu)": This analysis explores demand creation under the open architecture platform based on an empirical study of the smartphone industry. As theoretically shown in chapter "Evolution of Business Ecosystems", when products adopt an open architecture, barriers to entry become lower. Hence, many firms enter the market and develop many new products through trial and error. As a result, products evolve rapidly, customers can easily find items that meet their requirements, and demand creation is accomplished. The smartphone industry is such a case, so we can clarify the mechanisms that enable demand creation by studying this industry. We underline the importance of technological foundations (i.e., system knowledge and relevant technologies), including standard essential patents (SEPs). We also investigate the influence of a specific firm (Qualcomm) that develops and diffuses technological foundations throughout the industry.

Chapter "The Impact of Platform Providers' Knowledge on Interfirm Division of Labor: The Case of the Mobile Phone Industry (Shiu and Yasumoto)": Here, we examine capability building under open architecture platforms based on an empirical study of the mobile phone industry. Mobile phones, including smartphones, are open-modular products whose development and manufacturing require collaboration among three types of firms. These are the brand firms, which release the products under their own brand; the contract manufactures, which undertake development and manufacturing for brand firms; and the platform providers, which provide technical information and core components to both. By analyzing the results of a questionnaire survey, we show that both the brand firms and the contract manufactures derive the necessary information from the platform provider and utilize it to build their organizational capabilities.

Chapter "Conclusion (Fujimoto and Ikuine)": This concluding chapter summarizes the previous chapters and presents an augmented evolutionary framework of capabilities and architectures based on our research results. Furthermore, we argue that our framework and concepts may rather consistently explain both capability evolution in response to global competition and architectural evolution in digitalized industries, the two main industrial phenomena of the period in question.

Summary

To sum up, this book aims to present theoretical frameworks and empirical research results on how Japanese (as well as Asian and Western) firms and sites have built manufacturing capabilities and effectively created demand under the pressure of intensifying global competition. It also looks at the cases of certain digital industries in which the architectures of both products and platforms have changed significantly, causing rapid and cumulative demand creation.

Let us now proceed to the main discussion. Part I deals with our theoretical frameworks and models; Part II is mainly about empirical research on capability building at manufacturing sites in physical goods industries; and Part III primarily addresses architectural evolution and platform formation in digital goods industries. Note, however, that each chapter in Part II and Part III also includes some important theoretical implications derived from our empirical studies.

Bunkyo-Ku, Tokyo, Japan Takahiro Fujimoto
Tsukuba, Ibaraki, Japan Fumihiko Ikuine

Acknowledgments

The present book contains large amounts of empirical research results, and its publication was made possible thanks to the support of numerous individuals in industrial, academic, and government sectors. We would like to express our sincere gratitude to them for their kind cooperation, although we cannot include all of their names due to space limitations.

We first wish to thank the practitioners (the people at the genba) who kindly contributed to our field research. They respectfully answered our questions in interviews, participated in our questionnaire surveys, gave us approval to observe the factories and other industrial sites, and even let some of the authors work temporarily with them. Without their support, we would not have been able to collect the valuable factual data that became the robust foundations of our empirical research results.

We also wish to thank the Japan Association for Evolutionary Economics. Most of the authors had the chance to present some of the preliminary results of our empirical research at the annual meeting of the Japan Association for Evolutionary Economics in March 2015. All the comments and questions from the floor and other members of the society were precious and helped us to enrich the papers that were developed into the chapters of this book.

Above all, we would like to express our sincere gratitude to Ms. Francesca Viarengo for the editing and proofreading of our manuscripts. She devoted considerable effort in supporting our publication project. Note here that almost all the authors of this book are non-native English speakers. We also appreciate the strenuous effort of the staff at Springer Verlag to finish this thick book. In addition, we thank the Evolutionary and Institutional Economic Review (EIER) for granting permission to reprint chapters "The Nature of International Competition Among Firms". "Product Variety for Effective Demand Creation", and "The Diversity and Reality of Kaizen in Toyota".

Some of the researchers that contributed to this book are financially supported by the Japan Society for the Promotion of Science (JSPS), Grant-Aid for Scientific Research (KAKENHI) (A) 15H01960, (B) 15H03376, (C) 25380458,

(C) 18K01788, Grant-in-Aid for Challenging Exploratory Research 15K13032, and Grant-in-Aid for Young Scientists 16K17158, 18K12839, 18K12849.

Finally, we wish to express our gratitude to both Professor Yuji Aruka, the chief editor of this series, and Mr. Yutaka Hirachi at Springer Verlag, who gave us the opportunity to publish this book. Participating in this challenging project was indeed a great experience for all of the authors.

Takahiro Fujimoto and Fumihiko Ikuine, co-editors

Contents

About the Contributors

Takahiro Fujimoto (Chapters "A Design-Information-Flow View of Industries, Firms, and Sites", "The Nature of International Competition Among Firms", "Product Variety for Effective Demand Creation", "Capability Building and Demand Creation in Genba-Oriented Firms", "Evolution of Organizational Capabilities in Manufacturing – The Case of the Toyota Motor Corporation", "Balancing Standardization and Integration in ICT Systems: An Architectural Perspective", and "Conclusion"; Editor) is a Professor at the Graduate School of Economics, University of Tokyo, Japan. He specializes in technology and operations management (TOM) and evolutionary economics. His publications include *Product Development Performance* (with Kim Clark, Harvard Business School Press) *Evolution of a Manufacturing System at Toyota* (Oxford University Press), and *Industries and Disasters* (co-edited with Daniel Heller, NOVA).

Mitsuhiro Fukuzawa (Chapters "The Nature of Surviving Japanese Factories in the Global Competition: An Empirical Analysis of Electrical and Electronics Factories" and "The Effectiveness of Group Leaders in the Lean Production System: Time Study and Agent-Based Model of Leaders' Behavior") is an Associate Professor at the Faculty of Economics, Seikei University, Japan. His research interests include strategy formation process, development process of organizational capabilities, and operations management of Japanese factories. His publications include "Dynamic capability as fashion," *Annals of Business Administrative Science, 14*(2): 83–96, 2015.

Wei Huang (Chapter "Investigating the Creation and Diffusion of Knowledge for Demand Creation: The Case of the Telecommunications Industry") is a Doctoral Student at the University of Tokyo, Japan. His research areas are product innovation and development of new product in turbulent environments.

Fumihiko Ikuine (Chapters "Creating New Demand: The Impact of Competition, the Formation of Submarkets and Platforms", "Decline in Demand Creation: The Development Productivity Dilemma and Its Consequences", and "Conclusion"; Editor) is an Associate Professor at the Faculty of Engineering, Information and Systems, University of Tsukuba, Japan. His research interests are management of technology (MOT) and innovation. He is the author of *Development productivity dilemma: The innovation patterns in the digital age,* Yuhikaku (in Japanese), which won the Takamiya Award (prize for books) from the Academic Association for Organizational Science (AAOS) of Japan.

Nobuyuki Inamizu (Chapters "The Nature of Surviving Japanese Factories in the Global Competition: An Empirical Analysis of Electrical and Electronics Factories" and "The Effectiveness of Group Leaders in the Lean Production System: Time Study and Agent-Based Model of Leaders' Behavior") is an Associate Professor at the Graduate School of Economics, University of Tokyo, Japan. He is investigating the dynamics of the workplace organization within Japanese firms from both quantitative and qualitative aspects, and also working on modeling these dynamics using computer simulation. His publications include *Communication and decision-making in a fluid organization: Agent-based approach,* University of Tokyo Press (in Japanese), which won the Takamiya Award (prize for books) from the Academic Association for Organizational Science (AAOS) of Japan.

Shumpei Iwao (Chapter "The Diversity and Reality of Kaizen in Toyota") is a junior Associate Professor of the Faculty of Economics, Meiji Gakuin University, Japan. His research areas are management science, technology and operations management, and Kaizen (continuous improvement activities). He recently wrote "Linking Continuous Improvement to Manufacturing Performance," *Benchmarking: An International Journal,* 25(5) (with M. Marinov), and "Revisiting the existing notion of continuous improvement (Kaizen): literature review and field research of Toyota from a perspective of innovation," *Evolutionary and Institutional Economics Review,* 14(1): 29–59.

Young Won Park (Chapter "Balancing Standardization and Integration in ICT Systems: An Architectural Perspective") is a Professor at the Graduate School of Humanities and Social Sciences, Saitama University, Japan. His research interests are in technology management, Global Strategy and IT strategy, and global supply chain management. His publications include *Business Architecture Strategy and Platform-Based Ecosystem* (Springer) and *Building a Sustainable Global Strategy: A Framework of core competence, Product Architecture, Supply Chain Management and IT Strategy* (NOVA). His articles have been published in journals including *Management Decision, International Journal of Production Economics, International Journal of Technology Management, International Journal of Information Management, Business Horizons, Journal of Business Research, Journal of Purchasing and Supply Management, Benchmarking: An International Journal.*

Junjiro Shintaku (Chapter "The Nature of Surviving Japanese Factories in the Global Competition: An Empirical Analysis of Electrical and Electronics Factories") is a Professor at the Graduate School of Economics, University of Tokyo, Japan. His current research interests are global strategies of manufacturing companies, standardization strategy, and global supply chain management. His publications include "Demand Fluctuation and Supply Chain Integration," *Perspectives on Global Development and Technology,* 16: 564–586, 2017, "Flexible Supply Capability Driving Total Inventory Reduction," *International Journal of Productivity and Quality Management,* 15(2): 268–284, 2015, "What Panasonic Learned in China," *Harvard Business Review,* December 2012: 109–113, and "Architecture-based Approaches to International Standardization and Evolution of Business Models," *International Standardization as a Strategic Tool: Commended Papers from the IEC Century Challenge*, 2006.

Yoshinori Shiozawa (Chapters "The Nature of International Competition Among Firms" and "Product Variety for Effective Demand Creation") is Professor Emeritus at Osaka City University, Japan, and Fellow of Japan Association of Evolutionary Economics. He specializes in evolutionary economics and its theoretical foundations. On the extension of classical theory of value, he has succeeded in developing a new theory of international values, which provides a framework of technical change and the emergence of global value chains. His papers include *Guided Tour of the Backside of Agent-Based Simulation* (Springer, 2016), *The New Theory of International Values: An Overview* (Springer, 2017), and *A Large Economic System with Minimally Rational Agents* (forthcoming).

Jing-Ming Shiu (Chapters "Investigating the Creation and Diffusion of Knowledge for Demand Creation: The Case of the Telecommunications Industry" and "The Impact of Platform Providers' Knowledge on Interfirm Division of Labor: The Case of the Mobile Phone Industry") is an Assistant Professor at the National Cheng Kung University, Taiwan. His research areas are management of technology (MOT), international standardization, and new product development management in ICT (Information Communication Technologies) industry.

Nobutaka Suzuki (Chapter "The Nature of Surviving Japanese Factories in the Global Competition: An Empirical Analysis of Electrical and Electronics Factories") is an Associate Professor at the Graduate School of Information and Management Systems Engineering, Nagaoka University of Technology, Japan. His research interest is strategic management, management of technology (MOT), and emerging market strategy. He wrote "The Development of Manufacturing Industry and Economic Growth in India-Japan Relations," *Asia Pacific Journal of Social Sciences,* 3: 79–89, 2012.

Hirofumi Tatsumoto (Chapter "Evolution of Business Ecosystems") is a Professor at the Graduate School of Business Sciences, University of Tsukuba, Japan. His main area of research is strategic management in platform business in electronics, software, and semiconductor industry in the global market. His publications include "Platforms and the international division of labor: A case study on Intel's platform business in the PC industry,"(with Ogawa, K. and Fujimoto, T.) in Gawer, A. (eds.) *Platforms, Markets and Innovation*, 345–369 (Edward Elgar, 2009) and *Platform strategy for global markets: The strategic use of open standards and the management of business ecosystem* (Yuhikaku 2017), which won the Takamiya Award from the Academic Association for Organizational Science (AAOS) of Japan, the Irie Itaro Award from The Academy of Multinational Enterprises, the book award from Japan Academy of International Business Studies, and the academic book award in Management Consulting and Service area (MCS) from the Japanese Institute of Certified Public Accountants.

Masanori Yasumoto (Chapters "Investigating the Creation and Diffusion of Knowledge for Demand Creation: The Case of the Telecommunications Industry" and "The Impact of Platform Providers' Knowledge on Interfirm Division of Labor: The Case of the Mobile Phone Industry") is a Professor of Graduate School of Environment and Information Sciences, the Yokohama National University, Japan. His current research interests are knowledge network within and across firms, strategy to cope with collaborative standardization, and innovation management in open environments. His publications include *Management of Technology and Innovation in Japan, Springer* (co-author) and "Does cross-functional integration lead to adaptive capabilities?," *International Journal of Technology Management*, 30(3/4): 265–298, 2005 (co-author).

Kodo Yokozawa (Chapter "The Nature of Surviving Japanese Factories in the Global Competition: An Empirical Analysis of Electrical and Electronics Factories") is an Associate Professor at the Graduate School of International Social Sciences, Yokohama National University, Japan. He received a Ph.D. degree at the University of Twente in the School of Management and Governance. His Ph.D. research was in international Japanese management systems transfer, focusing on transfer of Kaizen (continuous improvement) activities. He received his Master's degree in Business Administration from the Eastern Washington University, USA. His current research focuses on the transfer of best production management systems to organization community, based on a case study or transfer of process oriented management system to Indian car component suppliers' community.

Part I
Capability-Architecture Frameworks to Explain Globalization and Digitalization

Part I of the present book discusses the theoretical frameworks and concepts adopted here to analyze the evolution of industries and firms. We pay special attention to their applicability to the main industrial trends of the late twentieth century and early twenty-first century, such as global competition between advanced and emerging countries and global digitalization of industries.

Chapter "A Design-Information-Flow View of Industries, Firms, and Sites" is an introductory chapter that presents a capability-architecture-performance framework to analyze the evolution of manufacturing industries and firms. Here, manufacturing capability is a system of organizational routines governing the flows of value-carrying design information to customers, and a manufacturing site is the place where such flows exist. Product architecture indicates the correspondence between a product's functional and structural design elements. We predict that the dynamic fit between a manufacturing site's capability and a product's architecture will result in higher productive performance (theory of design-based comparative advantage). This capability-architecture view of industrial performance is expected to explain with reasonable accuracy the industrial phenomena studied in this book, namely globalization and digitalization. When applying this framework to empirical cases, the two main themes are (a) capability building of firms and sites for productivity improvements in response to global competition, and (b) architectural strategies for demand creation in the digital industries, which we discuss in Part II and Part III respectively.

Chapter "The Nature of International Competition Among Firms", focusing on the issue of firms' capability building, explores the modern Ricardian theory of international values and comparative advantages and its application within the dynamic setting of global competition. After clarifying that Ricardian international values (relative prices) can be determined in certain N-commodity-M-county cases, we regard the global cost competition of the post-Cold-War era as inter-site competition in terms of physical productivity (i.e., Ricardian labor input coefficient) with international wage rates as handicaps. The capability building efforts for drastic productivity improvements made by many factories in Japan, the handicapped high-wage country, are explained through this dynamic model of Ricardian comparative advantage.

Chapter "Product Variety for Effective Demand Creation" turns to the design-architecture side of our framework and discusses the issue of creating demand by expanding product variety. Based on the past economic literature on the variety of goods and services as a source of economic growth, this chapter argues that, in the case of firms developing sets of functionally similar products, effective demand can be generated by increasing the variety of product designs to a certain extent. By proposing the concept of *expected coverage function*, we show how optimal variety can be estimated for a set of firm products with similar designs throughout their product cycle. We also explain that the cost of developing additional product variations balances the estimated additional sales. Thus, additional product variety creates demand, but firms aiming to maximize their long-term profits must be aware that there is a limit, i.e., optimal variety.

Chapter 'Capability Building and Demand Creation in "Genba-Oriented Firms"' focuses on firms that combine the two key elements of capability building and demand creation. These are the so-called genba-oriented firms (or community-oriented firms), which simultaneously pursue both target profit (mark-up) rates for themselves through capability building and stable employment for the local communities through demand creation. More specifically, when they face intense global cost competition vis-à-vis their rivals in lower-wage countries, firms of this type in higher-wage countries, like Japan, have to increase their physical labor productivity to ensure their own survival and, at the same time, generate effective demand to maintain adequate levels of employment. In order to explain the behavior of genba-oriented firms, we propose an economic model that incorporates Ricardo's cost function, product differentiation, employment function, and full-cost principle, i.e., the PXNW model. This chapter also presents several actual cases of Japanese manufacturing firms between the 1990s and the 2010s whose behavior is generally consistent with what the PXNW model predicts.

Chapter "Evolution of Business Ecosystems" presents newer approaches for explaining the dynamics of the industries with many network effects by using the theoretical lens of business ecosystem, a special form of industrial cluster characterized by diversity of firms, complex relationship among them and the presence of platform firms. Such characteristics come from network effects. Firms often set up open standards to trigger network effects that are advantageous for their business models. Among them, platform firms, which strategically use the open standards and exploit the network effects at most for their business models, play a special role in the business ecosystem because their strategic behaviors contribute not only to the increase in their market presence but also to the expansion of business ecosystem with stimulating entries of newcomer firms.

Compared with the industrial models proposed in the previous chapters, "Evolu tion of Business Ecosystems" illustrates a completely new pattern of industrial dynamics. In a business ecosystem, products interact not only by competing with but also by complementing one another, as opposed to traditional product-to-product competition. Many open standards, which are a source of network effects, allow firms to collaborate in an autonomous, decentralized manner. Recent two phenomena, digitalization and globalization, push forward the growth of business ecosystems up to global scale.

A Design-Information-Flow View of Industries, Firms, and Sites

Takahiro Fujimoto

Abstract This introductory chapter discusses the research scope, framework, and key concepts of this book. Regarding the scope of our empirical research, we focus on two main industrial phenomena occurring in the period between the 1990s and the 2010s, i.e., global competition and digitization, and apply certain evolutionary frameworks and concepts to explain them.

As for the research framework to analyze the evolution of manufacturing industries and firms, we propose a capability-architecture-performance framework, which is derived from our broad concept of manufacturing as managing flows of value-carrying design information to customers, as well as that of manufacturing site (*genba*, in Japanese) as the place where such flows exist. We adopt a genba-based view of the economy, in which the most basic units of industrial-economic analysis on the supply side are the manufacturing sites, where value flows exist, rather than the firm, which is composed of the former.

In this context, manufacturing capability is a system of organizational routines that governs and improves the flows of value-carrying design information to customers. Product architecture, on the other hand, refers to the correspondence between a product's functional and structural design elements.

The concept of design-based comparative advantage derives from this framework, which predicts that certain dynamic fits between a manufacturing site's capability and a product's architecture result in higher productive performances and lower unit design costs. For example, a country whose manufacturing sites have higher coordination capabilities tends to have a design-based comparative advantage in products with coordination-intensive design, i.e., those with relatively integral architecture.

The original version of this chapter was revised: This chapter was previously published non-open access and has now been changed to open access. The correction to this chapter is available at https://doi.org/10.1007/978-4-431-55145-4_16.

T. Fujimoto (✉)
Graduate School of Economics, The University of Tokyo, Bunkyo-Ku, Tokyo, Japan
e-mail: fujimoto@e.u-tokyo.ac.jp

T. Fujimoto, F. Ikuine (eds.), *Industrial Competitiveness and Design Evolution*, Evolutionary Economics and Social Complexity Science 12,
https://doi.org/10.1007/978-4-431-55145-4_1

1 Introduction

1.1 Purpose and Scope of This Book

This book is about the evolution of firms, industries, industrial sites, and products. These are the foundations of national or global economies on their supply side. They are all artifacts (designed things) which evolve overtime.

In this introductory chapter, we will discuss the purpose and scope of this book, as well as a basic framework for concepts such as economy, industry, firm, product, and site. Then, we will present a design-based view of manufacturing and industrial competitiveness, whose main components include organizational capability and product/process architecture. We will also provide a historical description of firms, industries, and sites in postwar Japan, the main research field of the present book. In the appendix, we will explore the possibility of adopting a dynamic and design-based version of the Ricardian (i.e., classical economic) model of international trade as its twenty-first-century reinterpretation.

As for the scope of this book, we will pay special attention to the period between the 1990s and the 2010s, characterized by unusually rapid evolutionary changes in industries, firms, sites, and products, when two major transformations of the world economy, namely, *globalization* and *digitization*, occurred almost at the same time.

By globalization we mean post-Cold War integration of the Western and Eastern bloc economies, which led to intense international cost competition between lower-wage emerging countries (e.g., China, India) and higher-wage advanced countries (e.g., USA, EU, Japan). By digitization we mean major industrial changes driven by innovations in digital information and computer technologies, including the Internet, mutually networked digital devices (e.g., personal computers and smartphones), and software assets that are often complementary to one another.

This historical coincidence resulted in major shifts of digital products toward open-modular architectures, massive location shifts of industrial sites toward low-wage countries, uneven growth of national economies and industries, emergence of large platform-leading firms, and increasing income inequalities. We will focus on changes in the patterns of international industrial competition during this period.

Geographically, we will look at an advanced country and its industries that were significantly affected by the abovementioned globalization and digitization, i.e., Japan and its industries, firms, and sites. As we will discuss in this book, after the end of the Cold War and China's entrance into the global market, export industries and their manufacturing sites in Japan—a neighboring country whose average wages for factory workers were over 20 times those of China in the 1990s—faced difficult-to-survive cost competition with the huge international wage handicap.

Japan is also endowed with coordination-rich manufacturing sites that tend to have competitive advantages in coordination-intensive products, including fuel-efficient automobiles, analog TV sets, and functional materials (Fujimoto 2007, 2014; Appendix). Thus, when digital technological innovations triggered a rapid growth in products and platforms with coordination-saving open-modular architectures and open interfaces, many leading Japanese firms in the electronics industry found it difficult to formulate appropriate strategies vis-à-vis the American platform-leading firms (e.g., Intel, Microsoft, Apple, Google, Amazon). Besides, their coordination-rich industrial sites lost much of their design-based competitive advantage in unit costs in various digital products with coordination-saving open architectures vis-à-vis rival factories in low-wage emerging countries, such as China.

Thus, Japan's economy, industries, firms, and manufacturing sites were significantly and negatively affected by post-Cold War global cost competition and industrial digitalization. As a result, the national economy suffered from extremely low growth rates between the 1990s and the 2010s due partly to the aforementioned globalization and digitization, as well as to the post-bubble financial crisis, chronic deflation, and population aging. Japan's industrial and trade structure also shifted toward a higher share of fuel-efficient automobiles, high-performance industrial machineries, functional chemicals, and other integral products, whereas digital electronics products experienced a decline. Also, average profitability across the country's firms decreased during the same period.

In this situation, Japan's domestic manufacturing sites struggled to survive both global competition and digital transformation. Many of them disappeared, particularly in the consumer appliances and digital devices/equipment industries, but many survived in industries with relatively integral and complex products. As of the mid-2010s, Japan's manufacturing industries still accounted for roughly 20% of its gross domestic products (GDP), a relatively high share for a larger advanced nation.[1] Moreover, physical productivity increased significantly in many of the surviving sites manufacturing tradable goods.

Based on the above observations, this book will focus mainly on the evolution of firms, industries, and industrial sites in Japan during the post-Cold War period between the 1990s and the 2010s. It will propose some theoretical frameworks that may be appropriate for analyzing such evolutionary industrial phenomena.

1.2 Some Basic Ideas on the Evolution of Firms and Industries

1.2.1 Product, Site, Industry, and Firm

Let us start by presenting some basic concepts to investigate the phenomena mentioned above. First, as an analytical framework for dealing with the research theme of the present book, we adopt a multilayer evolutionary framework encompassing economies, industries, firms, sites, and products (Fig. 1).

[1]According to the IBRD and other statistics, the share of the manufacturing sector in 2012 was about 18% in Japan, 22% in Germany, 13% in the USA, and 10% in the UK and France.

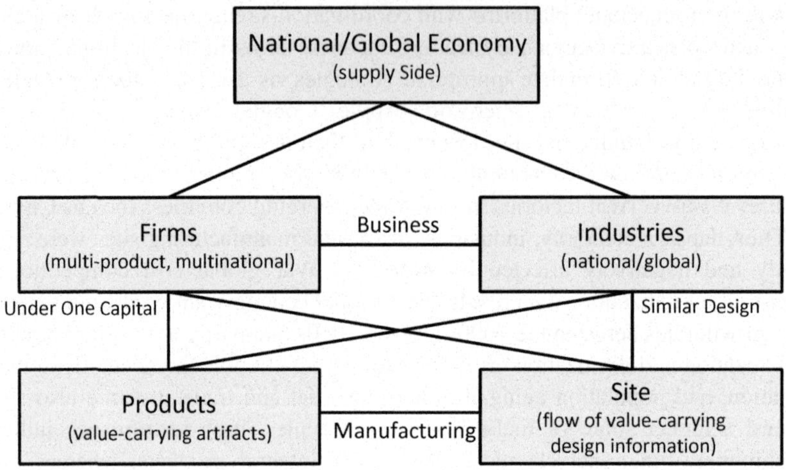

Fig. 1 Products, sites, industries, and firms

In this framework, a *product* is a value-carrying artifact that can be exchanged. More specifically, we assume that a product's value added dwells in its design information, i.e., information on its functions, structures, and their relations.

A *site* (i.e., industrial/manufacturing site; *genba* in Japanese) is the place where the value-carrying design information of certain products flows toward their markets, and it includes the people who collectively govern such flows. We regard an industrial site, or genba, as the most basic component of a firm, an industry, as well as a national economy. As such, a manufacturing site is the common element of both a firm and an industry.

An *industry* is a collection of sites that develop and produce functionally similar products and their components—this being the conventional definition of an industry. Besides, we also adopt a newer concept called *platform*, which is a collection or network of products that are functionally/structurally complementary to each other. Thus, we use a newer and broader definition of an industry as a collection of functionally similar products, platforms, and their components. Competition, collaboration, and transactions routinely occur among such sites, products, components, and platforms within an industry.

A *firm* can be seen as a collection of industrial sites that are under the control of a single capital, but they may produce very different products and be located in different countries. In other words, based on today's stylized facts, we assume that modern firms are mostly multiproduct and/or multinational firms. Thus, we distinguish between firms and industries, as well as firms and sites, based on our observations of actual economic activities today.

In any case, the fundamental value-adding units of the economy are industrial sites. Based on this multilayer framework, we analyze the evolution of the products, manufacturing sites, industries, and firms of a national/global economy. In other words, we argue that all four components evolve over time with dynamic and emergent interactions among them.

1.2.2 Manufacturing as Flows of Design Information

This book also adopts a broad and *design-based view of manufacturing* to analyze the evolution of firm, industries, and sites. That is, we argue that the value added of a product resides in its *design* or information/knowledge about the artifact's functions, structures, and their relations (Simon 1969; Suh 1990; Ulrich 1995).

Therefore, we use another basic framework to investigate industries, firms, and sites, i.e., a design-based view of manufacturing (Fujimoto 2007, 2012b). That is, we define *manufacturing*—or *monozukuri* in Japanese—broadly as all the activities of firms, industries, and sites that control and improve flows of value-carrying design information to the customers. As such, our design-based concept of manufacturing covers not only manufacturing industries but also services and other non-manufacturing sectors, as long as they involve flows of value-carrying design information to the customers.

Design here refers to knowledge or information about an artifact's functions, structures, and their relations (Suh 1990) that is created prior to its production (i.e., design realization).[2] As pointed out earlier, we argue that the source of a product's value added for customers lies in its design information, which is able to attract and satisfy them.

To the extent that the concept of manufacturing is defined broadly in this way, *manufacturing sites* (genba) are also defined broadly, including not only production factories but also product development projects, resale/wholesale stores, service facilities, farming fields, and so on, as long as value-carrying design information flows to the customers there.

In this context, an *industry* is conceived as a collection of manufacturing sites (e.g., factories, product development centers, service facilities), in which design information of a similar kind, combined with its media (i.e., materials and energies), flows to the customers in the market. This implies that an industry can be seen as a set of flows of functionally similar design information, or as a collection of mutually competing/cooperating/transacting sites that govern such value-carrying design flows.

---

[2]Design here means not only industrial design but also all kinds of engineering design, including both hardware and software as well as service design, work design, organizational design, system design, and so on, as long as it represents an artifact's functions and structures and their relations. A product is an exchangeable or tradable artifact with value added (i.e., the difference between its price and material cost). Thus, a product, like any other artifact, can be seen as a combination of design information and its medium, in the form of direct materials and energy—a modern interpretation of Aristotle's being as a combination of form and matter.

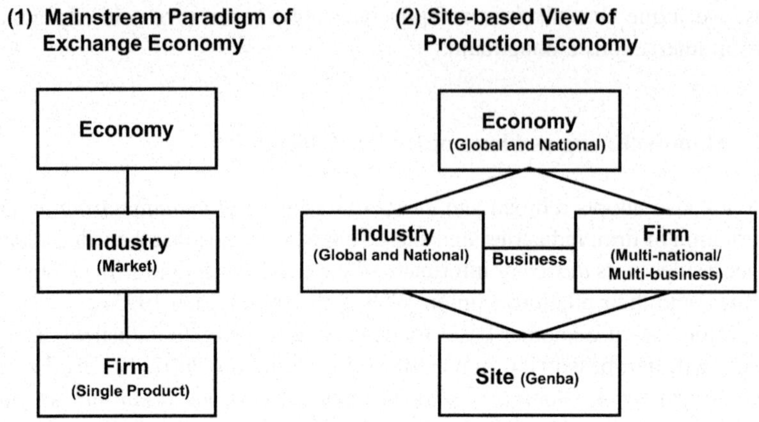

Fig. 2 Supply side of the economy

1.2.3 Manufacturing Site as Economic Agency

Next, let us discuss the nature of a manufacturing site (genba) as a semi-independent economic entity that evolves over time. In other words, we regard an industrial site as a multifaceted socioeconomic entity, since it belongs to a firm, an industry, and a community at the same time. When a firm is not simply profit-oriented but also site-oriented, and if a site (genba) is community-conscious and the community is employment-conscious, we may conjecture that a *site-oriented firm* behaves differently from a conventional profit-maximizing firm, in that the former pursues both profits (e.g., target markup ratios) and stable employment (e.g., a fixed number of regular employees) at the same time. We will discuss the nature of such site-oriented or genba-oriented firms later in this book (Chapter "Evolution of Business Ecosystems").

In a sense, recognizing a manufacturing site as a semi-independent socioeconomic agency means deviating from the standard economic model of a profit-maximizing firm. In the standard neoclassical microeconomic model (e.g., partial equilibrium analysis), the assumption is that a firm produces only one kind of product with identical design, so the distinction between firms and sites is not essential. Besides, an industry's market supply curve is regarded as a simple sum of the individual supply curves of the firms in this industry. Thus, in the prevalent neoclassical microeconomic model, the supply side of a national economy simply consists of three layers, i.e., firms, industries, and national economies (see Fig. 2, discussed later). No detailed analyses of industrial sites and products as artifacts are included in this standard model because a site is regarded merely as a dependent part of a profit-maximizing firm and a product as a non-differentiated commodity.

In this book, by contrast, we treat both firms and sites as mutually interdependent economic agencies, each of which has its own objectives, such as survival, growth, profit, and employment. Nowadays, a multiproduct and multinational firm selects its products, sites, and their locations, whereas an industrial site, which is a part of the firm and the community at the same time, tries to survive with stable employment as a semi-independent socioeconomic entity in itself.

As suggested above, when international industrial competition is intense, the simultaneous pursuit of (i) productivity improvements through the sites' capability-building and (ii) effective demand creation by means of product variety and design improvements is critical for a site-oriented firm that aims to achieve minimum profit for survival and stable employment. Thus, the present book analyzes dynamic interactions among firms and sites regarding profit, growth, survival, and employment in the current age of intense global competition.

1.2.4 Competition and Competitiveness as Driving Forces

We have so far argued that a firm and an industry can be regarded as a set of certain manufacturing sites or products and that the sites and products themselves evolve over time, which is the basic logic of the evolution of firms and industries. Then, our next question is: what is the major driving force behind such evolution? Our tentative answer is as follows: one of the main driving forces behind the evolution of firms and industries is *industrial competition*, or the competition between functionally similar products, components and platforms, as well as their manufacturing sites.

Generally speaking, we define *competition* as a socioeconomic entity's efforts to be selected by some other entities (i.e., the selectors) under the conditions of fairness and free choice by the latter and *competitiveness* (or competitive performance) as the selectee's ability to be selected by the selectors.

For example, in standard economic textbooks, it is assumed that products of a given quality compete on price in order to be selected by customers in the product market—this is *price competition*, which happens at the surface level of the economic system. Likewise, modern firms compete on profitability in order to be selected by investors and bankers in the capital market.

In the present book, on the other hand, we pay special attention to another type of competition among manufacturing sites at a more basic level of the economic system—*capability-building competition*—in which manufacturing sites compete to be selected by the firms to which they belong by improving their productive performance, such as production lead times, physical productivity, and manufacturing quality (Womack et al. 1990; Clark and Fujimoto 1991; Fujimoto 1999).

Facing today's intense global competition, many of the sites mentioned above as community-employment-conscious may collectively try to survive by improving their capability and productive performance, achieving minimum acceptable profits for the firms to which they belong, and thereby managing to be selected by said firms' top managers as the ones that can continue operations.

Hence, competitions and competitiveness are multilayered phenomena, since factories (or genba), products (or their business units), and firms all compete to be selected and to ultimately survive. In addition, because an industry, as illustrated earlier, consists of certain products of similar design as well as the manufacturing sites that produce them, we can regard price competition among the products and capability-building competition among the sites as *industrial competitions* at the surface and deeper level, respectively.[3]

We also predict that the competitiveness of products and sites in the same industry and in the same country or region (e.g., German automobile industry) will tend to converge, since said products and sites face similar competitive environments and are therefore forced to build similar capabilities over time. In these cases, it is meaningful to compare the average competitiveness performance (e.g., unit cost, productivity, quality, lead times, etc.) of an industry's products and sites among counties or regions (Womack et al. 1990; Clark and Fujimoto 1991). The comparison of physical productivity and unit production costs (proxy variables of prices) among countries and industries first originated with David Ricardo's trade theory of *comparative advantages* or *comparative costs* (Ricardo 1817), which we will discuss later in this book (Chapter "The Nature of International Competition Among Firms").

1.2.5 Evolution of Capabilities and Architectures

We have so far argued that the main driving forces behind the evolution of firms and industries are industrial competitions occurring among manufacturing sites, products, platforms, and the like. We may then ask: which characteristics of sites and products evolve over time? A brief answer may be as follows: *organizational capability* evolves on the side of the manufacturing sites, whereas *architecture* (in addition to technology) evolves on the side of the products.

First, we define a manufacturing site's *organizational capability* (i.e., manufacturing capability) as a set of organizational routines that control and improve the flows of design information in that manufacturing site (Fujimoto 1999). For example, the so-called Toyota production system (TPS) may be regarded as a manufacturing capability or a system of interconnected organizational routines (e.g., *kanban*, small-lot delivery, *jidoka*, multi-skilling, building-in quality, levelization, continuous improvements, etc.), each of which was generated through a combination of deliberate choices and emergent processes (Fujimoto 1999; Chapter "Evolution of Organizational Capabilities in Manufacturing: The Case of the Toyota Motor Corporation"). In other words, both organizational capabilities and routines may evolve over time as a result of capability-building capabilities, competition, and environments related to the manufacturing sites in question.

Second, we explore the nature of the *architectures* of products, components, platforms, processes, and other economic artifacts. Generally speaking, a product's design information has two aspects: technology and architecture. A product's

[3]Note here that competitions among firms regarding their profitability on the capital market are not the same as industrial competition to the extent that modern large firms tend to be engaged in multiple industries or businesses.

specific *technology* refers to concrete causal relations among its functions and structures, whereas its *architecture* describes the abstract graphical correspondence among them. To the extent that architecture is an abstract mathematical concept, product architectures may be compared across different industries, whereas many product technologies are industry-specific. Thus, in the present book, we explore how changes in a product's technologies and market requirements affect the evolution of product architectures.

Having sketched out some basic concepts to analyze the evolution of firms and industries—including a site-based framework for firms and industries, the notion of design information as a source of value added, a design flow-based view of manufacturing, the site (genba) seen as a socioeconomic entity, the idea of industrial competition as an evolutionary driving force, as well as the evolution of a site's capabilities and a product's architectures—we will try to construct a systematic framework to explore industrial competitiveness and evolution next.

2 A Framework for Analyzing Industrial Competitiveness

2.1 Industrial Competitive Analysis: A Missing Link in Modern Economics

Let us now look at the theoretical background of the evolutionary framework adopted in this book. The worldwide industry in the early part of the twenty-first century may be described by emphasizing various aspects—intensifying post-Cold War global competition involving both advanced and emerging nations, trends toward freer trade through various bilateral and multilateral agreements among nations, explosive growth of goods and services that use digital networking technologies, stricter constraints regarding environmental protection and energy conservation, increased complexity of artifacts that deal with such constraints, coexistence of fiercer price competition in commodity-type goods and product differentiation in brand-conscious goods, greater technology and market uncertainties faced by firms, and growing instability and influence of global financial networks on industries.

One of the propositions derived from the above description is that international *industrial performance* (*competitiveness*)—based on the concept of comparative advantage, devised in the nineteenth century by David Ricardo and other classical economists (Ricardo 1817)—still matters in this century. It is also important to note that improvements in physical labor productivity (i.e., labor input coefficients), the ultimate generator of industrial comparative advantage and national standards of living (Smith 1776), occur at industrial sites, or genba in Japanese, including factories, development centers, retailers, service facilities, and farming fields. An industry is nothing but a collection of industrial sites (genba) of a similar kind.

In the eighteenth and nineteenth century, major works by classical economists, including Adam Smith and David Ricardo, used to provide rich accounts of "industries" based on field observation (e.g., Smith's famous analysis of pin making). In the past one hundred years, however, after Alfred Marshall's *Industry and Trade* in particular (Marshall 1919), mainstream economics (i.e., the neoclassical school) tended to deemphasize the concepts of "industry" and "sites" while pursuing mathematical-theoretical sophistications such as the general equilibrium theory, which assumes profit maximization at the firm level. This meant that mainstream economics mostly neglected the field-based concept of *industrial performance*, even though it continued to be an empirically important notion for understanding the nature of today's world economy (Womack et al. 1990; Clark and Fujimoto 1991).

Indeed, the concept of comparative advantage of industries, both Ricardian and neoclassical, continued to be key to understanding the freer trade systems of the twenty-first century. Newer approaches—like the product life cycle (flying geese) theory, the new trade theory, and the new-new trade theory—certainly provided additional explanatory power to better understand today's trade phenomena involving emerging nations, foreign direct investment, product differentiation, and economies of scale (Akamatsu 1962; Vernon 1966; Helpman and Krugman 1985; Melitz 2003). Without introducing the concept of *design* of traded goods and services, however, we may not be able to capture the essential characteristics of today's international trade—*intra-industrial trade at minute levels*, such as sheet steel for inner automobile panels exported from Korea to Japan and that for outer panels exported in the opposite direction.

Besides, some 20 years after the end of the Cold War and the abrupt entrance of gigantic low-wage countries like China into the global market, the average wage in such emerging countries has finally started to soar, as the period of "unlimited supply of labor" (Lewis 1954) has come to an end. This may be a good time to introduce a somehow dynamic and field-based version of the Ricardian-Sraffian trade theory to examine how international differences in productivity and wage increases between advanced and emerging countries affect changes in global trade structures (Ricardo 1817; Sraffa 1960; Shiozawa 2007; Fujimoto and Shiozawa 2011–2012).

Against the above background, this section sketches out an evolutionary framework for the analysis of industrial performance by introducing such concepts as *manufacturing (monozukuri)* such as *design information flow*, *genba as value-flowing site*, *evolution of organizational capabilities*, *evolution of product-process architectures*, *dynamic fit between capabilities and architectures*, and *multilayer concepts of industrial performance* (see Fujimoto 2007, 2012b for details).

2.2 Economy, Industry, Firm, and Site Revisited

Here we revisit the multilayer framework concerning industries described earlier in this chapter. In order for us to conduct empirical research on industrial performance, we must modify our analytical framework for industrial analysis from the interpretation adopted by today's mainstream economics, or the "firm → industry → economy" paradigm of the "exchange economy" or catallactics as defined by Sir John

Hicks (Hicks 1976; Fig. 2 (1)), to a seemingly more realistic one, or the "site→firm/industry→economy framework of the "production economy" or plutology (Hicks 1976; Fig. 2 (2)).

Since the main goal of the present book is to empirically explore dynamic changes, or evolutions, on the production and development side of the economy, it would be natural for us to adopt a *site-based view of the production economy*, which is illustrated in Fig. 2 (2).

To sum up, the most basic value-adding units of the economy are the industrial sites or genba. In addition, economies, industries, firms, and sites all evolve over time (Nelson and Winter 1982; Fujimoto 1999, 2007, 2012a, b). Hence, many of the empirical studies presented in this book will start from field observation and perform an evolutionary analysis of firms and industries by focusing on their common components, i.e., industrial sites (genba), including factories, and development projects.

2.3 The Capability-Architecture-Performance Framework of Industrial Evolution

The field-based framework for the analysis of industrial performance and trade structures proposed in this chapter relies on the evolutionary framework of *design-based (architecture-based) comparative advantage*, as explained earlier, which predicts that certain dynamic fits between manufacturing capabilities and product-process architecture will result in an industry's international competitive advantage (Fig. 3).

This framework includes the following elements:

1. The design-based concept of *manufacturing* in a broad sense (monozukuri, in Japanese), which reinterprets development-production-sales activities as creation and transfer of value-carrying design information flowing from firms/sites to customers
2. The generic logic of *comparative advantage*, which assumes that a fit between a country's characteristics and a product's attributes results in the competitive advantage of a given product in a given country (Ricardo 1817; Fujimoto and Shiozawa 2011–2012)
3. The *evolutionary theory of organizational capabilities*, which explains ex post rational objects without fully depending upon ex ante rational reasoning (Fujimoto 1999; Chapter "Evolution of Organizational Capabilities in Manufacturing: The Case of the Toyota Motor Corporation)
4. The concept of *product-process architecture*, originating from the theory of axiomatic design in engineering (Suh 1990; Ulrich 1995)

Both the organizational capabilities of the manufacturing sites (genba) and the architecture of the artifacts (products and processes) collectively and dynamically

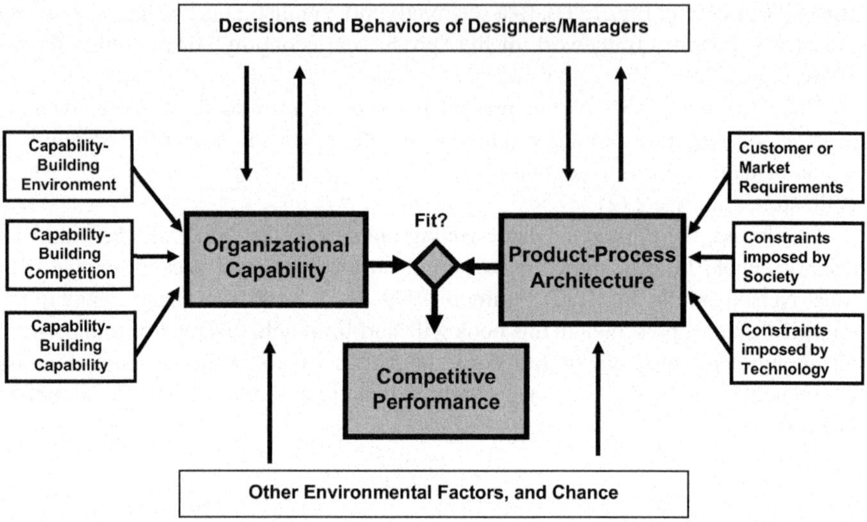

Fig. 3 Field-based view of industrial performance—design-based comparative advantage

influence the performance (competitiveness) of the industry in question, as illustrated by the shaded boxes in Fig. 2.

Moreover, both capabilities and architectures are treated here as endogenous rather than exogenous factors. This implies that they can change as their interactions with environmental and other factors change. There is no such thing as Japan-specific capability or automobile-specific architecture in a static sense, since capabilities and architectures are a result of the path-dependent historical evolution of the entire industrial system.

Having described the overall evolutionary framework of capability-architecture-performance, let us now look at its components in more detail. The following sections will briefly illustrate the design information view of manufacturing (monozukuri), industrial performance, organizational capability, product-process architecture, and design-based comparative advantage, in this order.

2.3.1 Manufacturing as Design Information Flows Among Productive Resources

Starting from the abovementioned framework to analyze industrial performance, our next question is: what are the key concepts that all industrial sites (genba) have in common? Here we adopt a broad and design-based concept of manufacturing (Fujimoto 1999, 2007) or monozukuri. According to this view, a genba is a place where *flows* of value added to the market exist and the value added in question ultimately resides in *design information*. A productive resource (Penrose 1959) or an

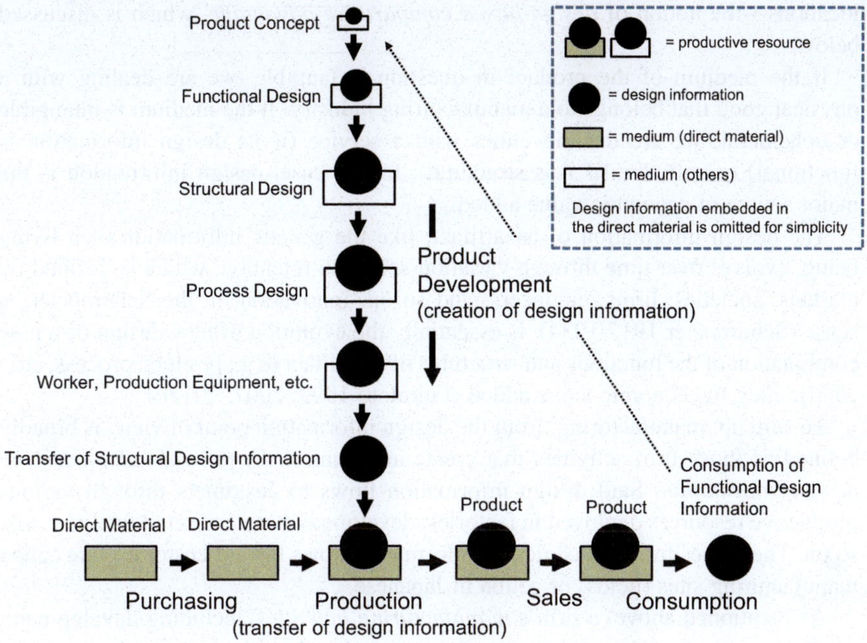

Fig. 4 Manufacturing as flows of design information

artifact (Simon 1969) existing inside the manufacturing site in question is nothing but a combination of value-carrying design information and its medium (Fig. 4).[4]

Thus, the common factors that can be observed in all industrial sites are (i) flows of design information to customers; (ii) artifacts (productive resources), each of which is a combination of design information and its medium; and (iii) the site's performance, measured as effectiveness of flows.

In this context, as mentioned earlier, *design* means information or coordination that interconnects an artifact's functional and structural elements (Suh 1990). A product (a good or service) is a tradable artifact consisting of design information and its medium, following Aristotle's logic of form and matter. Production is nothing but the transmission of a product's design information to its medium. Thus, design precedes production for a given product. In the context of trade theories, this implies that international selection of design locations tends to precede that of production

[4]A *productive resource* (Penrose 1968)—such as workers, equipment, dies, tools, standard operating procedures, digitized design files, raw materials, work in process, prototypes, or engineering drawings—is also an artifact, or a combination of partial design information and medium. In the production process, a part of the structural design information of a firm's products is embodied in workers, machine hardware, software, or other media. Raw materials and work in process are also productive resources that embody partial design information. In this sense, the design-information view regards a firm as a set of productive resources, which is nothing but design information assets deployed and stored in labor or capital stocks as their media.

locations—the notion of *design-based comparative advantage*, which is discussed below.

If the medium of the product in question is tangible, we are dealing with a physical good that belongs to a manufacturing industry. If the medium is intangible or ephemeral, we are dealing either with a service (if its design information is functional) or software (if it is structural). In any case, design information is the major source of economic value added.

The design information of an artifact, like the genetic information of a living being, evolves over time through variation-selection-retention, which is decided by markets, societies, firms, engineers, and so on. Innovation, in the Schumpeterian sense (Schumpeter 1912/1934), is essentially the evolution of new design or a new combination of the functions and structures of an artifact (e.g., product, process, etc.) contributing to economic value added (Fujimoto 1999, 2007, 2012b).

To sum up, manufacturing, from the design information point of view, is broadly defined as those firm activities that create and control the *flow of value-carrying design information*. Said design information flows to customers through various productive resources deployed in factories, development centers, retail facilities, and so on. The places from which design information flows toward customers are called manufacturing sites (fields) or genba in Japanese.

As mentioned above, a firm's *manufacturing activities*, including development, production, purchasing, and sales, can be regarded as flows (creation and transfer) of value-carrying design information among productive resources (Fig. 4).

Within this framework, *product development* is the creation and verification of value-carrying design information. It is essentially a process of translation, going from the evaluation of future consumption processes and technological possibilities to product concept creation, product functional design, and product structural design, and ending with production process design and preparation. Each stage consists of repetitive problem-solving cycles involving designing-prototyping-testing steps (Clark and Fujimoto 1991; Thomke and Fujimoto 2000).

Production, in this context, is the repetitive transfer of product design information from the production process to the materials or work in process (i.e., the medium of the product). At each stage of the process, a fraction of the product's design information, stored in workers, tools, equipment, manuals, and other productive resources, is transferred to the materials or work in process and transformed into an actual product.

Purchasing means obtaining media (i.e., materials) for the product from outside firms. In many cases, the materials already embody partial design information, so purchasing activities often involve design information flows (Asanuma 1989; Clark and Fujimoto 1991). *Sales* is the transmission of the design information embodied in the products from the firms to the customers.

Consumption of physical goods is another information-creation process related to the customers themselves, in which they use or operate the products and thereby convert their structures into their functions, which is then translated into the customers' own satisfaction or dissatisfaction. We may see this as the customers' self-

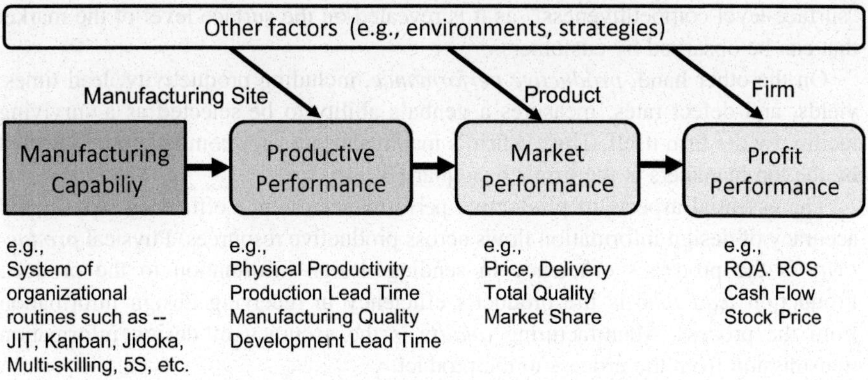

Fig. 5 Capability, competitiveness, and profitability

performed service activity, in which service means provision of value-carrying functions by the operators of certain structures.

2.3.2 Competitive Performance of Industries, Firms, and Sites

Let us focus on the manufacturing industries of trade goods for now. Generally speaking, *industrial competitiveness* refers to the productive or market performance of a certain set of products or sites belonging to the industry of a particular nation or region.

Competition here means a subject's efforts to be selected for a certain reward under either pre-determined rules and/or free choice on the part of the selector. Competition, in other words, is an interaction between mutually independent selectors and selectees. Thus, *competitiveness* (i.e., competitive performance) can be defined as *a selectee's ability to be selected by selectors* under the rule of independent choice.

In actual fact, there are at least three layers of competitive performance, depending upon what is selected by what: *profit performance of a firm, market performance of a product, and productive performance of a manufacturing field* (Fig. 5).

Profit performance refers to a firm's ability to be selected on the capital market (e.g., return on sales, return on assets, return on equity), or its attractiveness as a whole in the minds of investors. The level of *profit performance* is affected by the firm's productive and market performance, as well as by other environmental factors such as exchange rates, business cycles, and corporate strategic choices.

Market performance is a product's ability to be selected on the product market, or the attractiveness of the design information embodied in the product in question in the minds of customers. The product's ex ante market performance includes price, delivery time, and perceived product quality, whereas its ex post market performance is measured by its market share. We may also call market performance "surface-level competitiveness," as it is revealed on the surface level of the market that can be observed by customers.

On the other hand, *productive performance*, including productivity, lead times, yields, and defect rates, measures a genba's ability to be selected as a surviving facility by the firm itself. Thus, a firm's manufacturing sites compete to be selected by the top managers at the firm's headquarters.

The essential aspects of productive performance include efficiency, speed, and accuracy of design information flows across productive resources. Physical *productivity* is the process's efficiency in sending design information to the product. Production *lead time* is the product's efficiency in receiving design information from the process. Manufacturing *quality* is the accuracy of design information transmission from the process to the product.

Both productivity and lead time improve in proportion to *value-adding time ratios*, other things being equal. That is, physical productivity (i.e., units produced per person-hour) increases by N times when the ratio on the design-information-sending side (i.e., the percentage of design-information-sending time over the total operation time of a day) increases by N times, given the amount of the product's design information and the speed of its transmission. Likewise, production lead time (i.e., time elapsed between reception of the direct materials and completion of the product) is reduced to 1/N when the ratio on the design-information-receiving side (i.e., the percentage of design-information-receiving time over the same lead time) increases by N times. In both cases, the time during which information is *not* transferred from the process (e.g., workers, production equipment) to the product is called *muda* (waste) at the Toyota Motor Corporation (Ohno 1978).

In any case, the abovementioned productive performances are measured by the effectiveness of the flow of design information in the manufacturing sites. As indicated above in Fig. 5, the causal connection between a site's organizational capability and its productive performance is more direct than that between its capability and its products' market performance, as well as the firm's profit performance (Monden 1983; Shoenberger 1982; Womack et al. 1990; Fujimoto 1999).

We have so far discussed competitive performance at the level of firms (profit performance), products (market performance), and sites (productive performance). What about performance at the industry level? As mentioned earlier, an industry is a collection of manufacturing sites or their products, but not a collection of firms, which can be multi-industrial and/or multinational. Accordingly, it is not relevant to aggregate firms' profit performance at the industry level. An industry's ex post *market performance*, measured by market share, can be aggregated as a country's market share on the global market. As for ex ante market performance, such as price, total quality, and delivery, their distribution or average levels vis-à-vis rival countries may be used as summary indicators. An industry's *productive performance* indicators, such as productivity, lead time, and defect ratios, may also be captured as their distribution or average levels vis-à-vis rival countries (Womack et al. 1990; Clark and Fujimoto 1991).

2.3.3 Organizational Capabilities of Manufacturing Sites

Organizational capability is a concept developed in evolutionary economics and in the resource-based view (RBV) of firms in strategic management (Penrose 1959; Nelson and Winter 1982; Grant 2005; Fujimoto 1999). According to this view, a firm or one of its manufacturing sites is seen as a holder of firm-specific (or site-specific) organizational capabilities and managerial resources. Given the level of a firm's resources, its capabilities affect its competitive performance (e.g., productivity).

Organizational capability is an attribute of an organization, in that it is more than the simple sum of individual skills, and it affects interfirm differences in competitiveness and profitability in the long run. It influences, if not determines, the long-term survival rate of competing sites and firms. The organizational capability of a best-practice firm is difficult for other firms to imitate, so interfirm differences in competitiveness stemming from organizational capability tend to be sustainable over a long period of time. Organizational capability tends to be built up cumulatively by a firm rather than established through one major investment or acquisition. The process of capability-building is not always based on a deliberate planning process and may well be emergent (Mintzberg and Waters 1985) or evolutionary (Fujimoto 1999).

A firm's organizational capability may be found at the level of its headquarters (e.g., strategy formulation capability for creating a platform) or its manufacturing sites (genba). Although we discuss both types of capabilities in this book, let us focus on the latter for now. When a firm's organizational capability for controlling and improving the flow of value-carrying design information is found at the level of its manufacturing sites, we call it organizational capability in manufacturing, or simply *manufacturing capability*. It is a firm-specific or site-specific system of *organizational routines* that govern the design information flows among the site's productive resources.

As mentioned before, the physical productivity (the inverse of the labor input coefficient) of a manufacturing site is the efficiency of its design information flow to the market (Fujimoto 1999). It follows that the productivity of various factories or development projects differs depending upon the firm's technological choices and/or the sites' manufacturing capability (Womack et al. 1990; Clark and Fujimoto 1991). Our empirical analysis starts from the recognition of such interfirm and international differences in productivity within a global industry. Note that the standard (neoclassical) trade theories tend to assume that production functions (i.e., physical productivity) are identical across firms and national borders within an industry, which does not seem to be a realistic assumption in today's global competition.

The above view of industries can be seen as a dynamic interpretation of Ricardo's comparative advantage. Through their evolution, manufacturing routines and capabilities (e.g., the Ford system or the Toyota system) create international productivity differences across manufacturing sites (e.g., factories and projects) within the same industry (Fujimoto 1999). Indeed, we often find the productivity of a factory in one

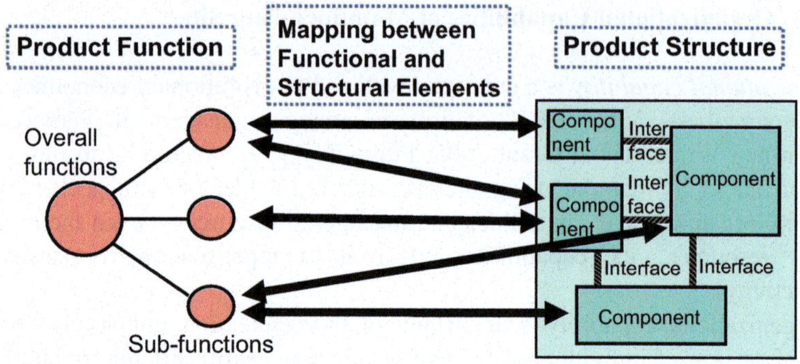

Fig. 6 Product architecture

country to be three or more times higher than that of a competing factory in another country that has adopted similar production technologies.

2.3.4 Architectures of Products and Processes

Let us now turn to the design attributes of value-carrying artifacts, including architecture and technology. *Architecture* is defined for any given artificial system (Simon 1969), including a product, use system, production process, platform, or business model. It refers to a formal pattern to link an artificial system's functional elements with its structural elements (Langlois and Roberstson 1992; Ulrich 1995). Thus, *product architecture* has to do with the engineers' basic way of thinking when they design the functions and structures of a new product. They may start from the product's overall functional requirements, derived from its concept, and deconstruct them into a set of subfunctions or functional elements. They then conceive the product's components, or structural elements, and map the relation between functional and structural elements. Thus, a product's architecture refers to a formal pattern of correspondence between its functional and structural elements (Fig. 6).

To the extent that the product's functional and/or structural elements are interdependent, the components (i.e., structural elements) need *interfaces* with other components, through which signals and energy flow for mutual adjustment. After completing a basic design of this sort, engineers can move on to the detailed design of each component.

Likewise, *process architecture* refers to the correspondence between the functional/structural elements of a product and its production process's structural elements. The concept of process architecture is important particularly in non-assembly-type industries, such as chemicals, steel, and other material industries, whose products are monolithic and difficult to deconstruct into discrete components.

The overall picture of product-process architecture may be illustrated by a matrix of product functions, product structures, and production process structures (Fig. 7).

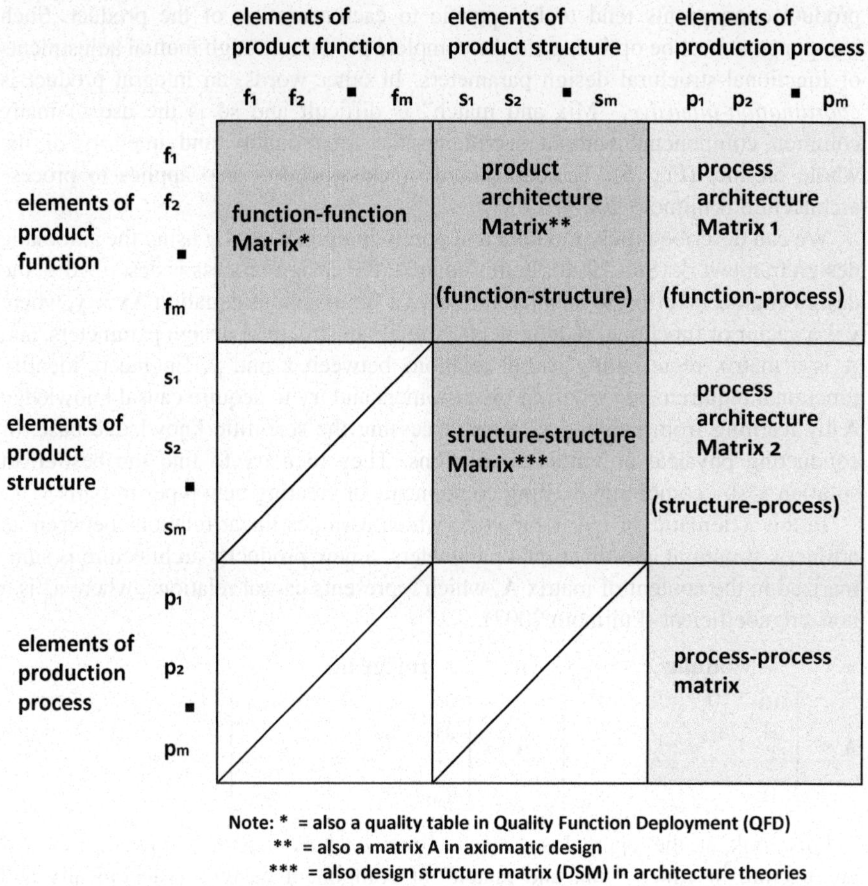

Fig. 7 Overall picture of product-process architecture

2.3.5 Basic Types of Architectures: Modular, Integral, Open, and Closed

There are certain basic types of architecture: modular versus integral and open versus closed (Ulrich 1995; Fine 1998; Baldwin and Clark 2000; Fujimoto 2007). *Modular architecture*, in its pure form, refers to a one-to-one correspondence between functional and structural elements. The parameters for components or production processes can be designed and operated relatively independently from one another, with less coordination among them. The *interfaces* among such components can be simplified and standardized, so "mix and match" of structural elements can generate variety within the total system (e.g., product) without sacrificing functionality. In other words, a modular product is *coordination-saving*.

Integral architecture, by contrast, is characterized by a many-to-many correspondence between a product's functional and structural elements. The designs of product components tend to be specific to each variation of the product. Such components must be optimized to the complete product through mutual adjustments of functional-structural design parameters. In other words, an integral product is *coordination-intensive*. "Mix and match" is difficult and so is the use of many common components without sacrificing the functionality and integrity of the whole product (Fig. 5). The same kind of classification also applies to process architecture (Fujimoto 2007).

We can describe purely modular and purely integral cases by using the axiomatic design framework (Suh 1990). In this context, the design process is described as the design engineers' effort to identify and solve a simultaneous equation $\mathbf{Ax} = \mathbf{y}$, where $\mathbf{y}$ is a vector of functional requirements, $\mathbf{x}$ refers to structural design parameters, and $\mathbf{A}$ is a matrix representing causal relations between $\mathbf{x}$ and $\mathbf{y}$. Engineers identify functional requirements $\mathbf{y}^*$ given by customers and try to acquire causal knowledge $\mathbf{A}$ by learning from existing systems, accessing the scientific knowledge base, or conducting physical or virtual simulations. They then try to find the best-effort solution $\mathbf{x}^*$ by combining existing components or creating new types of parts.

In this axiomatic design framework, which assumes linear relations between an artifact's structural and functional parameters, a new product's architecture is summarized in the content of matrix $\mathbf{A}$, which represents causal relations, where a_{ij} is a nonzero coefficient (Fujimoto 2007).

Modular **Integral**

$$\mathbf{A} = \begin{bmatrix} a_{11} & 0 & \cdots & 0 \\ 0 & a_{22} & & \\ \vdots & & \ddots & \\ 0 & & & a_{mm} \end{bmatrix} \qquad \mathbf{A} = \begin{bmatrix} a_{11} & a_{12} & \cdots & a_{1m} \\ a_{21} & a_{22} & & \\ \vdots & & \ddots & \\ a_{m1} & & & a_{mm} \end{bmatrix}$$

Let us look at the open/closed axes. *Open architecture* is a type of modular architecture in which "mix and match" of component designs is technically and commercially feasible not only within a firm but also across firms because certain industry-standard (i.e., open) interfaces are established among them. *Closed architecture*, on the other hand, is the case where a given component is functionally/ structurally connectable to other components only within a certain firm's boundaries because its inter-component interfaces are firm-specific.

By combining the modular-integral axis and the open-closed axis described above, we can identify three basic types of product architecture (Fig. 8): (1) open-modular (open), (2) closed-modular, and (3) closed-integral (integral).

As an ideal type, *open-modular* architecture is characterized by industry-standard interfaces that are shared by relatively functionally complete components (with one-to-one correspondence between their functions and structures) designed by different firms; *closed-modular* architecture features firm-specific common components that can be connected only to other components whose interfaces are designed by the same firm; lastly, *closed-integral* architecture entails a collection of optimally designed product-specific (i.e., customized) components with complex many-to-many connections among their functions and structures.

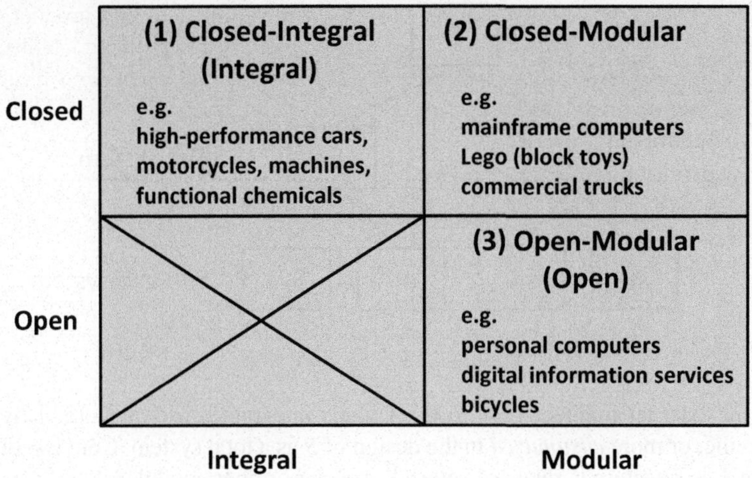

Fig. 8 Basic types of product architecture

The above design-information view of products, processes, sites, and industries naturally leads us to adopt an architectural approach to industrial classification based on architectures rather than a more traditional one relying on specific technologies. This architectural framework may provide additional insights into matters concerning intra-industry trade and a reinterpretation of the theory of comparative advantage, which is discussed next.

2.4 Architectural Positioning Strategy

2.4.1 Internal and External Architectures

The issue of strategic choices with regard to architectural positioning is another key aspect in our evolutionary analysis of firms and industries. When we analyze a firm's architectural positioning strategy for certain products, components, or platforms, we have to start from the notion that a complex artifact can be described as the hierarchy of a system with subsystems, sub-sub systems, and so on (Simon 1969; Langlois and Robertson 1992). Let us assume that the firm in question is engaged in a certain product category, which can be described as a three-layer hierarchy: a system S, subsystems S_i, and sub-sub systems s_{ij}. Let us also assume that this company develops and produces subsystem S_I by manufacturing or purchasing its sub-sub systems s_{Ij} and selling S_I to the assembler of the total system, S (Fig. 9).

In this case, the architecture of subsystem S_I regarding its functions and structures (i.e., sub-sub systems s_{Ij}) may be called the *internal architecture* of S_I, whereas the internal architecture of the total system S (in its part corresponding to S_I) may be regarded as the *external architecture* of S_I.

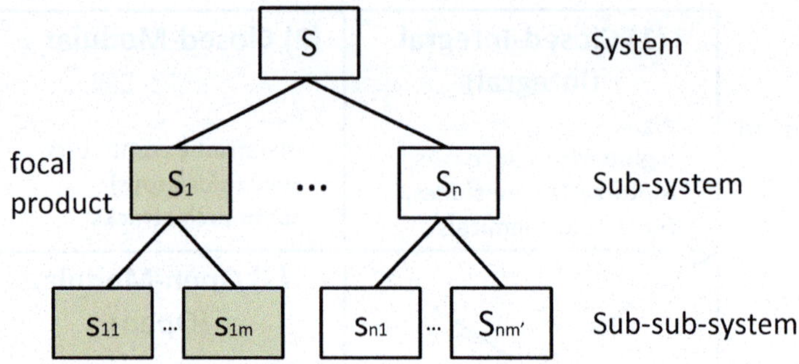

Fig. 9 The hierarchy of artifacts

If the external architecture of S_1 is relatively integral, S_1 will be more likely to be S-specific, or more *customized* to the design of S as a total system. Conversely, if its external architecture is relatively closed-modular or open-modular, S_1 will be more likely to be a customer-firm-specific *common part* or *industry-standard part*, respectively.

2.4.2 Architectural Positioning Matrix

Based on this distinction between the internal and external architecture of focal system S_1, we may analyze the integral/modular architectural positioning of S_1 by using a simple 2x2 matrix with four basic cells regarding strategic positioning: *integral-inside-integral-outside* (I-I), *integral-inside-modular-outside* (I-M), *modular-inside-integral-outside* (M-I), and *modular-inside-modular-outside* (M-M).

Different innovation/production/pricing/sales strategies may be applied to different cells. For example, price leadership or aggressive innovations to compensate for high unit costs will be crucial in the I-I strategy; sales and cost leadership with scale and learning effects will be key in the I-M strategy; solution business by sales engineers and direct selling will be effective in the M-I strategy; and simple production scale or profit-seeking of remaining players will be the best approach in the M-M strategy (Fig. 10).

Similarly, we can create a matrix to analyze internal/external and closed/open architectures, with the following architectural positioning strategies: *closed-inside-closed-outside* (C-C), *closed-inside-open-outside* (C-O), *open-inside-closed-outside* (O-C), and *open-inside-open-outside* (O-O) (Fig. 11).

Among these choices, the strategy most often adopted by the so-called platform leaders (e.g., Apple, Google, Amazon, Facebook, Intel, Microsoft, etc.) is *closed-inside-open-outside* (C-O). In order to understand the considerable impact of this C-O strategy, we need to explore the mechanisms of the cumulative network effect (network externality; Katz and Shapiro 1985) with complementary goods in the open-inside-closed-outside platform, which consists of closed-inside-open-outside core products.

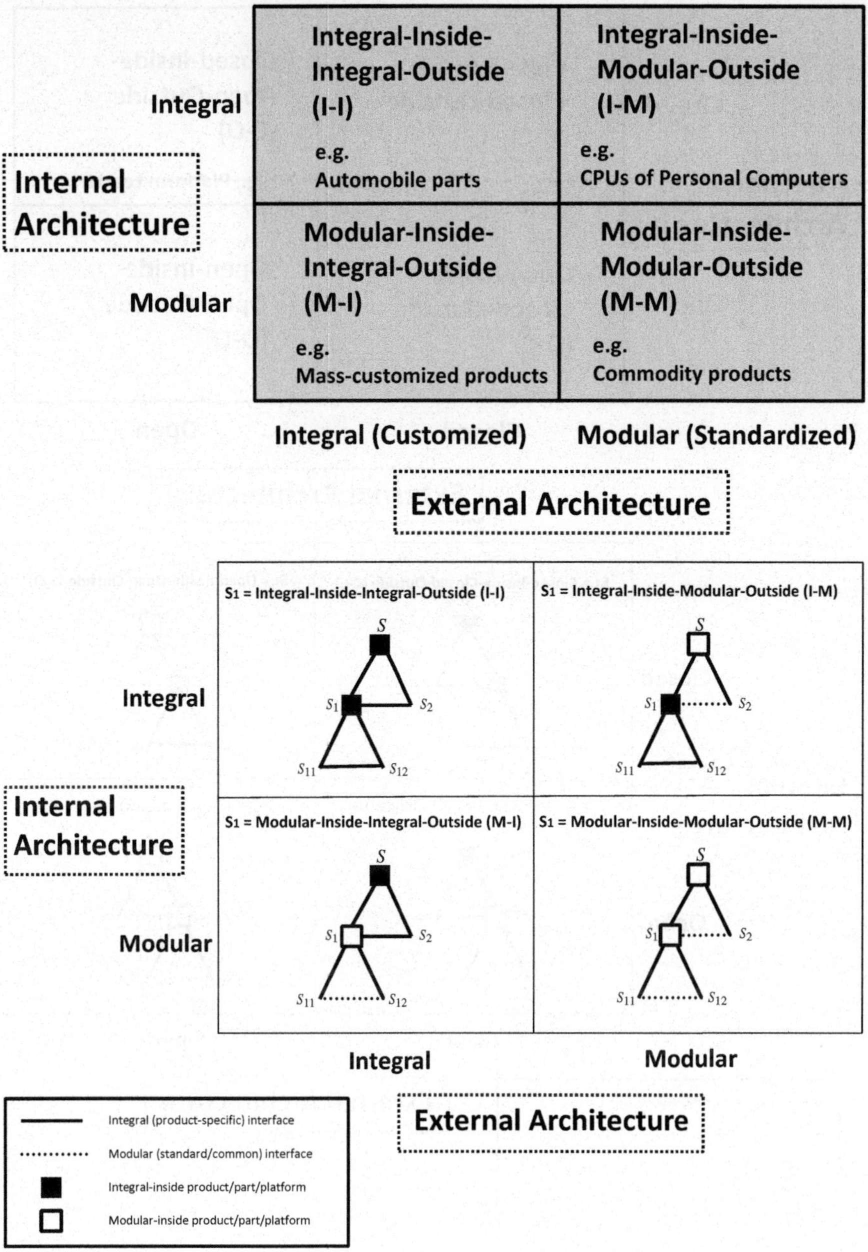

Fig. 10 Architectural positioning strategy (1) (integral/modular)

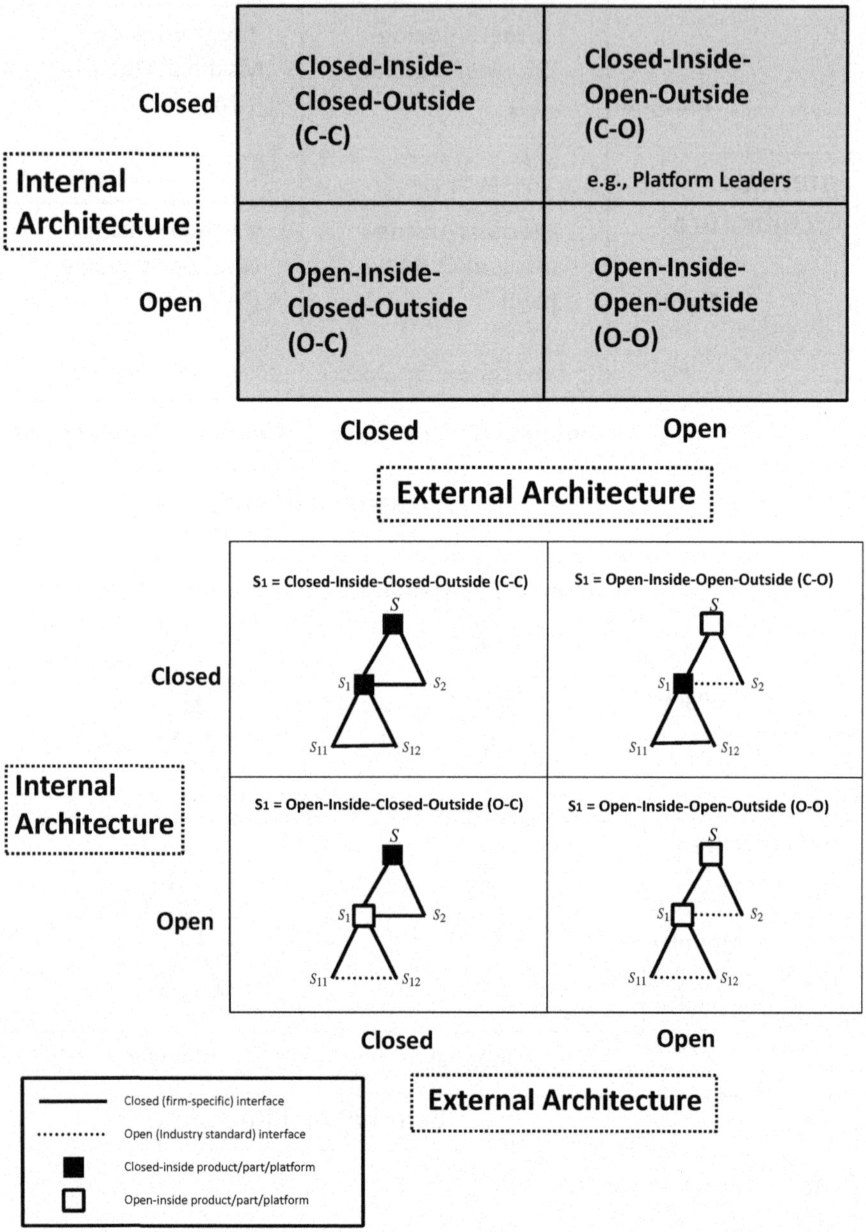

Fig. 11 Architectural positioning strategy (2) (closed/open)

2.4.3 Competition Among Open-Inside Platforms

Suppose that there is a technologically superb firm that has a broad range of deep systemic and component knowledge. In conventional inter-product competition, the high-tech firm in question will try to enclose all of its proprietary technologies and knowledge and maximize its product's market share by making the most of its technology-driven product competitiveness.

However, when the firm aims to become a platform leader, its strategy can be very different (Gawer and Cusumano 2002; Chapter "Capability Building and Demand Creation in 'Genba-Oriented Firms'"), as explained below:

(i) When the firm in question (from now on, the "core firm") finds that it possesses a core technology for an artifact with potentially open-modular architecture and network externality with complementary goods and promising markets, it tries to create an industry-standard interface, or open interface, around this key technology, which other firms will adopt.

(ii) The core firm further tries to make the "core area" within the newly established standard interface closed-inside-open-outside while also rendering the design information and technological knowledge inaccessible from outside the core firm by simply hiding them or protecting them by means of patents.

(iii) The core firm capsules its core technology in its "closed" area in the form of key components (e.g., CPUs for personal computers), core complementary goods (e.g., smartphone terminals), fundamental operating software (e.g., OSs for personal computers or smartphones), communication equipment (e.g., base stations for mobile phones), developmental services, communication/transaction services, etc., from which it earns revenues.

(iv) On the other hand, the core firm deliberately makes the design information and technological knowledge of the "peripheral area" outside the industry-standard interface open to other firms, so that the latter can develop complementary goods with open-outside architectures relatively easily by using the now accessible design/technological information. In this way, technological barriers to entry in the peripheral area become significantly lower, and many firms may enter this area.

(v) The core firm tries to preserve its technological leadership by monopolizing the design information in the core area, managing and maintaining the open interface, and making the design information in the peripheral area open and attractive to many existing or potential manufacturers/developers of complementary goods.

(vi) If the above efforts by the core firm are successful, cumulative network effects among core products/components/software and complementary goods and services may bring about rapid market growth for this network of products, components, software, and services, which we may call *platform* (Katz and Shapiro 1985; Brandenburger and Nalebuff 1996; Gawer and Cusumano 2002).

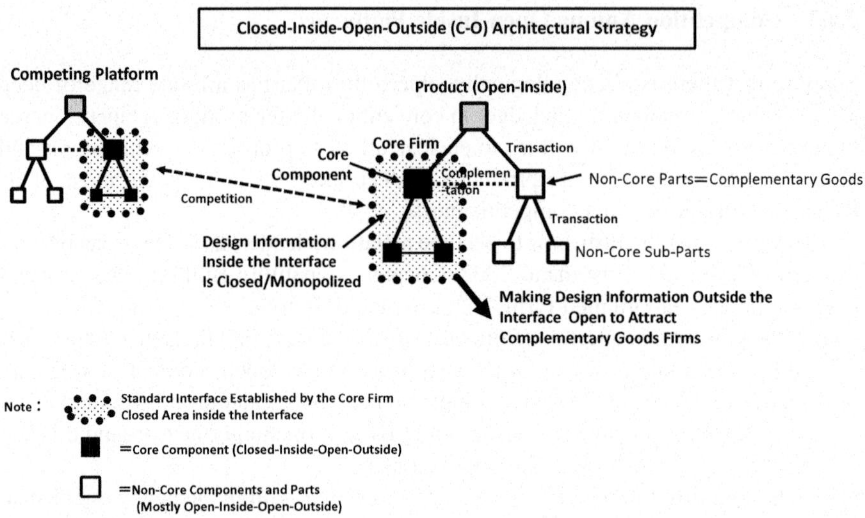

Fig. 12 Platform strategy by the key component firm

(vii) If the platform in question monopolizes the industry, we may regard this platform with open-inside architecture as an industry in a broad sense. If there are competing platforms that are functionally similar (e.g., Google versus Apple in smartphones), we may regard them as an industry with multiple and mutually competing platforms with open-inside architectures. Thus, an industry, in the broad sense of the word, consists of either competing products or competing platforms. Figures 12 and 13 illustrate the cases of core component and core product.

Thus, by introducing the concepts of hierarchies of artifacts, internal/external architectures, and open/closed architectures, our framework of competitiveness/capability/architecture can cover both conventional inter-product competition and newer inter-platform competition. In the former case, capability-building capabilities and design-based comparative advantages are the key elements to understand the dynamics of industrial competition. In the latter case, capabilities for creating and managing industry-standard interfaces and open-inside platforms, as well as cumulative network effects among complementary goods and services, are the main success factors for the platform-leading firms and the platforms themselves.

2.4.4 Product-Based Industry and Platform-Based Industry

When a set of products and components evolves into a platform created by the core firms (i.e., the platform-leading firms), which establish their industry-standard interfaces, a question arises as to how we should redefine an industry comprising the said platform. Conventionally, an industry consists of a set of products of similar design

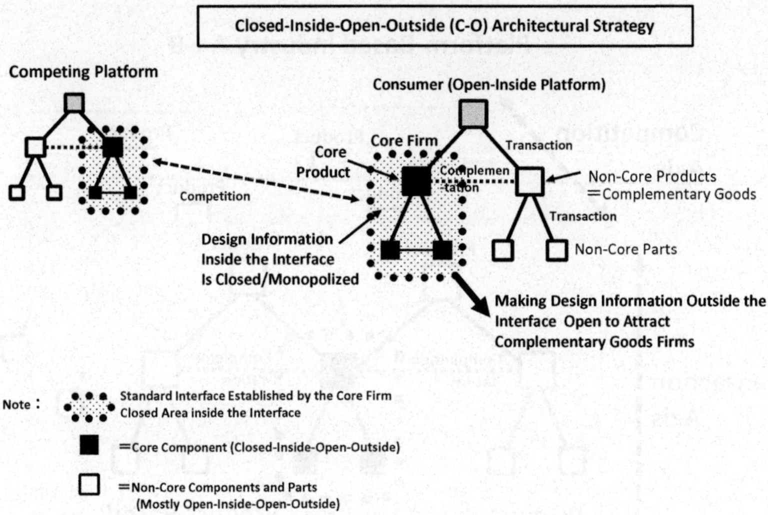

Fig. 13 Platform strategy by the key product firm

or functions competing with each other, as well as of the components and materials that their producers purchase. In other words, the products and components in an industry are mutually connected through competition or transaction.

On the other hand, a platform with open-modular architecture and network effects among complementary goods consists of products and components that are interconnected not only through competition and transaction but also through complementation (Brandenburger and Nalebuff 1996). Therefore, when a platform extends beyond a conventional industry, or a collection of functionally similar products and components, should we also call the set of industries interconnected via said platform "an industry"?

We believe that there should be two different definitions, i.e., the conventional *product-based definition of an industry* and the newer *platform-based definition of an industry*. That is, when an open-modular platform with standard interfaces encompasses multiple product-based industries, for instance, A and B in Fig. 14, we may redefine them as a single platform-based industry that includes both product-based industries A and B.

In the present book, an *industry* is understood as a collection of design information assets (e.g., products, components, services) that are networked or interconnected along three axes, i.e., transaction, competition, and complementation (see Fig. 14). When the products and components are not only competing with but also complementary to each other, we call such a network of products a *platform*. An industry can also be defined in terms of manufacturing sites and firms that are buying/selling, competing, and cooperating with each other. When such relations include "symbiosis" between complementary goods and their producers, we may call this industry an *ecosystem* (Iansiti and Lavien 2004). Hence, the definition of industry adopted here is rather broad.

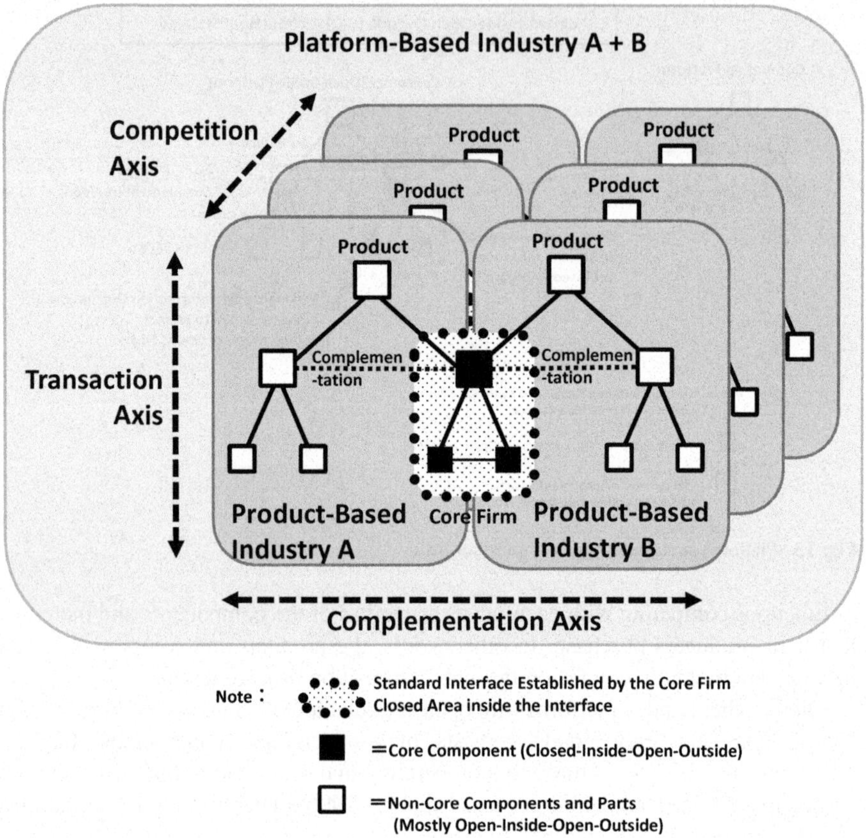

Fig. 14 Two definitions of an industry: product-based and platform-based

3 Implications and Conclusions

3.1 Summary of the Evolutionary Framework

The present chapter illustrated a field-based or site-based evolutionary framework for analyzing competition dynamics among sites, industries, and firms. This *field-based evolutionary framework of capability-architecture fit* may give us additional insights to better understand the general industrial dynamics of the early twenty-first century on a global scale.

After defining the competitive performance of industries, the capability of manufacturing sites (genba), and the architectures of products and processes, we illustrated the basic logic of *design-based comparative advantage* by connecting these factors (Fujimoto 2007, 2012b; see Fig. 2 again): the dynamic fit between a certain type of genba organizational capability, which has emerged in a given country, and a certain type of product architecture, which has evolved over time, tends to result in higher competitive performance of design locations in terms of comparative design costs.

Organizational capability in manufacturing is defined as a system of organizational routines that collectively control and improve the flow of design information to customers (Nelson and Winter 1982; Clark and Fujimoto 1991). To the extent that an organization is a system of coordinated activities (Barnard 1938), key dimensions of its capability will naturally include degrees and types of *coordination*. The evolutionary logic is also introduced here to explain why different types of organizational capabilities are unevenly accumulated in different countries and regions (Fujimoto 1999, 2007, 2012b, 2014).

The concept of *architecture* is defined as a formal pattern for coordinating the functional and structural design elements of an artifact, including product and process (Ulrich 1995; Fujimoto 2007). A product/process with integral architecture is coordination-intensive, whereas a product/process with modular architecture is coordination-saving, as mentioned earlier.

It follows from the above argument that a country's patterns of comparative advantage in design may be influenced by a certain fit between the *coordination capabilities* of its manufacturing sites (genba) and the *coordination intensities* of products and processes, both of which evolve over time. Specifically, a country whose industrial sites are relatively rich in coordination capabilities for evolutionary reasons, such as postwar Japan, might have a comparative advantage in design in relatively coordination-intensive products or those with integral architectures. Conversely, a country whose industrial sites have historically emphasized specialization-standardization-simplification of their products, processes, components, and their interfaces—such as the USA, whose industries grew rapidly, thanks to a massive inflow of immigrants—might have a comparative advantage in design in relatively coordination-saving products or those with modular architectures.

The present framework follows the general logic of comparative advantage theories, which emphasize country-industry fit and relative productivity advantage across countries. In addition, it adopts the design-based concepts of comparative advantage by integrating the design view of manufacturing into existing trade theories. In this context, both capabilities and architectures are treated as endogenous and dynamic. Hence, our approach assumes that a certain evolutionary process will result in the uneven distribution (i.e., endowment) of a given organizational capability across countries and firms. History does indeed matter.

The view of design-based comparative advantage also postulates that organizational capabilities are more difficult to move across borders than capital, goods, and services, even in our age of globalization, and that they tend to become country-specific. A country's capability-building environment (e.g., resource scarcity), the intensity of its industry's capability-building competition, and its firms' capability-building capability (i.e., evolutionary capability; Fujimoto 1999) all affect the prevalent nature of the capabilities of its manufacturing sites or genba.

The evolutionary view of architectures also argues that a product's overall *macro-architecture* is selected ex post by markets and society, whereas its *micro-*

architectures tend to be generated ex ante by engineers (Fujimoto 2012b). When a product must meet demanding functional requirements and/or strict constraints (e.g., safety and environmental regulations), its macro-architecture tends to become integral, other things being equal. By contrast, when the requirements and constraints are less strict, it tends to become more modular. Thus, a product's architecture is not a given—it evolves through micro-macro loops of design selections by engineers and markets.

Thus, the framework of design-based comparative advantage tries to explain why certain products are imported or exported within a global system of intra-industry trade of differentiated products, which is the overall trend of the twenty-first century.

3.2 Comparative Advantage in Production/Design Cost

In this introductory chapter, we argued that industries and firms are both a collection of sites engaged in providing certain products and that the products' market performance and the firms' profitability are both affected by the productive performance (i.e., deep-level competitiveness) of such manufacturing sites. We also pointed out that the value added of a product dwells in its design information.

The capability-architecture view of industrial performance adopted here is, in a sense, a reinterpretation of the Ricardian theory of comparative advantage. It extends the concept of comparative cost from production to design, so that it becomes a theory of comparative design cost. This implies that David Ricardo's classical theory of international trade, with certain modifications by modern economists such as Sraffa and Shiozawa (e.g., the multifactor multi-country version of Ricardo's comparative cost analyses), may be realistic enough to explain twenty-first-century trade phenomena in Japan and across the world (Sraffa 1960; Shiozawa 2007).

We predict that industries and industrial sites are more likely to be selected and thus survive when they build certain organizational capabilities and produce or develop products with appropriate architectures. Similarly, firms will be more likely to survive and grow when they choose a profitable mix of products and their architectures, as well as competent sites and their locations. In the following chapters, we will often use the abovementioned keywords: *industrial competitiveness*, *organizational capabilities*, and *product architectures*.

Theoretically, the framework of international industrial competitiveness put forward in this book may be regarded as a dynamic application of the Ricardian comparative advantage theory to design costs and locations, i.e., *design-based comparative advantage*. This approach starts from field observations of *industrial sites*, in which value-carrying design information flows toward the markets, evaluates the *organizational capabilities* that control or improve the value flows, identifies the *architectures* of products and processes, and analyzes the dynamic fits between architectures and capabilities and their impact on the *competitive performance* of sites, products, firms, and industries.

Thus, by dynamically reinterpreting the classical concepts of comparative advantage and by applying them not only to production costs but also to design costs, this volume will try to explain what has happened to various industrial sectors in postwar Japan, in terms of economic growth, labor shortage, yen appreciation, capability-building, international competition during the Cold War, global competition after the Cold War, and relative wage/productivity divergence and convergence vis-à-vis other advanced/emerging nations.

3.3 Evolution of Industries, Firms, and Sites

The evolutionary framework adopted here will also illustrate the growth and changes of national and global economies, industries, firms, and sites seen as interrelated dynamic processes.

First, the evolution of the capabilities of manufacturing sites (genba) causes productivity growth and differences, which influence country A's average productivity in industry X through market selection of high-productivity sites within the country.

Second, different industries in country A, with different patterns of capability-building processes and design architectures, display different levels of relative productivity vis-à-vis the industries in competing countries B and C.

Third, the resulting profile of the relative productivity ratios of industries X, Y, and Z between competing countries A and B affects relative wage ratios between the two countries (Fujimoto and Shiozawa 2011–2012). In other words, the profile of all industries' relative productivity ratios vis-à-vis competing countries affects the relative wage ratio.

Fourth, as a result of the relative productivity and wages mentioned above, the relative costs and prices of industries X, Y, and Z in the competing countries are revealed. In the long run, following the logic of Ricardian comparative advantage, the industrial portfolios of countries A, B, and C emerge through selection by the global markets of "comparatively advantageous industries," which have higher relative productivity ratios vis-à-vis rival countries rather than other domestic industries.

However, the industrial structures of trading countries may constantly change to the extent that, as *capability-building competition* among sites and firms continues, their products' design attributes (e.g., architecture) change. For instance, while the relative wage ratio between two countries may change, as their all-industry profiles of relative productivity ratios mentioned above change, further changes in relative productivity ratios may, in turn, change the patterns of comparative advantage. If the relative productivity ratios of industry X in countries A and B converge faster than the relative wage ratios in the same two countries, due to technological standardization, lower-wage country B may gain a comparative advantage vis-à-vis country A, thereby shifting its status from importer to exporter in industry X, as Akamatsu's *flying geese* theory or Vernon's product cycle theory suggest (Akamatsu

1962; Vernon 1966). This is not always the case, though, as the international trade situation in the 2010s indicates—international wage gaps may decrease faster than physical productivity gaps between two countries (e.g., China and Japan).

Fifth, in order to secure profits and growth, manufacturing firms worldwide will try to select certain advantageous combinations of products and locations by moving and expanding across countries and industries. Multinational firms may find overseas locations for their manufacturing sites by balancing two principles, i.e., physical proximity to markets and comparative advantage. Therefore, a firm's business structure will evolve through selection of advantageous industries, products, architectures, technologies, site locations, and so forth.

It ought to be noted that the above account follows the Ricardian logic of classical economics, which argues that normal (natural) prices are determined by unit labor cost, or by the combination of labor productivity (labor input coefficients) and hourly wages. Additionally, the evolutionary analysis of firms and industries presented here essentially follows the assumption of classical economists (or their prominent successor P. Sraffa) that prices and volumes of a given product are determined separately (Sraffa 1926, 1960; Shiozawa 2007).

3.4 Between New and Old Theories

The early twenty-first century is the age of truly global competition, environmental and energy constraints, rapid changes in digital and other technologies, increasing numbers of demanding customers, parallel advancements in product commoditization and differentiation, rapid changes in relative wages and productivity across borders, and high levels of socioeconomic uncertainty. To the extent that the above is true, the concepts of comparative advantage and industrial performance and their evolution will continue to be key to sustaining or improving our children's living standards and quality of life.

Some may argue that we need newer theories able to handle the new realities of the twenty-first century. In fact, in analyzing open architecture platforms in digital industries, we will apply newer economic concepts, such as complementary goods, network externality, and inter-platform competition. In this particular area, we certainly need to introduce new economic concepts for new realities.

At the same time, we have also explored an alternative idea—returning to the older theories of the nineteenth century, namely, classical economic theories, and modifying them to fit the realities of the current global economy. More specifically, our work tries to start from the classical (Ricardian) trade theory and then reinterprets it dynamically to explain the rapid changes in relative wages and productivity (Fujimoto and Shiozawa 2011) and to incorporate the concept of product design and architecture into it (Fujimoto 2007, 2012b). Consequently, this chapter proposes a field-based evolutionary framework of capability-architecture fits to analyze industrial performance, i.e., the concept of design-based comparative advantage (Fig. 2).

It is worth noting that the present framework of dynamic and design-based comparative advantage is, in many ways, complementary to existing trade theories. Indeed, the neoclassical (Heckscher-Ohlin-Samuelson) theory, the product-cycle (flying geese) theory (Akamatsu 1962; Vernon 1966), the new trade theory (Helpman and Krugman 1985), and the new-new trade theory (Melitz 2003) all capture certain important aspects of today's industrial competition and trade dynamics.

However, in order to understand the trade phenomena of the post-Cold War era, when minute-level intra-industrial trade of highly differentiated goods is common and relative wages and productivity of individual sites are changing rapidly worldwide, we would need additional insights that may borrow ideas from other schools and disciplines, including the concept of comparative cost from classical economic theories, of system emergence from evolutionary economics, of organizational capabilities from strategic management, of product architecture from design-artifact theories, and of monozukuri (manufacturing as design flow) from technology and operations management (Fujimoto 1999, 2007, 2012a; Fujimoto and Shiozawa 2011–2012; Mintzberg and Waters 1985; Nelson and Winter 1982; Penrose 1959; Shiozawa 2007; Simon 1969; Sraffa 1960; Ulrich 1995). What we need in this context seems to be a dynamic and interdisciplinary framework that can capture the essence of the multifaceted entity called *genba.*

Appendix: The Case of Postwar Japan—Capability-Building and Architectural Fit

By applying the capability-architecture framework to the case of Japanese industries in a dynamic way, this book will argue that postwar Japanese industries tended to possess a rich endowment of coordinative capabilities (e.g., teamwork of multi-skilled engineers/workers), mostly due to certain historical reasons, and that Japan's *coordination-rich* industries typically enjoyed design-based comparative advantage in *coordination-intensive* products (Fujimoto 2014). In other words, we assume that Japan's industrial innovations mainly developed in industries with relatively *integral* (i.e., coordination-intensive) architectures, including automobiles and functional chemicals, rather than those with modular (i.e., coordination-saving) architectures, such as digital products and software.

Using the above evolutionary framework of industry performance, let us look at the history of Japanese manufacturing industries (i.e., trade goods) after World War II. To briefly describe the *postwar history of Japan's genba*, we have divided this period into roughly 20-year spans (1950–1970, 1970–1990, 1990–2010, 2010–; see Table 1).

(i) Following a period of turmoil immediately after World War II, the beginning of the Cold War and Japan's strategic geographical position brought about opportunities for rapid economic growth at an unexpectedly early stage. In the 1950s

Table 1 A postwar history of genba (manufacturing sites) in Japan

Period (roughly)	Characteristics	Consequences in genba (manufacturing sites)
1945–1950	Beginning of the Cold War	Capability rebuilding of genba starts (QC, etc.)
1950–1970	High growth without immigrants	Emergence of genba coordination capabilities through "economy of scarcity" (e.g., Toyota system)
1970–1990	International competition during the Cold War (among advanced nations)	Productivity/quality improvements overcoming yen appreciation and oil crises, trade friction
1990–2010	Global competition after the Cold War (with low-wage emerging nations)	The Dark Ages of Japan's genba in trade goods, capability-building unable to overcome wage handicap
2010–prediction	Global competition with richer emerging nations	Japan's capability-building genba may have better chances of survival as wage gaps narrow

and 1960s, the "economy of scarcity" forced many Japanese factories and sites to develop coordination-rich manufacturing capabilities based on the teamwork of multi-skilled employees. This historical imperative subsequently brought about Japan's comparative advantage in coordination-intensive (i.e., integral architecture) goods, such as small cars and analog consumer appliances.

(ii) In the 1970s and 1980s, internal and international competition became tougher due to yen appreciation and slower economic growth, but many of Japan's manufacturing sites (monozukuri genba) accelerated their efforts in capability-building and productivity increases to overcome these handicaps. As a result, many Japanese manufacturing industries enjoyed competitive advantage and Japan's trade surplus expanded, which created trade friction and a boom in the Toyota production (lean production) system in and outside Japan. This was the era of international competition between advanced nations during the Cold War.

(iii) In the 1990s and 2000s, however, the competitive environment surrounding Japan's industrial sites changed drastically. As the Cold War ended, highly populated low-wage countries like China started to enter the global market as major exporters. China's typical wage rate in the 1990s was, roughly speaking, one-twentieth that of Japan, due partly to what A. Lewis might describe as "unlimited supplies of labor" from China's agricultural inland provinces to industrializing coastal areas (Lewis 1954). Thus, many of Japan's manufacturing sites in trade goods sectors found that their advantage in physical labor productivities (i.e., labor input coefficients) did not help them maintain their Ricardian comparative advantages in production costs. Besides, the wave of digital innovations since the mid-1990s created major market shifts from coordination-intensive analog products to coordination-saving (i.e., architecturally open-modular) digital products, where Japan's relatively coordination-rich sites could not maintain their design-based (i.e., architecture-based) comparative advantage (Fujimoto 2007, 2012b).

As a result of the abovementioned changes in global competitive environment and product technologies in the electronics industry, and continuing post-bubble recession and yen appreciation, Japan's manufacturing sites (genba) faced a "Dark Ages" period during most of the 1990s and 2000s. This was particularly true in the digital electronics sector, in which China, Korea, and Taiwan became major exporters, whereas some of Japan's most prominent factories were closed down despite severalfold increases in labor productivity. However, Japanese factories in the trade goods sector continued their capability-building efforts, further improved productivity, and many of them survived the "Dark Ages," particularly in coordination-intensive (i.e., integral architecture) products, including fuel-efficient cars, sophisticated industrial machinery, highly functional chemicals, steel, and so forth. Indeed, Japan maintained a trade surplus for much of this period.

Nonetheless, the competitive situation is now changing worldwide. As we enter the next 20 years (2010–), we can already observe significant changes in the shape of global competition, including a rapid increase in wage rates in China and other emerging countries (Thailand, Indonesia, India, and others). The wage-related handicap that Japan's trade goods sectors labored under for many years has diminished somewhat since the beginning of the 2010s. Accordingly, although no nation can escape country-level changes in industrial structure driven by dynamic comparative advantage, the chances that Japan's domestic factories persevering in their capability-building efforts can survive will increase in the coming years. In this sense, many of Japan's high-productivity manufacturing sites are gradually moving past their worst period, i.e., the end of the post-Cold War era.

Yet, this may be the time of darkness before the dawn, in which observers and decision makers may make major mistakes, as they overlook the abovementioned changes and assume an overly pessimistic view that the gloom of manufacturing industries in high-wage advanced nations will continue forever. Some managers in Japan's large enterprises have considered only short-term cost data, de-emphasized their domestic factories' productivity advantages, and ignored their potential for further productivity increases, erroneously closing down factories that could have survived with proper measures. Some media sources and scholars are also responsible for spreading unduly pessimistic views, predicting the hollowing out of Japan's manufacturing sector as a whole. These are erroneous interpretations that ignore both the genba's realities and the theoretical principles of comparative advantage. If all managers in Japan's trading sectors were to follow the advice of such commentators, the result might indeed be the hollowing out of the entire manufacturing sector as a self-fulfilling prophecy—the result of human error rather than of global competition.

This is why we need a solid theoretical-analytical framework for evaluating industrial performance and its potential in each sector or product. The author believes that this framework should be based on field observations and industrial data, as well as the logical integration of various disciplines, including classical trade theories, design-architecture theories in engineering, and evolutionary views of capability-building.

References

Akamatsu K (1962) A historical pattern of economic growth in developing countries. Dev Econ 1 (Supplement s1):3–25

Asanuma B (1989) Manufacturer-supplier relationships in Japan and the concept of relation specific skill. J Jpn Int Econ 3:1–30

Baldwin CY, Clark KB (2000) Design rules: the power of modularity. MIT Press, Cambridge, MA

Barnard CI (1938) The functions of the executive. Harvard University Press, Cambridge, MA

Brandenburger AM, Nalebuff BJ (1996) Co-opetition. Doubleday, New York

Clark KB, Fujimoto T (1991) Product development performance. Harvard Business School Press, Boston

Fine CH (1998) Clockspeed: winning industry control in the age of temporary advantage. Perseus Books, Reading

Fujimoto T (1999) The evolution of a manufacturing system at Toyota. Oxford University Press, New York

Fujimoto T (2007) Architecture-based comparative advantage: a design information view of manufacturing. Evol Inst Econ Rev 4:55–112

Fujimoto T (2012a) Evolution of firms and industries. Evol Inst Econ Rev 9(1):1–10

Fujimoto T (2012b) An economic analysis of architecture and coordination: applying Ricardian comparative advantage to design costs and locations. Evol Inst Econ Rev 9(1):51–124

Fujimoto T (2014) Innovation management in Japan. In: Dodgson M et al (eds) The Oxford handbook of innovation management. Oxford University Press, Oxford

Fujimoto T, Shiozawa Y (2011–2012) Inter and intra company competition in the age of global competition: a micro and macro interpretation of Ricardian trade theory. Evolu Inst Econ Rev 8 (1–2):193–231

Gawer A, Cusumano MA (2002) Platform leadership. Harvard Business School Press, Boston

Grant RM (2005) Contemporary strategy analysis, 5th edn. Blackwell, Oxford

Helpman E, Krugman P (1985) Market structure and foreign trade: increasing returns, imperfect competition and the international economy. MIT Press, Cambridge, MA

Hicks JR (1976) 'Revolutions' in economics. In: Latsis S (ed) Method and appraisal in economics. Cambridge University Press, Cambridge, pp 207–218

Iansiti M, Lavien R (2004) The keystone advantage. Harvard Business School Press, Boston

Katz ML, Shapiro C (1985) Network externalities, competition, and compatibility. Am Econ Rev 75(3):424–440

Langlois RN, Robertson PL (1992) Networks and innovation in a modular system: lessons from the microcomputer and stereo component industries. Res Policy 21:297–313

Lewis WA (1954) Economic development with unlimited supplies of labour. Manch Sch 22:139–191

Marshall A (1919) Industry and trade. Macmillan, London

Melitz MJ (2003) The impact of trade on intra-industry reallocations and aggregate industry productivity. Econometrica 7:695–725

Mintzberg H, Waters JA (1985) Of strategies, deliberate and emergent. Strateg Manag J 6 (3):257–272

Monden Y (1983) Toyota production system: practical approach to production management. Industrial Engineering and Management Press, Norcross

Nelson RR, Winter SG (1982) An evolutionary theory of economic change. Harvard University Press, Cambridge, MA

Ohno T (1978) Toyota seisan hoshiki (The Toyota production system). Daiyamondo-sha, Tokyo

Penrose ET (1959) The theory of the growth of the firm. John Wiley and Sons, New York

Penrose ET (1968) The theory of the growth of the firm. Basil Blackwell, Oxford

Ricardo D (1817) On the principles of political economy and taxation. John Murray, London

Schumpeter JA (1912/1934) The theory of economic development. Harvard University Press, Cambridge, MA

Shiozawa Y (2007) A new construction of Ricardian trade theory: a many-country, many-commodity case with intermediate goods and choice of production techniques. Evol Inst Econ Rev 3 (2):141–187

Shoenberger R (1982) Japanese manufacturing techniques: nine hidden lessons in simplicity. Free Press, New York

Simon HA (1969) The science of the artificial. MIT Press, Cambridge, MA

Smith A (1776) An inquiry into the nature and causes of the wealth of nations. In: Cannan E (ed) Methuen, London, 1904, in 2 vols

Sraffa P (1926) The laws of returns under competitive conditions. Econ J 36(144):535–550

Sraffa P (1960) Production of commodities by means of commodities: prelude to a critique of economic theory. Cambridge University Press, Cambridge

Suh NP (1990) The principles of design. Oxford University Press, New York

Thomke S, Fujimoto T (2000) The effect of "front-loading" problem-solving on product development performance. J Prod Innov Manag 17:128–142

Ulrich KT (1995) The role of product architecture in the manufacturing firm. Res Policy 24 (3):419–440

Vernon R (1966) International investment and international trade in the product cycle. Q J Econ 80 (2):190–207

Womack J et al (1990) The machine that changed the world. Rawson Associates, New York

The Nature of International Competition Among Firms

Yoshinori Shiozawa and Takahiro Fujimoto

Abstract International competition among firms and units of a multinational enterprise is a game with wage rates as handicaps. This simple truth is commonly known among managers of firms competing in the international arena. Although it is instinctively trivial and evident, this image of international competition is quite different from the traditional comparative advantage argument. Supporting the new view requires a new theory. Sections 2, 3, 4, 5, and 6 are an introduction to the new theory of international values. The second part of the chapter (Sects. 7, 8, and 9) offers some preliminary remarks on how to use it. The third part (Sects. 10, 11, and 12) is a demonstration of how the new theory can be used to analyze international competition among firms. The last part (Sects. 13 and 14) illustrates how the new theory can be applied to a more realistic situation and provides some remarks regarding the wider scope of application of the new theory.

1 Introduction

International competition among firms is a game with wage rates as handicaps (Fujimoto 2011; Fujimoto and Shiozawa 2011–2012 Sect. 6). This proposition is true also among production sites (or production units, like plants and factories) of a multinational enterprise located in different countries. The purpose of this analysis is to show that the above proposition can be justified theoretically.

This simple truth is commonly known among managers of firms that are competing in the international arena. Japanese firms have had to play this game with Chinese firms with a big handicap. The game started with China's reform and opening-up (1978), but a real impact began to be felt in the 1990s. In the first

Y. Shiozawa (✉)
Osaka City University, Osaka, Japan
e-mail: y@shiozawa.net

T. Fujimoto
Graduate School of Economics, The University of Tokyo, Bunkyo-Ku, Tokyo, Japan

T. Fujimoto, F. Ikuine (eds.), *Industrial Competitiveness and Design Evolution*,
Evolutionary Economics and Social Complexity Science 12,
https://doi.org/10.1007/978-4-431-55145-4_2

phase of this competition, Japanese wage rates were at least 20 times higher than those of Chinese urban workers, but the handicap became less severe after two decades of hardship for the Japan-based firms and plants of multi-nationalized Japanese firms, mainly thanks to the considerable rise in wage rates in China. Now the Japan/China wage rate ratio among urban workers is around 5 to 1.

In Fujimoto and Shiozawa (2011–2012), we argued extensively how this simple point of view, symbolized by the handicap proposition, can explain the broader issue of international competition among firms and among units of a multinational enterprise. Indeed, one major difference between domestic and international competition lies in the fact that, in the latter case, firms are affected by serious handicaps that are impossible to observe within a country. Of course, international competition is not limited to this kind of competition, as it also occurs among firms of advanced countries with similar wage levels. Although competition is fierce, there are no major differences in the features of international and domestic competition. Wage rate handicap is a proper aspect of international competition.

We can easily observe considerable differences in wage rates between developed and underdeveloped countries. It may sound strange to common people, but for a long time, wage rate differences among countries were not a typical theme of international trade theory. There were two main reasons which may explain this situation. One was the absence of real competition within the same industry between developed and underdeveloped countries. The emergence of China after the reform and opening-up, for example, lent a new feature to international competition. A country with low-wage rates and appreciable levels of technology suddenly appeared on the scene of the free world—a rare, if not unique, experience for the world economy. The other reason was the theoretical structure of the trade theory. A big wage difference was excluded by the assumption from the most popular and standard theory of international trade, i.e., the Heckscher-Ohlin-Samuelson (HOS) model. One of the model's core theorems is the factor price equalization theorem. In this model, wage rate is the price of a factor called labor. When some of its conditions are satisfied (i.e., if factor proportions lie in a cone of equalization), the factor price equalization theorem holds and the wage rates become the same for all countries. The multi-cone situation is rarely studied due to greater difficulties in its analysis.[1] The new generations of trade theories, including the new and new-new trade theory, did not bring wage differences into focus. An exception was Dorbusch et al. (1977), which set the relative wage rate as a main variable for adjustment.[2] In these circumstances, the situation of large wage rate differences was not often studied in trade theory.

---

[1]A greater problem of the HOS model is that it assumes the same technology for all countries. This excludes firm-level competition, as its main aspect is competition in technology for better products and production techniques.

[2]Another important and much more radical exception is Pasinetti (1993 Chap. 9). This paper focuses on wage rate differences at the very basis of international economic relations. Its exposition is complementary to the new theory of international values, since it offers a good illustration of why wage rate differences are crucial in the examination of international trade.

While the economic situation in the real world has changed tremendously, the theoretical situation has also changed substantially since the beginning of the twenty-first century. A new interpretation of Ricardo's trade theory has become influential (Faccarello 2017; Tabuchi 2017a, b; see Sect. 2 of this chapter for a more detailed account). A new theory in the Ricardian tradition started to be developed in 2007 (Shiozawa 2007, 2014, 2017a). At the time when Fujimoto and Shiozawa (2011–2012) was written, this theory was not well developed and many points had been left theoretically unsolved. Shiozawa wrote a new book in 2014 and the overall features of the new theory became clearer (Shiozawa 2014). A first English introduction was published in 2017 (Shiozawa 2017a). Even after Shiozawa (2014), there was a big theoretical advancement inspired by the discussions had in the Workshop on the Theory of International Values. With the new theorems, Theorem 5-1 and Theorem 6-1, it became much easier to talk about the circumstances in which international value remains invariable in the situation that is out of full employment. The present analysis is the first attempt to explain international competition among firms and among production units (plants) of a multinational enterprise based on the new theory of international values. Readers of this study, we believe, will admit that it is much easier to examine and discuss international competition based on the new theory.

Competition among countries with big wage rate differences is of special significance for firms in both low- and high-wage countries. As for low-wage countries, the big discrepancy in wage rates provides a chance to overcome various disadvantages in technology, human skills, and infrastructures. In the third quarter of the twentieth century, the dependency theory argued that the terms of trade are unfavorable for developing countries. In the twenty-first century, after the East Asian Miracle and the resurgence of the BRICs, a diametrically opposite proposition has been argued. Low-wage rates might be a strong weapon in competing with firms in developed countries, and learning by doing (Arrow 1962) may help low-wage countries to accumulate the necessary experience for production. A new strategy for economic development may be designed with this new vision. However, this chapter does not treat this aspect of international competition. We concentrate on the actions and reactions of firms and production units in countries with high-wage rates when they face fierce competition with firms/units in countries with low-wage rates. In this sense, the present analysis is a summary note of our long paper Fujimoto-Shiozawa (2011–2012). The main difference between the two studies is that we show the theoretical basis more systematically in the present study.

Alongside our main explanation, we mentioned that the traditional image is far from what Ricardo thought and hinted at why this misleading image may have emerged. The new theory of international values is a part of the classical theory of values and may foreshadow its revival.

The remainder of this chapter is organized as follows. The first part, Sects. 2, 3, 4, 5, and 6, is an introduction to the new theory of international values and focuses mainly on the new result explained above, which does not appear in Shiozawa

(2017a). In this part, we explain step by step how the new theory of international values is constructed, starting from the classical four magic numbers and moving on to the pure labor input economy and then to the general case where input goods are freely traded. The second part, Sects. 7, 8, and 9, provides some preliminary remarks for the use of the new theory. The third part, Sects. 10, 11, and 12, is a demonstration of how the new theory can be used to analyze international competition among firms. Section 11 is a recapitulation of our results from a historical perspective, while the other two sections examine new aspects of international competition, including its economic effects. The last part, Sects. 13 and 14, illustrates how the new theory can be applied to a more realistic situation. Lastly, Sect. 15 concludes the chapter. A special warning should be made here. The matrix A has different meanings in different sections. Given the limited space, this was inevitable to make references to older works accessible to readers. They are requested to pay special attention in what meaning matrix A is used.

2 A Brief Account of Traditional Trade Theories

Almost all textbooks on international trade contain discussions of Ricardo's four magic numbers (Samuelson's denomination). For the convenience of later arguments, let us start from Table 1.[3]

These four numbers represent labor requirements to produce certain quantities of goods in countries J and C. In Ricardo's original text (Ricardo 1951, p 134–135), cloth and wine are exchanged between England and Portugal. Specifically, a certain amount of cloth is exchanged for another amount of wine of the same value. The four numbers correspond to the labor requirement for these quantities of goods (Ruffin 2002; Maneschi 2004; Tabuchi 2017b). Many textbooks today explain that these four numbers represent labor input coefficients, i.e., labor per unit for each good, with no reference to exchange ratio. A different interpretation requires a different type of explanation logic with regard to gains from trade and patterns of specialization (or patterns of trade). Chipman (1965) made a typical mistake by confusing the differences in the meanings of the four numbers.[4]

Another common misunderstanding is that Ricardo assumed a pure labor input economy. This is wrong because he understood labor as the total amount of labor

Table 1 Trade economy of two countries and two goods

	Good 1	Good 2
Country J	a_{J1}	a_{J2}
Country C	a_{C1}	a_{C2}

[3]Following the convention of Fujimoto-Shiozawa (2011–12), we attribute the symbols J and C to high-wage countries and low-wage countries, respectively. Readers can draw a parallel to the symbols A and U attributed to advanced and underdeveloped countries in Pasinetti (1993 Chap. 9).

[4]Chipman considered Ricardo's explanation a "non sequitur" (Chipman 1965, p. 479).

directly and indirectly necessary for the production of a product. In a modern context, we should think of Ricardo's labor as representing the full cost of production. Thus, the standard textbook explanation that Ricardo's model assumes labor as the only factor of production is highly misleading. It can well include inputs of parts and materials. In other words, we can regard those numbers as integrated coefficients (see McKenzie 1953; Jones 1961; Shiozawa 2017a Sect. 2). Of course, there have been limits to the theory in the Ricardian tradition, among which the greatest was that the traditional theory could not deal with input trade. Although McKenzie (1953) emphasized the importance of input trade (or "trade in intermediate goods" in his expression[5]), calls for an extension of the theory to incorporate input trade went unheard for a long time. As we observe in today's world economy, global supply chains are changing the patterns of specialization among nations, and it becomes ever more important to build a theory by which we can deal with input trade. However, difficulties are overcome at last thanks to the new theory of international values, which is an extension of Ricardo's theory of international trade.

The third misunderstanding, which is more serious, is this: The traditional trade theory that is ordinarily called Ricardian theory is in fact a descendant of John Stuart Mill's theory and not Ricardo's own theory. This misunderstanding is important for the present chapter, because the new theory of international values is constructed, as indicated above, on the extension of Ricardo's original theory and not in the vein of John Stuart Mill (Faccarello 2017; Tabuchi 2017a). The first person to protest about this misunderstanding was Graham (1923). With no mention of Ricardo, Graham opposed Mill's problem settings and his solution, which is commonly known as the theory of reciprocal demand.[6] As we shall see, Graham's idea plays an important role in Sect. 8. Unfortunately, he called the theory by Mill and his followers *classical theory*, and this became another source of confusion, because Mill's theory was in fact the origin of neoclassical economics (Shiozawa 2017b).

The main aim of this chapter is to prove, by means of the new theory of international values, that international competition among firms is a game with wage rates as handicaps. Sadly, the hurdles that we face are a little too big to deal with the general result directly. We must take a short detour in order to approach it more easily. So, let us return to the four magic numbers. To determine the pattern of specialization, it does not matter which of the two interpretations we follow.[7] For convenience of later arguments, we assume that the four numbers represent (conventional) coefficients per units of goods.

[5]Jones (1961) claims that he explored the case in which intermediate goods are freely traded, but his theory was restricted to RII in my classification (Shiozawa 2017a Sect. 2). Jones and Kierzkowski (1990, 2001) and Jones (2000) are about a theory of fragmentation but not a general theory of input trade.

[6]In later works, Graham makes a clear distinction between Ricardo's classical theory of international trade and the theory developed by John Stuart Mill and his successors (Graham 1932, p. 616).

[7]It does matter in the arguments of gains from trade. See Chipman's mistake, mentioned above.

The textbook explanation takes two ratios of the four numbers and then determines which one is bigger than the other. In taking ratios, two options are possible: (1) to divide a coefficient by another from the same line, i.e., interindustrial ratio in a country, and (2) to divide a coefficient by another from the same column, i.e., intraindustrial ratio between countries. As we explained in Fujimoto and Shiozawa (2011–2012), the two options give the same result if we draw a good comparison, because the following two conditions are mathematically equivalent:

$$a_{J1}/a_{J2} < a_{C1}/a_{C2} \Leftrightarrow a_{J1}/a_{C1} < a_{J2}/a_{C2}. \qquad (2-1)$$

However, it is sometimes argued that the interindustrial ratio comparison is the only right method to find the pattern of specialization.

The comparison of the two ratios teaches us about the pattern of specialization, but we contend that observations based on these ratios lack a crucial point, i.e., the proportion of the wage rates of the two countries. Wage rate ratios are hidden in the background. This might be another reason why the simple proposition that international competition is a game with wage rates as handicaps was not noticed much earlier.

The first condition of (2-1) is in reality a comparison between the price ratios of two countries if profit rate r, or more exactly the markup rate, is equal in the two countries.[8] In fact, for the product of each country, we have

$$p_{J1} = (1+r)w_J\, a_{J1},\ p_{J2} = (1+r)w_J\, a_{J2},\ p_{C1} = (1+r)w_C\, a_{C1},\ p_{C2} = (1+r)w_C\, a_{C2},$$

where p_{J1}, p_{J2}, p_{C1}, and p_{C2} indicate prices and w_J and w_C the wage rates for the two countries. This implies

$$p_{J1}/p_{J2} = a_{J1}/a_{J2} \text{ and } p_{C1}/p_{C2} = a_{C1}/a_{C2}. \qquad (2-2)$$

Then, the first condition of (2-1) signifies that the ratio of labor inputs for good 1 with respect to good 2 in country J is smaller than the same ratio in country C. Intuitive logic may imply that country J exports a relatively cheaper good and imports the other good. In this logic, the wage rate ratio is irrelevant to how specialization occurs. We can determine the specialization pattern with no reference to the wage rate ratio.

This has been a typical explanation of how two countries specialize in which goods. As long as we examine the pattern from a macroscopic point of view, this logic is perfect. Yet, it is not enough when we want to observe how international

[8]We use the symbol r in order to ensure continuity with the older discussions. This symbol represents a markup, as clarified in Sect. 6. The assumption of the same markups can be easily generalized to the case where each industry of each country has its own markup rate. See Shiozawa (2017a).

competition unfolds among firms and even among units of production of a multinational enterprise (MNE). In other words, if we observe international competition at the microscopic level, the wage rate ratio becomes one of the most important variables that function as determining factors of competition. In fact, the wage rate ratio turns out to be the key variable that dictates competitiveness and specialization patterns.

As for the microscopic analysis, the second condition of (2-1) is useful. In fact, it is clear that a ratio w_J / w_C exists such that

$$a_{J1}/a_{C1} < w_C/w_J < a_{J2}/a_{C2}. \qquad (2-3)$$

If w_J and w_C are wage rates expressed by a common international currency for countries J and C, (2-3) means that

$$\begin{aligned} p_{J1} &= (1+r)w_J\, a_{J1} < (1+r)w_C\, a_{C1} = p_{C1}. \\ p_{J2} &= (1+r)w_J\, a_{J2} > (1+r)w_C\, a_{C2} = p_{C2}. \end{aligned} \qquad (2-4)$$

The meaning of inequalities (2-4) is simple and direct. Country J exports good 1 to country C and imports good 2 from country C, because the cost of production (including profit) of good 1 in J is lower than that in C and the cost of production of good 2 in C is lower than that in J. In expressing this in form (2-4), we obtain estimations that are useful for firms producing each good in each country.

Although these relations in money terms are not usually mentioned, Ricardo's four magic numbers do indeed imply this comparison in terms of value. Faccarello (2017) and Tabuchi (2017a) emphasize that Ricardo's chapter on foreign trade is in fact constructed in money terms and should be interpreted as such. As Tabuchi (2017a p 268) put it, summarizing Faccarello, "there are no specific international prices, and every exchange is a monetary exchange, micro-agents act in their self-interest, and prices they pay tend to be natural prices." Nevertheless, the international trade theory was analyzed in "real terms" until the 1930s, when Ohlin's book (1933) first appeared (Tabuchi 2017a Sects. 3 and 4).

The first question that this change in viewpoint raises is: Is this viewpoint extendable to more general cases? The answer is quite simple. It is extendable to practically all plausible situations, whereas the old approach based on a comparison of physical coefficients is not. We will show how this can be done in the following two sections. As the logic of the general case is not very visible, in Sect. 3 we will illustrate the basic logic through a simple example with no input trade. Although it requires a little mathematics, we will present more general results in Sect. 4, which can be skipped if one is mainly interested in international competition at the firm and unit level. All the information needed for the subsequent discussion is summarized at the beginning of Sect. 5, which, along with the subsequent sections, will show how these results can be conveniently used for firm- and unit-level analyses of competition.

3 How Relative Wage Rates Are Determined (Elementary Discussion)

As a simple situation, let us examine a two-country, three-commodity case. Here, complete specialization is impossible[9]. If each country produces only one good, there is no country that produces the third good. Then, there are two possibilities: (1) the case where there is no common good produced by two countries and (2) the case where both countries produce a good in common.[10]

Let a_{ij} be the input coefficient for producing good j in country i. In case (1), the relative wage rates can move within a certain range. For example, assume that country 1 produces goods 1 and 2, whereas country 2 produces good 3. In order for this case to be possible, the following constraints must be satisfied:

$$w_1 a_{11} < w_2 a_{21}, \quad w_1 a_{12} < w_2 a_{22}, \quad w_2 a_{23} < w_1 a_{13}. \qquad (3-1)$$

From this section, we omit the coefficients $1+r$ that appear in both sides of the inuqualities. (3-1) means that w_2 / w_1 satisfies inequalities

$$a_{11}/a_{21}, \quad a_{12}/a_{22} < w_2/w_1 < a_{13}/a_{23}. \qquad (3-2)$$

There exists a real number w_2 / w_1 that satisfies (3-2) if the coefficients satisfy the conditions:

$$a_{11}/a_{21}, \quad a_{13}/a_{23} < a_{12}/a_{22}. \qquad (3-3)$$

For any given set of coefficients a_{ij}, relations like (3-3) hold when we renumber goods appropriately for each country. In fact, for any given set of a_{ij}, the three ratios

$$a_{11}/a_{21}, \quad a_{12}/a_{22} \text{ and } a_{13}/a_{23} \qquad (3-4)$$

can be arranged in an increasing order. Suppose that all ratios of (3-4) are different from one another.[11] If the order is, for example, such that

$$a_{11}/a_{21} < a_{12}/a_{22} < a_{13}/a_{23}, \qquad (3-5)$$

we have two possibilities: (1a) when $a_{11} / a_{21} < w_2 / w_1 < a_{13} / a_{23} < a_{12} / a_{22}$, country 1 produces good 1 and country 2 produces goods 2 and 3 competitively, and

[9]The term "complete specialization" is in fact ambiguous. It may signify two different situations: [A] Each product is produced only within a country. [B] There are no products that are produced in more than two countries. In this chapter we use the term complete specialization only in the meaning [A].

[10]There are other possibilities, such as the case of two countries producing all three goods simultaneously. This case should be excluded from the general situation, because it is possible only when the input coefficients satisfy special constraints.

[11]We exclude the case where (3-5) contains equality, as it is not the generative case.

(1b) when $a_{11} / a_{21} < a_{13} / a_{23} < w_2 / w_1 < a_{12} / a_{21}$, country 1 produces goods 1 and 3 and country 2 produces good 2. In both cases, one country produces two goods and the other country produces the third good.

The cases in which w_2 / w_1 coincides with one of terms of (3-5) are excluded from case (1), because they produce case (2). This is easily seen when we reason from the opposite side. Suppose that country 1 produces goods 1 and 3 and country 2 produces good 2 and 3. In a competitive condition, each country produces only those goods which have no greater cost than those of the other country. Then, we have inequalities:

$$\begin{array}{ll} w_1 a_{11} < w_2 a_{21}, & w_1 a_{13} \leqq w_2 a_{23}, \\ w_2 a_{22} < w_1 a_{12}, & w_2 a_{23} \leqq w_1 a_{13}. \end{array} \qquad (3-6)$$

In (3-6) we cannot take strong inequality < for left inequalities. If we arrange (3-6), it means that

$$a_{11}/a_{21},\ a_{12}/a_{22} < a_{13}/a_{23} = w_2/w_1. \qquad (3-7)$$

We can obtain these inequalities for a general set of a_{ij} when we renumber goods appropriately. Indeed, just like in case (1), we can arrange the three ratios (3-4) in an increasing order. Suppose that we obtain inequalities (3-5). Then, three cases are possible: (2a) $a_{11} / a_{21} = w_2 / w_1 < a_{13} / a_{23} < a_{12} / a_{22}$, (2b) $a_{11} / a_{21} < a_{13} / a_{23} = w_2 / w_1 < a_{12} / a_{22}$, and (2c) $a_{11} / a_{21} < a_{13} / a_{23} < a_{12} / a_{22} = w_2 / w_1$. These cases correspond, respectively, to the case when (2a) country 1 produces good 1 and country 2 produces all three goods, (2b) country 1 produces goods 1 and 3 and country 2 produces goods 2 and 3, and (2c) country 1 produces all three goods and country 2 produces good 2. The cases in which w_2 / w_1 is smaller than a_{11} / a_{21} or greater than a_{13} / a_{23} are excluded, because they produce no relveant situation.

The above tedious examination can be easily illustrated by Fig. 1.[12] This represents the set of production possibilities of a two-country, three-good trade economy. In fact, each case (1a), (1b), (2a), (2b), and (2c) corresponds, respectively, to an economy in which the world final demand lies on Ridge 1 or Ridge 2 or in Domain 1, Domain 2, or Domain 3. Details of these correspondences are omitted here. A more detailed explantions is given in Sect. 5 after Definition 5.2.

When we observe Fig. 1, we recognize three features: (i) there is no internal vertex (a vertex in the interior of positive orthant), (ii) the ridges correspond to cases (1) when there is no commonly produced good, and (iii) most of the points on the frontier lie somewhere in a two-dimensional facet, which corresponds to cases (2). There is a commonly produced good and wage rates and prices remain constant as long as demand stays in the same facet.

In a more general case, i.e., with a greater number of countries and commodities, the situation does not change much. The set of maximal points of a production possibility set is called *frontier* of the production possibility set. When the number of

[12]Figure 2 in Shiozawa (2017b, c)

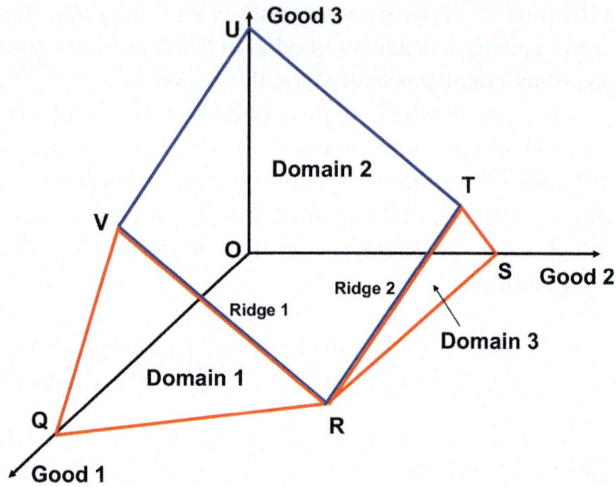

Fig. 1 Set of production possibilities for a two-country, three-good trade economy

commodities N is greater than the number of countries M, there is no internal vertex. In the majority of instances, the world final demand, if it is on the frontier, lies in a *facet*, i.e., in a face of codimension 1,[13] and wage rates and prices remain constant as long as demand remains within the same facet. When the world final demand lies on a ridge (face of codimension 2) or smaller dimensional face, relative wage rates have a degree of freedom that is equal to the codimension minus one. As almost all points on the frontier lie in a facet (face of codimension 1), except the points of a set of measure 0, we can safely say that wage rates and prices are uniquely determined (up to scalar multiplication) in a regular situation.

This presents a very different image from what John Stuart Mill and many of his followers, including Ronald Jones, believed to be a typical situation in international trade. It was the situation of complete specialization. Indeed, Ronald Jones wanted to find a pattern of complete specialization (Jones 1961).[14] Such a state is possible only on the internal vertex of the frontier. Internal vertices can exist only when $M \geqq N$. However, when $M < N$, there is no such point. In fact, since in the case of complete specialization each country produces only one product, the countries as a whole can produce at most M different products and it is impossible to produce all products.

Mill and Jones examined the complete specialization case and considered how prices are determined. When $M < N$, which is more realistic than the $M \geqq N$ case, this kind of question is invalid.[15] We should change our view of how the world economy in international trade works. Prices cannot be a main regulator of what happens in the economy. In a typical case, prices remain constant and quantities are adjusted

[13]Codimension of a face is the dimension of the space minus the dimension of the face. It can also be defined as the dimension of the normal cone to the face.

[14]Jones (1961) found a necessary and sufficient condition for the existence of complete specialization (see Shiozawa 2017a p. 13).

[15]There are about 200 countries in the world, but the number of commodities easily exceeds 10 million.

according to demand. When Mill examined the two-country, two-commodity case, the complete specialization point was the only one where the two countries could achieve gains from trade.[16] In the two-country, three-commodity case, we have a large parallelogram RTUV or Domain 2, in which two countries both enjoy gains from trade. We should think of this as a typical situation of international trade.

Suppose that the world final demand is situated somewhere near the center of the parallelogram. Country 1 produces goods 1 and 3 and country 2 produces goods 2 and 3. Wage rates and prices are fixed as long as demand stays inside the parallelogram. World demand can move within it and quantities are adjusted not through price movements but through changes in production scales. When examining complete specialization, Mill believed it necessary to abandon the cost-of-production theory of value and return to the law of demand and supply, which he deemed more fundamental and anterior to it. He thus paved the way for the shift of economics toward neoclassical economics (Shiozawa 2017b). However, the situation observed in the two-country, three-commodity trade economy is more similar to the classical theory of value (Shiozawa 2016). In fact, except for the fact that there are several possible values, the economic process taking place is almost identical to the closed economy case as long as the world demand remains in the same facet. Indeed, the latter can be seen as a special case of a world economy consisting of only one country.

Of course, there are some conspicuous differences between a closed economy and a world economy with many countries. The essential point in a trade economy is that we cannot treat workforces in different countries as the same labor. We have to treat them as different kinds of labor. We may say that there are as many factors of production as the number of countries.[17] The minimal price theory (or non-substitution theorem, after Paul Samuelson) does not hold because of this. But this does not imply that the Ricardian trade theory should be formulated as a neoclassical value theory. Once the wage rates of all countries are determined, there is no essential difference between a closed economy and a trade economy. As Faccarello (2017) and Tabuchi (2017b) claim, this must be the vision that Ricardo had when he wrote the *Principles* (Ricardo 1951[1817–21]). There was only one problem that Ricardo could not solve, i.e., to determine the relative wage rates of different countries. We contend in this chapter that, given the set of production techniques for each country and a competitive pattern, we can determine the relative wage rates in a proper way.[18]

[16]This situation inevitably led Mill to consider international trade as a problem of "pure" exchange economy (Shiozawa 2017b).

[17]In opposition to neoclassical economics, which considers capital a factor of production, we do not regard capital goods as the same factors of production as labor, because capital goods are products from production activities, whereas labor power is not produced inside the capitalist production system. At least, capital goods are not primary factors. It may be better to say that there are as many *primary* factors as the number of countries.

[18]This is the new result which does not appear in Shiozawa (2017a).

4 A Short Detour into Elementary Graph Theory

To attain a good characterization of competitive patterns, it is necessary to use some terms pertaining to graph theory. This requires elementary knowledge of graph theory and no difficult mathematics is involved. We simply have to procure some words to correctly and generally express the arguments of the previous section. If readers have a small amount of patience, they can easily read through this section. If they already have some basic notions, they can skip it and all they need to know is this: If sets of production techniques are given for all countries and a competitive pattern with certain properties is specified, a set of wage rates is uniquely determined (up to scalar multiplication).

Graph theory is a field of mathematics. *A graph* is a set of vertices and edges, an example of which is given in Fig. 2. An edge connects two vertices and the graphical expression only indicates these connecting relations. In the trade theory context, each edge represents a production technique. When an edge connects a country with a good, it represents a production technique which belongs to the country and produces the good. Hence, the places of the vertices are not important. In the same vein, edges that connect vertices can be line segments or curves. If there are more than two edges connecting the same two vertices, curves must inevitably be used in order to make them visible. If it is clear that an edge connects specific vertices, exactness is not required. More formally, a graph G includes a set of vertices V and a set of edges E. Each edge should be assigned two vertices (including the case when two edges coincide). Normally, we assume V and E to be finite.

A world economy is composed of countries and goods. Let V_1 be the set of countries and V_2 be the set of goods and V be the union of V_1 and V_2. A production technique is a process that produces a good j in a country h. For each production technique, we can assign an edge that connects an element of V_1 and an element of V_2. Therefore, a production technique can be expressed as an edge that connects V_1 and V_2. If $\sum$ is the set of all production techniques that is known and realizable, this state of technology can be represented by a graph $G = (V_1 \cup V_2, \sum)$, which we call a *technology graph*.

In representing a state of technology, the set of vertices is always composed of two disjoint parts V_1 and V_2. If a set of vertices is composed of two disjoint sets V_1

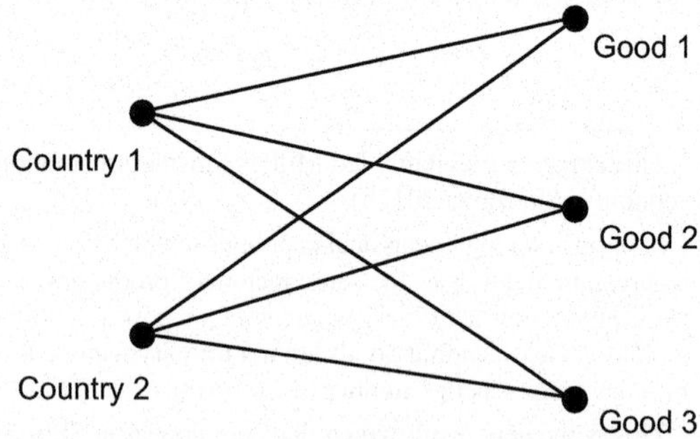

Fig. 2 Complete (2,3)-bipartite graph

Fig. 3 A spanning tree. This indicates how wage rates and prices are related

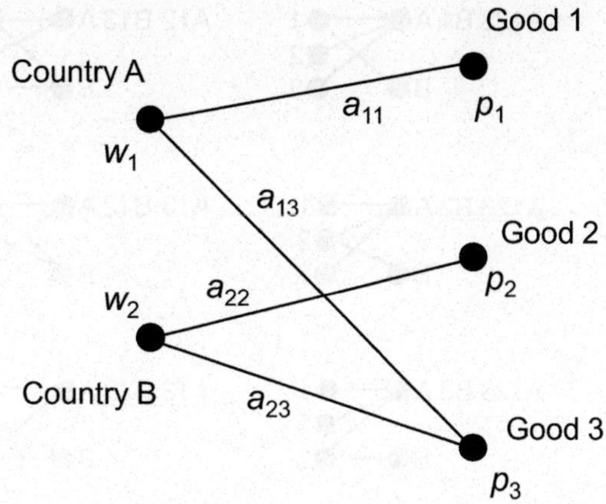

and V_2 and the edges connect an element of V_1 and an element of V_2, such a graph is called a *bipartite graph.*[19] If there exists one and only one production technique for each country/good pair, all the vertices of V_1 will be connected to one of the vertices of V_2 and vice versa. Such a bipartite graph is called a complete bipartite graph. Figure 3 is the complete bipartite graph for a two-country, three-good economy. If each country possesses one and only one production technique for each good, the technology graph is a complete bipartite graph.

A graph is said to be *connected* if any pair of vertices can be connected by a chain of edges. A vertex which is connected by only one edge is called a *leaf.* A graph is called *spanning* if all vertices are connected at least by one edge. It is called a *tree* if it is connected and contains no cycles. A *cycle* here means a chain of edges which starts from a vertex and returns to the same vertex without passing along the same edge. For a characterization of regular international values, we need a concept of spanning tree which is a subgraph of the technology graph.

Figure 3 is an example of a spanning tree for a two-country, three-good economy, for which there are 12 different spanning trees in total. Figure 4 presents all such spanning trees. For the sake of simplicity, a *character string* is used to express a subgraph of the complete graph. For example, string A123B1 refers to the graph in which edge A is connected to edges 1, 2, and 3 and edge B is connected to edge 1.

In Sect. 3, we observed five possible cases (1a), (1b), (2a), (2b), and (2c) for a two-country, three-commodity trade economy. Each case corresponds to faces of the maximal boundary (or frontier) of the production possibility set. A point in the relative interior of Ridge 1, which corresponds to (1a), has three competitive production techniques: country 1 produces good 1 and country 2 produces goods 2 and 3. If we write A and B instead of country 1 and 2, respectively, this gives

[19] Bipartite graphs are used implicitly and explicitly in product architecture theory in manufacturing design (Fujimoto 2007, 2017, Tatsumoto 2018). Note also that any bipartite graph can be represented by a matrix with 0–1 entries and vice versa.

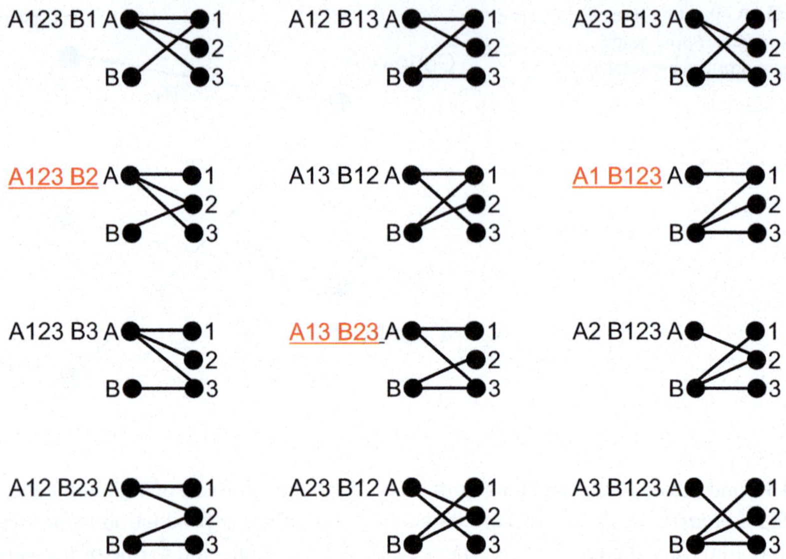

Fig. 4 List of all (2,3) spanning trees

pattern A1B23. Ridge 2, which corresponds to case (1b), has also three competitive production techniques, and the pattern of specialization can be written as A13B2. Graphs of these cases are not spanning trees, because they are not connected. In the same way, Domain 1 has string A1B123, Domain 2 A13B23, and Domain 3 A123B2, respectively. All three domains have four competitive production techniques and the corresponding graphs are all connected. Thus they are all spanning trees. This is all we need to know about graph theory.

5 Elementary Discussion Continued

It is convenient to treat wage rates and prices of goods as a set. We put $\mathbf{v} = (w_1, \ldots, w_M, p_1, \ldots, p_N)$ and call it an international value. When two values are proportional, we consider them to be different expressions of the same value. If we use this convention, we can simply say that the international value is determined uniquely, instead of saying that it is determined uniquely up to scalar multiplication.

Suppose that we have a Ricardian pure labor input economy (R0 economy in Shiozawa 2017a).[20] Production techniques are all of the pure labor input type. Thus, they are represented by an a_{ij} coefficient. Suppose that we are given a set T of production techniques that form the spanning tree A13B23 and all production

[20]In Pasinetti's terminology, *pure labor* (*production*) *economy* (Pasinetti 1993)

techniques of T are competitive.[21] Here, *competitive* means that the production cost (in the full cost meaning) of the technique is equal to the price of the product. In other words, if production technique A2 is competitive, it means that $w_1 a_{12} = p_2$. This equality is called *cost-price equality* or *value equation*. Note that in the expression above a_{ij} denotes the modified input coefficient $(1+ m_{ij})\, a^{\#}{}_{ij}$ in order to simplify the expressions, where m_{ij} is the markup rate for industry j in country i and $a^{\#}{}_{ij}$ is the physical input coefficient. See Shiozawa (2017a), Sect. 5.

Figure 3 expresses how wage rates and prices are related to each other when all production techniques of spanning tree A13 B23 are competitive. We can start from any vertex, say country A, and choose a positive wage rate w_1 arbitrarily. Vertex A is connected to vertices 1 and 3. As production technique A1 is competitive, we have $p_1 = w_1\, a_{11}$. In the same vein, as A3 is competitive, we have $p_3 = w_1\, a_{13}$. Thus, prices p_1 and p_3 are determined. Vertex 3 is connected to country B. If production technique a_{23} is competitive, then we have $w_2\, a_{23} = p_3$. Thus, wage rate w_2 is determined. As vertex B is connected to vertex 2, we have $w_2\, a_{22} = p_2$. In this way, if we choose a positive w_1 arbitrarily, we can determine all other wage rates and prices by tracking the paths starting from vertex A. If competitive production techniques form a spanning tree, all vertices are connected via a unique path from the starting vertex. By tracing the path and determining the values of all the vertices step by step, we can determine the value of the chosen vertex. This procedure does not produce any contradiction, since there are no cycles in the tree. Thus, we have Theorem 5-1.

Theorem 5-1 (Spanning Tree Determines an International Value, R0 Case)
Let A be a labor input coefficient matrix of a Ricardian pure labor input economy (i.e., R0 economy) with M countries and N goods. If competitive production techniques form a spanning tree, then there is a unique international value (up to scalar multiplication) that satisfies cost-price equalities.[22] □

An international value thus determined by a spanning tree T is called an international value *defined by*, or *associated to*, spanning tree T. On the other hand, T is called *supporting* spanning tree of the value. Let us reexamine Fig. 1. This represents a typical production possibility set for a two-country, three-good economy. It has three facets: two triangles QRV and RST and a parallelogram RTUV. They represent cases (2a), (2b), and (2c). Using the string expression, they correspond to A1B123, A13B23, and A123B2, respectively. Stringe codes of these three spanning trees are colored in red and underlined in Fig. 4.

What happens when we choose another spanning tree, say A123B1, as our set of competitive techniques? Theorem 3-1 tells us that there is an international value. What is the difference between the international values of the above three spanning trees and the international value $\mathbf{v}$(A123B1), which is defined by spanning tree A123B1? Remember that we obtain three spanning trees on the assumption that

[21]For character string expressions such as *A*13*B*23, see the explanation at the end of Sect. 4.

[22]This theorem can be easily extended to the RI and RII cases in Shiozawa (2017a).

inequalities (3-5) hold. Converting the first inequality, we have $a_{11} / a_{13} < a_{21} / a_{23}$ or equivalently $a_{23} < a_{21} \cdot a_{13}/a_{11}$. Then

$$w_2 a_{23} < w_2 a_{21} \cdot w_1 a_{13} / w_1 a_{11} = p_1 p_3 / p_1 = p_3.$$

To derive the second equation, we used the fact that production techniques B1, A3, and A1 are competitive. In the same way, we have

$$w_2 a_{22} < w_2 a_{21} \cdot w_1 a_{12} / w_1 a_{11} = p_1 \cdot p_2 / p_1 = p_2.$$

Thus, in the case of international value **v**(A123B1), we have

$$p_2 > w_2 a_{22} \text{ and } p_3 > w_2 a_{23}.$$

This means that production techniques B2 and B3 have positive excess returns when **v**(A123B1) holds. Let us see another spanning tree, say A13B13. When inequalities (3-5) hold, we have

$$w_1 a_{12} > w_1 a_{13} w_2 a_{22} / w_2 a_{23} = p_3 p_2 / p_3 = p_2.$$

Thus, production technique A2 has a positive excess return (we will omit the adjective "positive" from here on). In a similar way, all the spanning trees other than A1B123, A13B23, and A123B2 have at least one production technique that has an excess return. How about spanning trees A1B123, A13B23, and A123B2? Let us examine A13B23. We have the following two estimations:

$$w_1 a_{12} > w_1 a_{13} w_2 a_{22} / w_2 a_{23} = p_3 p_2 / p_3 = p_2,$$
$$w_2 a_{21} > w_1 a_{11} w_2 a_{23} / w_1 a_{13} = p_1 p_3 / p_3 = p_2.$$

There are no production techniques that have excess returns. The same is true for A1B123 and A123B2.

If the international value has some production techniques with positive excess returns, this value system will be destroyed, since firms possessing such production techniques will start to use them (at least when they know them) and will set lower product prices than the current product price. A stable international value must not admit production techniques that have positive excess returns. Let us call such an international value *admissible*. This leads us to the next definition.

Definition 5-2 (Admissible Value)
An international value is called admissible when it does not admit any production techniques with positive excess returns.[23] □

[23] In Sect. 6 we define the same concept using more explicit inequalities. The concept of admissible value assumes a set of production techniques, and, if the set changes, the same value may become non-admissible. Shiozawa (2017a Definition 3–7) provides a different definition. Concerning the equivalence of the two definitions, see the argument put forward later in this section.

Based on this definition, we can say that, for a labor input coefficient matrix *A* that satisfies inequalities (3-5), the international values defined by competitive A1B123, A13B23, and A123B2 are admissible, whereas no other international values defined by other spanning trees are admissible. Thus, we see that there is a one-to-one correspondence between three elements: (1) the facets of the production possibility frontier, (2) the admissible international values defined by a spanning tree, and (3) the spanning trees that have admissible international values.

The correspondence between (2) and (3) is trivial, as the two expressions simply offer a different focus on the same entity. The correspondence between (1) and (2) is more substantial. We cannot provide detailed proof here, but the general steps are as follows. Let us imagine a production frontier, which is a set of maximal points of a polytope like in Fig. 1. At any point in the interior of a facet of the frontier, we have a normal vector **p** that is normal (perpendicular) to the facet. Vector **p** is positive, in the sense that all components of the vector are positive. There is always a conjugate vector **w** for vector **p**, and the international value **v** = (**w**, **p**) satisfies the relations given by Theorem 3-4 in Shiozawa (2017a) (see also Theorem 6-2 in this chapter). The international value **v** is admissible, because the set of production techniques satisfies inequality (iii) of Theorem 3-4 mentioned above. The existence of such a value vector is ensured by Theorem 6-2 of this chapter. As the international value is uniquely determined, there are enough production techniques that form a connected technology graph. If the graph is not connected, the production techniques cannot determine the international value because, in that case, the international value has a degree of freedom as the number of connected components minus one.[24] Thus, the competitive technology graph must be a spanning tree in general cases, because the production techniques cannot satisfy all the cost-price equalities if the graph contains a cycle.

Conversely, assume that an admissible international value **v** = (**w**, **p**) is defined by a spanning tree *T*. Put $F = \{\mathbf{y} \mid \forall \mathbf{y} \geqq \mathbf{0},\ y_{ij} = 0 \text{ if } h(i, j) \notin T \text{ and } \mathbf{y} A = \mathbf{q}\}$ and $P = \{\mathbf{z} \mid \forall \mathbf{z} \geqq \mathbf{0},\ \mathbf{z} A \leqq \mathbf{q}\}$, where *A* is the labor input coefficient matrix, **q** is the vector of labor powers for each country, and **y** and **z** are activity level vectors. *P* is the production possibility set and we will prove that *F* is a facet of *P*. As **v** is admissible, we have $A\,\mathbf{w} \geqq \mathbf{p}$. Then, there is no point **z** of *P* that satisfies the inequality $\mathbf{y} \leq \mathbf{z}$ for any point y of F.[25] In fact, suppose that there exists a boundary point **y** of *P* that satisfies $\mathbf{y} \leq \mathbf{z}$, this implies

$$\langle \mathbf{z}, \mathbf{p} \rangle > \langle \mathbf{y}, \mathbf{p} \rangle = \langle \mathbf{y}, A\mathbf{w} \rangle = \langle \mathbf{y}A, \mathbf{w} \rangle = \langle \mathbf{q}, \mathbf{w} \rangle,$$

because **p** is strictly positive and $y_{ij} = 0$ for all *h*(i, j) not in *T*. This contradicts the fact that **z** is an element of *P*, because all points in *P* satisfy the inequality

[24]This fact is vaguely stated in McKenzie (1953 p 172–3).

[25]We distinguish between $\mathbf{y} \leq \mathbf{z}$ and $\mathbf{y} \leqq \mathbf{z}$. The former means $\mathbf{y} \leqq \mathbf{z}$ and $\mathbf{y} \neq \mathbf{z}$.

$$\langle \mathbf{z}, \mathbf{p} \rangle \leqq \langle \mathbf{z}, A\mathbf{w} \rangle = \langle \mathbf{z}A, \mathbf{w} \rangle \leqq \langle \mathbf{q}, \mathbf{w} \rangle.$$

Thus, any point of F is in the maximal frontier of *P*. The last step is to prove that set F forms a facet, i.e., a face that has codimension 1 or dimension *N*-1. This is a special case of Theorem 6-1 in the next section. To sum up, we have Theorem 5-3. It is also easy to see that the set F defined by a non-admissible spanning tree cannot be a facet because points of *F* cannot be maximal point of the set P.

Theorem 5-3 (Correspondence Between Regular Values and Admissible Spanning Trees, R0 Case)
Let *A* and **q** be a labor input coefficient matrix and a vector of labor powers of a Ricardian trade economy with *M* countries and *N* goods. In general, i.e., in the case where *A* is in a general position,[26] the regular international value on a facet of the production frontier is admissible and equal to the international value defined by an admissible spanning tree of production techniques and vice versa. □

Since in the above theorem, we use a term—namely, admissible spanning tree—that has not yet been defined, we now present Definition 5-4. According to Theorem 5-1, a spanning tree uniquely defines an international value. Hence, we need to determine whether this international value is admissible or not. We define a spanning tree as admissible when the international value is admissible in the sense of Definition 5-2.

Definition 5-4 (Admissible Spanning Trees, R0 Case)
A spanning tree *T* in an R0 trade economy is called *admissible* when the international value that is uniquely defined by *T* is admissible. □

Shiozawa (2017a) defines the international value as *admissible* when it satisfies conditions (i) to (iv) and *regular* when the price vector is normal to a facet of the production frontier (Shiozawa 2017a Definition 3-7). By economizing terms, we define the international value as admissible after Definition 5-2 or, equivalently, when it satisfies inequality (6-1) and iqualities (6-4) and regular if such admissible value is supported by a spanning tree. Then, Theorem 5-3 contends that there is a one-to-one correspondence between facets of the production possibility frontier and admissible spanning trees of the production techniques, and they both define the same international value. Therefore we can *redefine* the concept of regular international value by the one associated with an admissible spanning tree. Note that whether a spanning tree is admissible or not depends on the labor input matrix *A*.

In this way, almost all examinations of the production frontier can be translated into examinations of admissible spanning trees. The latter examinations are more suited to many types of analyses, including competitive analyses of firms, which we will attempt in this chapter for the first time. Another important topic is involuntary unemployment. The value concept in Shiozawa (2017a) relied on world final

[26] This condition is necessary because, in a degenerated case, F may not be a facet of codimension 1. The meaning of "general position" will be explained in Sect. 6.

demand on a facet of the production possibility set, and, although it was possible to partially analyze involuntary unemployment, the status of the international value remained ambiguous.

There are many other interesting properties of the Ricardian trade economy (see for example Shiozawa 2015), but let us content ourselves with pointing out some results about the number of facets or admissible spanning trees. The number of spanning trees *s*(*M*, *N*) of the (*M*,*N*)-bipartite graph is known as Scoin's formula[27]:

$$s(M,N) = M^{N-1}N^{M-1}. \qquad (5-1)$$

As we saw in the case of a two-country, three-good trade economy, the number of all the different spanning trees is $2^2 \cdot 3 = 12$. In the same case, the number of facets or admissible spanning trees is 3. Let us denote the number of facets or admissible spanning trees of an *M*-country, *N*-commodity trade economy c(*M*, *N*). Is there a general formula for the number *c*(*M*, *N*)? In the case of an R0 economy,[28] this number c(*M*, *N*) is known for coefficient matrix *A* in a general position:

$$c(M,N) = (M + N - 2)!\,/(M - 1)!(N - 1)!. \qquad (5-2)$$

This is the number of different (assignment) classes after Ronald Jones (1961 p 164), although he gave no explanation as to why this notion is relevant. A class of assignments or competitive patterns is the set of technology graphs that have the same number of directly connected goods for all countries. As countries and goods are symmetric in technology graphs, this is the same as saying that a class is the set of technology graphs that have the same number of directly connected countries for all goods. The ratio *c*(*M*, *N*) /*s*(*M*, *N*) is a decreasing function. Although *c*(*M*, *N*) is a big number for slightly large *M* and *N*, the ratio decreases rapidly. For example, in a three-country, five-good economy, the ratio is 15/6075≒0.0025. In a five-country, ten-good case, the ratio is approximately 0.0.000000037 or 1 over 30 million. Thus, admissible spanning trees occupy a very small percentage of all spanning trees.

6 How Relative Wage Rates Are Determined (Input Trade Case)

In Sects. 3 and 5, we examined trade economies that are pure labor input economies R0. As explained in Shiozawa (2017a, Sect. 2 p 14), the results obtained for R0 can be extended to RI and RII, because they have in fact the same structure and we can

[27]See, for example, Berge (1971 p 108).

[28]The formula is true for RI and RII, because they have the same structure (Shiozawa 2017a Sect. 2). Little is known about the RS economy case.

apply all arguments of R0 to RI and RII. However, the RS-type trade economy requires a separate analysis. A Ricardian-type economy is called an *RS trade economy* (abbreviation of Ricardo-Sraffa trade economy) when input goods are traded and countries have possibly different technologies. Since it is often evident that we refer to trade economy, we can also say *RS economy* for short. An RS economy is structurally different from the R0, RI, and RII classes of economies even when they have the same number of countries and goods. In this section, we will present some key results concerning the RS economy and show that the same formulation presented in Sect. 5 is applicable to the RS economy too.

Let us start by recollecting several concepts from the new theory of international values (Shiozawa 2017a). The main target of this theory is to investigate an RS trade economy with *M*-countries and *N*-goods. Each country has a set of production techniques that are linear and operational within a certain limit of capacity. It is worth noting that, in contrast to many traditional trade theories, such as the Heckscher-Ohlin-Samuelson (HOS) and Heckscher-Ohlin-Vanek (HOV) models, no similarity assumptions were imposed with regard to the sets of production techniques of the different countries. An important consequence of this technological non-similarity is that the countries may have very different wage rates (measured by a common currency).[29] Our basic view is that major differences in wage rates exist across the world, and they stem mainly from significant differences in the technology sets that the countries have. In this regard, we share the opinion of many technological progress economists, like Dosi et al. (1990 Chap. 1).[30] As we stated in the introduction to this chapter, wage rate differences form a basic condition for international competition among firms and subsections of MNEs. Firms and production units compete with wage rate differences as a main handicap.

In contrast to Sects. 3 and 5, in this and later sections (except Sect. 9), we will adopt different notations for the symbols used. A production technique is represented by a set of two vectors ($\mathbf{u}$, $\mathbf{a}$). Vector $\mathbf{u}$, which is an *M*-row vector composed of labor input coefficients, has only one positive entry u_h among *M* entries, while all other entries are 0 (production in a single country).[31] Vector $\mathbf{a}$, which is an *N*-row vector composed of *N* entries representing the *net output* coefficient for good j, has only one positive entry, while all other entries are either 0 or negative (no joint production assumption). In addition, we assume in Sects. 6, 8, and 10 that the single positive element of $\mathbf{a}$ takes the value 1. A production technique ($\mathbf{u}$, $\mathbf{a}$) that has positive entries at the i-th entry in $\mathbf{u}$ and at the j-th entry in $\mathbf{a}$ is the

---

[29]Thus, the new theory of international values provides the theory that Pasinetti (1993 Chap. 9) thought necessary.

[30]Pasinetti (1993 Chap. 9) acknowledges the same fact but believes it to be ascribable to labor immobility. This is not correct, because, even if there is labor mobility, the wage rate difference remains when the set of production techniques is conserved through migration (Shiozawa 2017a Sect. 11).

[31]This does not exclude production by using global supply chains, because it is treated as using chains of trades and productions. In the RS economy, input goods are freely traded among countries.

production technique of country i and produces good j. Thus, a production technique can be represented by an edge of a bipartite graph, as in the case illustrated in Sect. 5, and all the terminology related to graph theory is also valid in this section.

As a typical case, we assume no transportation and transaction costs.[32] Thus, the location of the good in question is neglected. We also assume that the set of production techniques of each country is productive. These assumptions can be weakened in various ways. For example, the assumption that net output coefficient vector **a** has only one positive entry means that we exclude joint production. However, joint production, which arises from the existence of fixed capital goods, i.e., machines and installations, can be incorporated into the new theory (Shiozawa 2017a Sect. 8). For other details, see Shiozawa (2017a). The productivity assumption for each country can easily be weakened. In fact, it is sufficient that the set of production techniques for the world be productive. But this requires a longer description for each proposition and we wish to avoid it.

An international value vector $\mathbf{v} = (\mathbf{w}, \mathbf{p})$ is also a pair consisting of wage rate vector $\mathbf{w} = (w_1, \ldots, w_M)$ and price vector $\mathbf{p} = (p_1. \ldots, p_N)$. Vectors **w** and **p** are treated as column vectors and assumed to be elements of dual vector spaces. In the same way, vectors **u** and **a** are elements of two row vector spaces of dimensions M and N, respectively. We express the total sum of products by a bracket (linear form expression):

$$\langle \mathbf{u}, \mathbf{w} \rangle = u_1 w_1 + \cdots + u_M w_M \quad \text{and} \quad \langle \mathbf{a}, \mathbf{p} \rangle = a_1 p_1 + \cdots + a_N p_N.$$

The above notations may seem redundant especially in the case of labor input vector **u**, because there is only one positive entry. However, this is necessary in order to obtain a simple, uniform notation for all production techniques belonging to different countries.

The total set of production techniques is expressed by a pair of matrices J and A, where J is the matrix of labor input coefficients and A is the matrix of net material output coefficients (See Shiozawa 2017a, Sect. 5 for the treatment of these coefficients when markup rates are positive). If the set of production techniques Σ contains H elements, J is a matrix of H rows and M columns and A is a matrix of H rows and N columns. Each row of J and A is written, respectively, as $\mathbf{u}(h)$ and $\mathbf{a}(h)$ for production technique h. Symbol A has totally different meaning from A in Sect. 5. It is important not to cofuse the two.

There are two important concepts related to international values, i.e., that of *admissible* value and that of *regular* value. In order to define the regular value for an RS economy, the result of Theorem 5-1 is needed, whereas the admissible value can be defined with no preparation. An international value, $\mathbf{v} = (\mathbf{w}, \mathbf{p})$, is said to be *admissible* when it is positive and satisfies inequalities:

---

32 Regarding the case with positive transportation costs, see Shiozawa (2017a Sect. 9).

$$J\mathbf{w} \geqq A\mathbf{p} \tag{6 – 1}$$

If we extract the row for production technique *h*, this means

$$\langle \mathbf{u}(h), \mathbf{w} \rangle \geqq \langle \mathbf{a}(h), \mathbf{p} \rangle \quad \forall h \in \Sigma \tag{6 – 2}$$

If

$$\langle \mathbf{u}(h), \mathbf{w} \rangle = \langle \mathbf{a}(h), \mathbf{p} \rangle, \tag{6 – 3}$$

we say that production technique *h* satisfies the *cost-price equality* or *value equation*. In this case, we also say that production technique *h* is *competitive*.

In Sect. 2, we assumed that all countries have the same "profit rate," but this expression was actually inaccurate.[33] It would be more correct to say that each firm sets its product prices by using a fixed markup rate, i.e., according the full-cost principle. If the unit cost is *c* and the markup rate is *m*, the price will be (1 + *m*) *c*. There are many explanations for how and why this markup rate is determined. It is largely fixed following tradition, but, to a certain degree, it reflects the state of competition between the product and competing products (see Shiozawa 2014 Appendix (in Japanese) and Shiozawa 2016 for a short explanation). If markup rate *m* is determined for product j in country i, the coefficient vectors $\mathbf{u}(h)$ and $\mathbf{a}(h)$ should be multiplied by factor 1 + *m*, except for the coefficient of the main product j (see Shiozawa 2017a Sect. 5 for details). In the discussion below, all coefficient vectors are assumed to be expressed by this equivalent expression. When we say that a production technique is competitive, it means that the full cost of production using the said technique is equal to the product price. Value equations are expressed exactly through this method, but it is worth noting that quantity relations should be considered based on bare coefficients before adopting equivalent coefficients (Shiozawa 2017a Sect. 5, Oka 2017). We can now state two main theorems.

Theorem 6-1 (Uniqueness of International Values)
Let (*E*, Σ) be an RS trade economy of *M*-countries and *N*-goods. Suppose that the set of production techniques *T* in a general position as an RS trade economy satisfies the following conditions:

(a) *T* is productive.
(b) The technology graph *T* is a spanning tree.
(c) A positive international value $\mathbf{v} = (\mathbf{w}, \mathbf{p})$ exists and the value equation holds for any production technique *h* of *T*.

Then the following two propositions hold:

(i) Vectors $\{ (-\mathbf{u}(h), \mathbf{a}(h)) \mid \forall h \in T\}$ are linearly independent.

[33]"Profit rate" is normally defined as the ratio between gross profit and the total value of fixed capital. Thus, it is a function of demand or production volume and is not determined by unit cost and product price.

(ii) The international value **v**, which satisfies condition (c), is unique up to scalar multiplications. □

Note that proposition (ii) is only a consequence of proposition (i). This uniquely determined international value **v** = (**w**, **p**) is said to be *regular* when it is admissible with respect to Σ and satisfies Eq. (6-3) for all elements *h* of spanning tree *T*. We call this international value **v** the regular value defined by spanning tree *T*.

Theorem 6-2 (Existence of Regular International Values)
For any RS trade economy (*E*, Σ), there exists at least one regular international values **v** = (**w**, **p**). In other words, there exists a positive vector **v** = (**w**, **p**) and a spanning tree *T* chosen from Σ such that

$$\mathrm{Jw} \geqq \mathrm{Ap} \qquad (6-1)$$

and

$$\langle \mathbf{u}(h), \mathbf{w} \rangle = \langle \mathbf{a}(h), \mathbf{p} \rangle \quad \forall h \in T. \qquad (6-4)$$

□

A key concept for Theorem 6-1 is that the set of production techniques *T* is in a general position. This notion was also used for labor input coefficient matrix *A* with no explanation (Sect. 5). General position is a widely used mathematical concept, but it requires a definition for each case. The simplest definition of general position for the set of production techniques *T* whose technology graph forms a spanning tree is that the set of vectors $\{ (-\mathbf{u}(h), \mathbf{a}(h)) \mid \forall h \in T\}$ is linearly independent. By virtue of this definition, Theorem 6.1 becomes trivial. However, in this case, it is necessary to prove that the set of spanning trees in a general position covers a dense open set in a space of input coefficients that satisfy the three conditions (a), (b), and (c).[34]

Theorem 6-1 is completely new and does not appear in Shiozawa (2017a). Shiozawa has presented written proof of it and asked his colleagues to check it. As the check is still in progress, it would be more correct to say that it is still a conjecture. However, as we observed in Sect. 5, Theorem 6-1 holds for economies without trade of input goods (trade economies of type R0 or, equivalently, of types RI and RII), and a variant of it is most likely to hold too, provided that the conditions are suitably modified in case the present theorem is not entirely correct. This is why we present it as a theorem. Theorem 6-2 is a simple consequence of Theorem 3-4 in Shiozawa (2017a). Note also that the definition of regular international value is modified, but, like in the case of pure labor input economy in Sect. 5, the new and old definitions are equivalent.

[34]The set of $M + N-1$ vectors $\mathbf{a}(h)$ in $\mathbf{R}^N$ forms a densely open set in $\mathbf{R}^{N(M+N-1)}$, but we have to prove that the set of linearly independent *T* forms a densely open set in the set of $M + N-1$ vectors that satisfy (a), (b), and (c) by means of the topology induced by $\mathbf{R}^{N(M+N-1)}$.

The new definition of the concept of regular international value (regular value from here on) and Theorem 6-1 are needed to make the new theory of international values more adaptable to the analysis of Keynesian involuntary unemployment.[35] As noted in Sect. 5, in the old formulation (Shiozawa 2017a), the status of regular values was ambiguous with respect to the significance and uniqueness of the examined regular values.

Regular values whose existence is assured by Theorem 6-2 are not unique. For example, in a Ricardian trade economy (of type R0, RI, or RII) with two countries and three goods, there are normally three different regular values. The more general case, i.e., an (M, N) Ricardian economy, is analyzed through Formula (5-2), but the general formula for RS economies is not yet known. Ogawa's example (Ogawa 2017 Fig. 5) with three countries and three goods shows that there are seven regular values, whereas (5-2) gives six.

7 Historical Nature of Regular Values

Multiplicity of regular values does not pose serious problems, but we can truly understand them only when they are studied within their historical and evolutionary processes (David 2001; Malerba et al. 2016; see also Krugman 1991). A sequence of economic states is path dependent. If we can determine when the present regular value is conserved and when it changes, we may well be able to develop our analysis in an evolutionary way or in a history-friendly manner.[36]

Suppose that, in an economy, a regular value $\mathbf{v}$ and a set of competitive production techniques are given. We assume for the moment that the set of production techniques does not change. Historical analyses inevitably begin with an *always-already given structure* (Althusser 1965 Sect. 4). Suppose that productions are made only by using competitive production techniques. If the net production of the economy is equal to the world final demand, the economy reproduces its state and there is neither the need nor the possibility for the economic situation to change.[37] Suppose that the final demand changes for whatever reason. In the general equilibrium framework, one looks at how the prices change. Except for the cases of agricultural products, whose prices are affected by climatic conditions, and of other products whose production volumes are difficult to change rapidly, in the

[35]Shiozawa wishes to thank Taichi Tabuchi and is grateful for the discussions in the Workshop on the Theory of International Values. Without Tabuchi's criticism, he would have never attempted this theory reorientation.

[36]Although we essentially have the same idea, path dependence emerges in our case not through a multiplicity of institutions, but through the existence of mutually exclusive international values (i.e., pairs of wage rates and prices).

[37]See the argument below. Here we consider a simple reproduction case, while the steady growth case requires modifying the input coefficients. See Shiozawa (2017a Sect. 5) and Oka (2017).

Ricardian economy (including the RS economy), wage rates and prices do not change as long as the following four conditions are satisfied[38]:

1. World final demand can be produced as net product by using only competitive production techniques.
2. Each country has enough labor power for the operation of the above production.
3. Production volumes remain within the capacity limits.[39]
4. There is no exhaustion of underground resources, such as rare metals.

The simplest interpretation is to assume that each firm keeps sufficient product stocks. If changes in demand flow are slow, stocks can be adjusted to the new demand by increasing or decreasing the production volume according to the demand flow expressed to the firm. This is a complicated process, but evidence shows that this quantity adjustment process works well when firms have enough inventory stocks and use the sales averages of several periods to predict demand (see Shiozawa et al. in press).

As long as the above four conditions are satisfied, there is not only no need but also no possibility that production techniques will change. Take any production technique h in Σ that is not competitive. Then, as $\mathbf{v} = (\mathbf{w}, \mathbf{p})$ satisfies inequality (6-1), we have

$$\langle \mathbf{u}(h), \mathbf{w} \rangle \geqq \langle \mathbf{a}(h), \mathbf{p} \rangle.$$

If h is not competitive, this means that

$$\langle \mathbf{u}(h), \mathbf{w} \rangle > \langle \mathbf{a}(h), \mathbf{p} \rangle.$$

The left-hand side represents labor cost and the right-hand side corresponds to the net return from material input and output (in full-cost accounting). Adopting production technique h causes a loss, so no firm will want to use production technique h as long as international value $\mathbf{v} = (\mathbf{w}, \mathbf{p})$ prevails. Thus, we are *locked in* the state of the present value and system of competitive production techniques. No single firm can switch to another production technique, although another value system is perfectly possible if an entity, such as a central planner, orders firms to change their production techniques, wage rates, and prices all at once. Contrary to claims put forward by some neoclassical economists, in this context of international trade, lock-

[38]The case of agricultural products is excluded from our examination, because it requires a very different argument. Another important but implicit condition is that exchange rates are fixed.

[39]For the sake of simplicity, we do not refer to the capacity that a set of fixed capital determines, but we implicitly suppose that each firm has a set of equipment, machinery, buildings, facilities, and vehicles of proper proportion and suitably arranged. Within this framework, production is a proportional input-output relation which is possible within a capacity that is determined by capital equipment. If we take into account the depreciation of fixed costs, we are in fact considering an increasing-returns-to-scale situation. The approach based on the proportional cost assumption is retained if we suppose that the headquarters of the firm charge, as a part of input costs, a rate of capital usage which is proportional to the quantity of production. See Fujimoto (2012b).

in is not the result of menu costs or other frictional factors that prevent prices from moving freely.

What happens when one of above four conditions is not satisfied? Of course, it all depends on which one it is. If condition (3) is violated, it is possible that firms will invest to increase production and, after some time, there will be enough production capacity and there may be no price change. Of course, capacity increases will be decided before overcapacity is reached. The case of material shortage may unfold in a similar way. For example, firms supplying materials affected by shortages may ask buyers to wait for a while until production is increased and full demand is met.

The case in which condition (4) is not satisfied is difficult to analyze. Its effects are not limited to the producing countries. Unless new deposits are discovered or substitutes are invented, all firms that use the extracted materials are obliged to change their production techniques to other ones that require lower quantities of those materials or do not use them at all. This may induce a chain effect from the industries that use the materials directly to those that use them indirectly.

The case in which condition (1) is violated for a single product arises rarely. If there is greater demand for a product, the firms that provide it will increase production. If there are no restrictions for what concerns conditions (2), (3), and (4), a simple increase in production volume may ensure that demand is met. Nonetheless, in rare situations, demand for a product increases very rapidly and its manufacturers cannot keep up, but this is an uncommon and happy coincidence for firms.

The most interesting and relevant case arises when a country does not satisfy condition (2). Condition (2) may be somehow unclear. Labor supply is, in a sense, quite flexible. Even when all available workers are employed, they can work longer hours and produce larger quantities of products in a situation of strong demand. There is no specific point at which the economy achieves full employment, and, even before reaching that state, many firms find it difficult to employ enough staff. When this occurs in a given country and labor supply is not adjusted rapidly, the situation may continue for a long time and the wage rate of that country will gradually increase. This will change the international value and, as a consequence, trade patterns may also be modified.

Assume, for example, that country 1 experiences severe labor shortage and its wage rate, i.e., w_1, is driven up substantially. All industries in country 1 will face higher labor costs. If all products are made in country 1 without using imported products, the price hike will be uniform, because production price p is expressed by $w_1\mathbf{u}\ (I{-}A)^{-1}$ if $\mathbf{u}$ is the labor input coefficient vector and A is the square material input coefficient matrix in equivalent expression (Shiozawa 2017a, Sect. 5). However, this is a rare case, because almost all countries import various input goods. If the prices of these imported goods do not change, the wage rate hike will affect production costs differently for different products. The cost of some products will increase as a percentage of the wage rate hike. Other products will be less affected when the value composition of imported goods is high.

When they recognize this permanent change in wage rates, firms may be obliged to change their product prices. Some industries may be strongly affected and it may become difficult for them to continue production and exportation. Other industries may boost labor productivity, thereby succeeding in eliminating the effects of the wage rate hike. If the product is competitively produced in other countries, only two options are viable: to abandon production or to increase productivity so as to compete with the wage rate hike. A similar but more severe situation occurred in Japan, although for completely different reasons, when China opened up its economy in the 1990s. We will examine this situation in later sections.

A country's wage rate hike inevitably induces a change in the international value.[40] Suppose that a new value $\mathbf{v}(1) = (\mathbf{w}(1), \mathbf{p}(1))$ is established and a new pattern of specialization emerges after a while. If this new value is admissible, no problem will arise and firms will not change their production techniques. If the new value is not admissible, there must exist, among the previously competitive production techniques, at least one or more techniques that cannot be operated competitively. The firms that have been operating them will be obliged to abandon their former production techniques and replace them with new ones. The international value must change to a new value $\mathbf{v}(2) = (\mathbf{w}(2), \mathbf{p}(2))$ and will continue to do so until it becomes admissible. Furthermore, some production techniques will yield excess returns. The price of the product may remain constant for a while before competition forces firms to reduce it to a normal markup rate.

The case of admissible but not regular value is more difficult to examine, since uniqueness of value ceases to exist. Whatever the reason for the change in value may be, there is no pressure to return to the former value, so the admissible value can fluctuate randomly. In this case, demand pressure may cause the value to change. If the value remains admissible, the degree of freedom is rather modest. If the world final demand is on a ridge (face of codimension 1), it will be only one. If the value changes in one direction, it will soon reach a regular value. Thus, we can say that an admissible but not regular value is rather unstable and eventually moves toward one of the regular values "adjacent" to the admissible one. Yet, it may be difficult to tell which of several regular values will be chosen in this case, as this may be determined by chance and circumstances.

8 A Short Remark on Technical Change

So far, we have assumed that the set of production techniques Σ is given and fixed. However, in the real economy, production techniques change as time passes by. This section will present some comments in preparation for the analysis of how technical change affects both the international value and patterns of specialization.

Technical change is a complex process. We will not discuss it in detail, because it requires specific examination and also because a large body of literature already exists on the topic (see, for instance, a survey by Chris Freeman, Freeman 1997).

[40]The assumption is always that the wage rate is expressed by an international currency.

Technical change comprises innovation in both product and production techniques, but we will concentrate on the effects of new production techniques. The question of new products will be referred to only when we talk about demand creation.

When exploring technical change, the focus is usually on innovative commodities. Instead, we are here more interested in products that compete internationally. Such products are called *link commodities* or commodities produced in common by Graham (1948).[41] Graham (1932) called this special type of competition *linked competition.*[42] In the following discussion, the main target of our investigation will indeed be linked competition. In technology graphs, a link commodity is a good that has at least two edges connecting different countries.

Around the world, there are commodities that are produced only in one country. Such products are called *monopolistic commodities* even when they are produced competitively by many firms within that country. *Innovative commodities*, which are the result of product innovation, are, at least in the first phase of their existence, monopolistic commodities.[43] Traditional commodities may remain monopolistic if demand is limited to one specific country and transportation costs are high.

The evolution of a production technique takes the form of an enlarged set of production techniques, since the previous production techniques are not easily forgotten or lost. Suppose that a new production technique h is developed. If the quality and the normal unit fixed capital cost of the product remains the same, the criterion to determine whether the new production technique is better than the old technique $h^{\#}$ is simple: the new production technique must have a smaller unit production cost than the older one. If value $\mathbf{v} = (\mathbf{w}, \mathbf{p})$ is in place, the condition for the new production technique to be adoptable is

$$\langle \mathbf{u}(h), \mathbf{w} \rangle + \langle \mathbf{b}(h), \mathbf{p} \rangle < \langle \mathbf{u}(h^{\#}), \mathbf{w} \rangle + \langle \mathbf{b}(h^{\#}), \mathbf{p} \rangle. \qquad (8-1)$$

Differently from the previous sections, here we use notations which express inputs and outputs separately. Thus, $\mathbf{b}(h)$ and $\mathbf{b}(h^{\#})$ are material input coefficient vectors for production technique h and $h^{\#}$, respectively, when the output is supposed to be a unit. Vectors $\mathbf{u}(h)$ and $\mathbf{u}(h^{\#})$ express the labor inputs for the production of a unit. Thus, the net output coefficient vector a production technique which produces product i is expressed as

$$\mathbf{a}(h) = \mathbf{e}(\mathrm{i}) - \mathbf{b}(h). \qquad (8-2)$$

---

[41]Shiozawa wishes to thank Hideo Sato for providing details on Graham's usage of terms (mail of February 15, 2017). See also Sato 2017 Sect. 3.

[42]Graham avowed that he owed this naming to his former student C. R. Whittlesey (Graham 1932, p. 581, n.3).

[43]Dosi et al. (1990 p 200) distinguish between two groups of commodities: Ricardian commodities and innovative commodities. In our terminology, commodities in the first group are link commodities and those in the second group are monopolistic commodities.

It is superfluous to say that **e**(i) is a unit vector that has entry 1 for the i-th component and is 0 for all other entries. In this notation, the cost-price equality for $h^{\#}$ is given by

$$\langle \mathbf{u}(h^{\#}), \mathbf{w}\rangle + \langle \mathbf{b}(h^{\#}), \mathbf{p}\rangle = p_{\mathrm{i}}\,. \qquad (8-3)$$

Consequently, inequality (8-1) means that the net return of operating production technique h is greater than that of the formerly competitive technique $h^{\#}$ (in the full-cost sense).

If production technique h satisfies inequality (8-1), the former regular vector $\mathbf{v} = (\mathbf{w}, \mathbf{p})$ is no longer regular nor admissible with respect to the new set of production techniques $\Sigma^{\&} = \Sigma \cup \{h\}$, because inequality (6-1) ceases to hold. If a firm adopts the new technique h and the markup rate remains unchanged, the price of the product will decrease. If the product is used as input in the production of a second product, the cost of production of this second product will be reduced and its price will also change. Thus, the introduction of a new production technique causes a chain of price changes.

If the new set $\Sigma^{\&}$ remains unchanged for a certain period of time, a new regular international value will be found. As production techniques are improved continuously, the regular value will change incessantly. The new theory of international values contains within itself the logic that explains how a better production technique is chosen from among the set of production techniques. Changes in international values go hand in hand with changes in competitive production techniques. Coevolution of value and technology is a theme that the ambitious book by Dosi et al. (1990) tries to tackle. While its core theory is formulated in a static way (Σ is given), the new theory of international values can be applied to the analysis of more dynamic cases since it contains the logic of how production techniques are chosen over time. We will attempt such an analysis in the following sections. The main focus of this chapter is to investigate the dynamics of linked competition that ensue from competition among firms or units of a MNE situated in different countries.

In the case of an innovative commodity, condition (8-1) has no meaning, because there is no old production technique with which to compare the new production technique. Whatever the production cost may be, the production technique can exist as a "competitive" one. The success of the product entirely depends on how high its demand is on the market. If it is an input product, it may cause major changes in production techniques through selection of the new product as input. If it is a generic product, say blue LED, its influence will be strong and felt globally, but in this chapter we will not deal with this aspect of technical change. As explained in Sect. 14 below, new products are indispensable factors for the survival of firms and nations.

9 A Factor Behind the Comparative Advantage Theory

Our main purpose here is to explain how international competition—or, more specifically, linked competition—unfolds among firms and among units of a multinational enterprise. For this purpose, it is worth mentioning that the traditional trade theory had a strong bias toward the macroscopic viewpoint, whereas our investigation is performed from a more microscopic perspective, close to that of firm managers and factory leaders. Indeed, we wish to shed light on how they think and act in the situations in which they are thrown. What follows is essentially the same analysis that was presented in Fujimoto and Shiozawa (2011–2012), but we believe that we have made substantial improvements, in that we now have a more articulated theory of international values and can argue various aspects relying on a solid theoretical basis.

In explanations of comparative advantage, there are two different points of view, as we argued in Sect. 2 and more extensively in Fujimoto and Shiozawa (2011–2012). These points of view are best illustrated by the two equivalent inequalities in (2-1). They are mathematically equivalent, but many textbooks claim that the left-hand side inequality is the only relevant one when considering the comparative advantage of two countries. This inequality explains the relative difference in domestic product prices, and it may be an easier means to guess which industry in a country has a comparative advantage. Those economists who explain the comparative advantage theory in this way consider the question from a macroscopic perspective, i.e., from the standpoint of an economist or government official who looks at comparative advantage as competition among countries.

However, whether a commodity is exported or imported is decided by makers and merchants,[44] who specialize in specific commodities but know little about the state of production of other groups of commodities. The terms on the right-hand side (i.e., intraindustrial ratios) are what they use as reference to determine whether they can manufacture a product in their country at a cheaper price than potential makers of the product in competing countries. If they know the wage rates w_{J} and w_{C}, they can easily calculate which of the inequalities like those in (2-4) holds among the terms appearing in (2-1). If the upper inequality of (2-4) holds, country J can export commodity 1 to country C. If there are no other countries which produce the product, country J can export to all other countries. If the lower inequality holds, country C can export commodity 2 to country J and to other countries if there are no other countries manufacturing the product. Remember that, for the sake of simplicity, we assume no transport costs.

Although the wage rate of a country is often difficult to know, it is much more accessible than the productivities of other industries. The left-hand side condition of

[44]This is also one of three points that Faccarello (2017) identifies as Ricardo's basic view of international trade.

(2-1) is not an easy fact to deal with for the mangers of firms and factories. Yet, if we know the wage rate ratio between two countries, we can easily determine, for any product k, which of the following equalities or inequalities holds[45]:

$$\begin{aligned}(1+m_J)\, w_J\, a_{Jk} &< (1+m_C)\, w_C\, a_{Ck}\\ (1+m_J)\, w_J\, a_{Jk} &= (1+m_C)\, w_C\, a_{Ck}\\ (1+m_J)\, w_J\, a_{Jk} &> (1+m_C)\, w_C\, a_{Ck}\end{aligned} \qquad (9-1)$$

If the markups are similar in both countries, these relations are reduced to simple comparisons between $w_J\, a_{Jk}$ and $w_C\, a_{Ck}$. In these comparisons or estimates, the terms appearing in the right-hand side condition of (2-1) are the only necessary data, in addition to the wage rate ratio.

We can now understand why the macroscopic viewpoint expressed by the right-hand side inequality of (2-1) is considered to be the only valid explanation.[46] Ricardo, John Stuart Mill, and many other theorists of comparative advantage did not have a theory as to how the wage rates of different countries are determined. Pullen (2006) calls the left-hand side condition SNDG, or same-nation-different-goods ratio [comparison], and the right-hand side condition DNSG, or different-nations-same-good ratio [comparison]. Fujimoto and Shiozawa (2011–2012) names them interindustrial ratio comparison and intraindustrial ratio comparison. Through the left-hand side inequality, we can know the patterns of specialization without knowing the wage rates of trading countries. This is the main reason why the theorists preferred reasoning based on the left-hand side condition of (2-1), i.e., SNDG or interindustrial comparisons.

This situation has changed completely with the introduction of the new theory of international values, which explains why and how wage rates are determined for each country. As we have already pointed out, after Faccarello (2017) and Tabuchi (2017a, b), Ricardo's theory of international trade should be interpreted as a comparison of costs rather than in relation to comparative advantage. The textbook version of the Ricardian theory lacked a theory for wage rate determination. It was inevitable that the macroscopic viewpoint would prevail in the argument on comparative advantage.

Another long-standing tradition is to interpret the comparative cost theory and the comparative advantage theory as synonyms, but it would be best to abandon this practice, because we can now differentiate between the two. The former is about how international trade patterns are determined by comparative cost structures among countries. Conversely, the comparative advantage theory is a special theory that reveals trade patterns through a simple formula of input coefficients. Yet, this is applicable only to simple two-country, pure-labor-input cases or to low-dimensional

---

[45]Remember that a in Sect. 2 stands for the labor input coefficient. We follow the same convention in this section.

[46]According to Pullen (2006 p 63), this way of thinking goes back to Cairns. But it is also possible that John Stuart Mill's formulation determined the future path (Shiozawa 2017b).

economies. The left-hand side inequality of (2-1) is one of such fortunate cases. When there exists input trade, Alain V. Deardorff (2005) proposed nine versions of the definition of comparative advantage but failed to give any logical explanation as to why this or that definition can indicate that an industry has a comparative advantage over others (Deardorff 2005, see also Shiozawa 2017c Sect. 7 and Tabuchi 2017a Sect. 3).

The lack of an appropriate trade theory within the Ricardian framework created much confusion with regard to the dichotomy between absolute and comparative advantage. Respectable economists, like Giovanni Dosi and his collaborators, are also confused about absolute and comparative advantage (Dosi et al. 1990 Sect. 6.2). It is not right to claim that "absolute advantages dominate over comparative advantages" (ibid. p 151), as this violates well-established economic concepts without fair and sufficient reasons. We understand that they want to emphasize the importance of technological efficiency in physical terms. They introduce a new concept, *absolute competitiveness* (ibid. p 149–51). This is a good attempt but does not seem to succeed in constructing a new theory, which is what they wished to achieve. We believe this ambiguity to be caused by the lack of a good trade theory through which to examine technological competition in international trade situations. The main weakness appears to be the inability to grasp the role of wage rate differences. Dosi et al. (1990 Sect. 6.1 in particular) know well that a nation's income per capita and wage rate are closely related to its technology level but fail to realize that the wage rate of a nation is the most important factor for firms that are engaged in linked competition. Wage rate has been missing for a long time in the argument of comparative advantage.[47]

10 Main Features of Cost Competition

Suppose that we are given sets of production techniques for countries J and C and a regular international value $\mathbf{v} = (\mathbf{w}, \mathbf{p})$ supported by a spanning tree T. We express the production techniques using symbols $\mathbf{u}$, $\mathbf{b}$, and $\mathbf{e}$ following the notation of Sect. 8. Suppose that good 1 is a link commodity between countries J and C. As the firms of both countries compete with each other, we have the equation

$$w_C\, u_C + \langle \mathbf{b}_C, \mathbf{p} \rangle \fallingdotseq w_J\, u_J + \langle \mathbf{b}_J, \mathbf{p} \rangle. \qquad (10-1)$$

[47]The Marxian theory of international values tried to shed light on how the wage rate of each nation is determined. However, it failed because it relied too much on interpreting Marx's texts. A rare exception is Kinoshita (2003). In the 1960s, Kinoshita asserted that the value added that a worker of a nation produces at a given unit of time is proportional to the productivity of the nation (p 52). Although he could not give an exact formulation, what he claimed was essentially correct.

10.1 Three Meanings of the Near Equation

We use the symbol ≒ to express three meanings. The first meaning is the same as the mathematical one. When the products of two countries are identical in all aspects, it means that the values of the two sides are nearly identical. Theoretically, we can assume that the firm that sets a lower price will absorb the whole demand (Joseph Bertrand's principle). In the real world, this does not happen, since there are at least some differences in certain aspects (e.g., place and delivery time). A number of firms or consumers might prefer to buy the more highly priced product. Delivery, quality, and supply stability are possible reasons for this behavior. All products are intrinsically differentiated, a point which will be discussed when analyzing the third meaning. Whether Bertrand's principle holds or not, however, we can imagine a situation in which firms set the same price, even though production costs are not exactly equal. When production costs are nearly equal, firms can set the same price. Firms and business units (production sites or *genba*) are going concerns. When there is a small disadvantage in costs, they can overcome it by reducing their markups. In that difficult moment, workers may accept a lower wage rate than the standard one. Even when productivity increases, thanks to the efforts of the workers at the site, they may be content with the present wage rates if their jobs are secured. These attitudes were widely observed around the turn of century among Japanese firms trying to rival their Chinese competitors.

The second meaning of using the symbol ≒ is to refer implicitly to the case in which transportation costs, including transaction costs and tariffs, exist. For the sake of simplicity, transportation costs are usually assumed to be zero. This assumption is necessary in order to show the essential logic of international trade, but it is a gross simplification that affects the features of international trade itself. Therefore, it is important to consider (1) how the theory is reconstructed in the presence of transportation costs and (2) the effects of positive transportation costs. As for point (1), we explicitly showed how to deal with this situation (Shiozawa 2017a Sect. 9). The approach is to treat the same good situated in different countries as different goods. As for point (2), knowing the effects of positive transportation costs in a general case is by no means easy. The main point is if a good is imported from country A to country B, the firms in country B can compete within country B as long as the production costs of A plus transportation costs from A to B are comparable with the cost borne by firms in B. If the transportation costs are considerable, this gives the firms in country B an ample margin of cost differences. So, readers are requested to make the necessary modifications when they want to apply our arguments to a concrete situation involving at least a positive amount of transportation costs.

If we do not take the extent of this effect into account, we make a fundamental mistake. The Ricardian trade theory is often criticized because it seems strange that the model results in nearly complete specialization. Of course, in the real world we see that a huge number of goods are produced commonly in many advanced countries. The criticism claims that this contradicts the Ricardian model. However, critics forget that nearly complete specialization, as in the case of the spanning tree, is the result obtained when there are no transportation costs. In actual fact, when transportation costs are positive and constitute a substantial portion of total costs, it is quite possible that many products are commonly produced in many countries. Positive transportation costs drastically change the features of international specialization. We should become accustomed to reading equalities like (10-1) for what they really mean when there are positive transportation costs.

The third meaning is needed when products are differentiated. If a product is superior in quality to others, the price of an inferior product can be seen as equal to the price of the superior product discounted by a suitable constant. If this discount rate can be fixed, all difficulties caused by the differences in quality can be solved. This may sound like an excuse and, indeed, it is. But we do not enter into this question here, because it is the most vulnerable aspect of the present theory. The concept of near equality in the third meaning may make some sense if we observe that products of different qualities have different meanings for different customers.

We accept Armington's idea of treating products of different countries (or more precisely of different producers) as differentiated products (Armington 1969, Petri 1980, Blonigen and Wilson 1999). Even when prices are different, it is rare that a single product may absorb the whole demand, which will instead be spread across products with certain ratios among them. For example, a product of high quality but with a high price will not attract most of the demand but will occupy some positive ratio of the whole demand in its class of products. In these cases, we must say that prices are at the near equality point and we will be permitted to write equalities like (10-1). However, we cannot place this treatment at the very core of our theory, because the new theory of international values is based on the assumption that two or more countries competitively produce common commodities. A breakthrough to overcome this dilemma still needs to be achieved.

10.2 The Japanese Experience After the Chinese Reform and Opening-Up

In the previous sections (Sects. 7 and 9 in particular), we saw that the relative wage rate of each country is fixed as long as the pattern of specialization (indicated by the spanning tree T in Sect. 6) is given. We emphasized in Sect. 9 that these relative wage rates are the factor that hides behind the traditional comparative advantage theory. As noted in the introduction, international competition among countries is a game with wage rates as handicaps. If wage rates differ to a large extent, this handicap becomes the single factor that decides the major features of how competition takes place.

This point can be illustrated most simply by analyzing the Japanese experience in the two decades after Deng Xiaoping's *reform and opening-up* of China and the end of the Cold War around 1990. Before the end of the Cold War, many Japanese manufacturing industries enjoyed competitive advantage and trade surplus expanded. It was the age of strategic trade policies and *Japan bashing*. The end of the Cold War and the opening-up of China drastically changed the environment surrounding Japanese industries. A new competitor, with wage rates equal to one-twentieth those of Japan and a huge number of workers, appeared on the scene. The wage rate ratio between Japan and China is not easily determined, because there are great disparities within China. One-twentieth is the figure that the managers of Japanese firms that established production sites in China used to mention and refers to the ratio between coastal city workers in China and workers in Japan. The average ratio between the two countries is actually much greater because, according to foreign labor statistics elaborated by the Japanese Bureau of Labor Statistics, the hourly wage of Chinese production workers was 1/30 that of Japanese production workers as late as 2005 (cited from Table 1, Krugman and Obstfeld 2009).

The opening-up of China occurred rather suddenly. Before the 1990s, there was mutual trade between Japan and China, but it was tightly controlled so that no perverse political and economic effects occurred in either country.[48] Direct investment into China was not easy and Chinese firms were mostly publicly owned companies, i.e., state- and community-owned companies. Although China' s living standards were still very low, the country had a complete set of industries and some areas, such as aerospace and nuclear weapons, boasted a high level of engineering skills. For Japan, the opening-up of China represented the birth of a major industrial actor in its vicinity, with sufficiently trained workforce, low-wage rates, and commercially talented entrepreneurs. Thus, many Japanese firms suddenly faced competition from China. Initially, Chinese firms' labor productivity was not as high as that of Japanese firms. If we measure labor productivity as the inverse of the labor input coefficient and call it physical labor productivity, most Japanese companies had physical labor productivity that was 5–10 times greater than that of China. Moreover, the disparity in wage rates remained substantial. If we suppose for the moment that both countries' material input coefficients vectors are the same (i.e., trade economy of type RII; the general case, i.e., RS economy, will be discussed later) and the prices of the input goods are the same everywhere, near equality (10-1) means this:

$$w_C\ u_C \fallingdotseq w_J\ u_J. \qquad (10-2)$$

In our experience, this simple equation was a valid criterion to judge whether a Japanese firm or a plant of a multinational enterprise located in Japan could compete with Chinese firms. Takahiro Fujimoto, one of authors of this book, interviewed more than 100 presidents and top managers of Japanese firms and plants whose products are in competition with Chinese products. Almost all of them said that they wondered whether their labor productivity was high enough to effectively compete with Chinese firms, given the wage rate discrepancy between the two countries (one-twentieth in the first phase).

---

[48]Officially, the reform and opening-up started in 1978 but it remained limited until 1992.

According to these managers, when the Japanese wage rate was about 20 times greater than that of China, competition was very difficult. In some cases, the top management at the Tokyo head offices of multinational enterprises often decided to close down units in Japan and transfer all production to China. Some factory managers asked to postpone the closing down of their plants in deficit and, within 2 years, succeeded in raising productivity enough to be able to compete with Chinese firms or plants of multinational enterprises. The latter cases indicate that labor productivity can change substantially, even drastically, if workers and production managers cooperate to increase it. In recent years, mainly due to a considerable growth in Chinese wage rates, the wage rate discrepancy has decreased and approached the ratio of 5 to 1, and Japanese managers say that it is now much easier to compete with Chinese firms. Chinese productivity has indeed gone up, but it is easy for Japanese firms or plants, they say, to keep productivity five times higher than that of Chinese companies (for a more detailed description of the Japanese experience, see the Appendix to chapter "A Design-Information-Flow View of Industries, Firms, and Sites" in this book, i.e., Fujimoto 2018).

10.3 The Case Where Material Input Costs Are the Same

The above occurrences teach us that Eq. (10-2) is a key criterion for managers to assess their factory's competitiveness vis-à-vis Chinese firms or plants. If this is true, then the difference $\langle \mathbf{b}_C, \mathbf{p} \rangle - \langle \mathbf{b}_J, \mathbf{p} \rangle$ does not play a significant part in the production costs of both countries. In this case, the wage rate ratio w_J / w_C marks the tipping point in terms of whether the firms or plants of one country can compete with those of the other. Indeed, if

$$u_C/u_J > w_J/w_C, \qquad (10-3)$$

country J's production is more competitive than that of country C. If inequality (10-3) holds in the opposite direction, then it is difficult for the production of country J to compete competitively with the production of country C (for more details, see Fujimoto and Shiozawa 2011–2012).

In the following analysis, we use the abbreviated expressions *C firm* and *J firm*, instead of saying firm or site in country C or country J. Condition (10-3) can be changed to a more familiar expression by using the concept of labor productivity. Indeed, as labor productivity π is the inverse of labor input coefficients u, (10-3) can be expressed as

$$\pi_J > (w_J/w_C) \cdot \pi_C. \qquad (10-4)$$

Inequality (10-4) means that, to remain competitive, the productivity of J firms must be greater than the wage rate ratio times the productivity of C firms.

When dealing with relations like (10-3) or (10-4), for the sake of simplicity, we will say that a J firm is more competitive than a C firm. When an international value is known, it means that a production technique for the product in country J has lower costs than that of country C.

This criterion is valid even in the case when unit material cost exists as long as difference $\langle \mathbf{b}_C, \mathbf{p}\rangle - \langle \mathbf{b}_J, \mathbf{p}\rangle$is small in comparison with the unit labor cost.[49] In such cases, we obtain a justification for the traditional Ricardian trade theory that assumes a pure labor input economy (Viner 1937 Chap. 8, Dorbusch et al. 1977).

Our arguments may seem a long, needless detour if we come to the same conclusion of pure labor input economy even when we take material inputs into account. Yet, this is wrong because, thanks to our examination, we have come to know the condition when criterion (10-3) is valid. Such reflection is almost forgotten in economics, but knowing the range of validity is one of most important aspects of the laws of physics. The fallacy in forgetting this becomes clear when we examine the case of material costs in two countries being different.

10.4 The Case Where Material Costs Are Different

Now, let us consider cases where $\langle \mathbf{b}_C, \mathbf{p}\rangle - \langle \mathbf{b}_J, \mathbf{p}\rangle$ is not negligible with regard to direct labor costs. This covers a variety of situations, and we must content ourselves with discussing only some typical cases.

As an extreme case, we can imagine that

$$\langle \mathbf{b}_C, \mathbf{p}\rangle - \langle \mathbf{b}_J, \mathbf{p}\rangle > w_J u_J \qquad (10-5)$$

In this case, it is impossible for a C firm to compete with J firms, no matter how low the wage rate w_C in country C is. Indeed, in this case, we have

$$w_C\, u_C + \langle \mathbf{b}_C, \mathbf{p}\rangle \leqq w_J\, u_J + \langle \mathbf{b}_J, \mathbf{p}\rangle$$

even when $w_C = 0$. Thus, it is evident that differences in material input coefficients must be sufficiently small for a C firm to compete with J firms.

Let us now examine another case, the situation in which the material costs for C firms are 60% of the total costs and those for J firms are 40% of the total costs. When they compete with comparable costs, this means that the labor costs ratio is around 40 to 60 for C and J firms, respectively. The nearly equal cost condition (10-2) now becomes

[49]Ronald Jones (Jones 1961 p 167) assumed that material input coefficients are the same for all countries (i.e., trade economy of type RII) when examining the trade of "intermediate products" (in his expression of 1961) or more simply "input trade" (in his later expression). "As a partial justification of this assumption," Jones cited the fact that "the prices of intermediate (and final) commodities are the same in all countries with free trade and zero transport" (p. 167 note). This is an odd justification because, even when the prices of input products are the same, input coefficients may differ for various reasons: experience, skill of workers or engineers, product design, etc.

$$40 \text{ wC uC} \fallingdotseq 60 \text{ wJ uJ}.$$

This alleviates the burden of workers in J firms, because the condition for J firms to remain competitive against C firms is now

$$1.5 \quad \pi J \geqq (\text{wJ}/\text{wC}) \cdot \pi C. \qquad (10-6)$$

Compare this result with (10-4). We can see how the difference in material input costs alleviates the labor productivity constraints. In the same vein, we can consider an example in which the material costs are 50% for C firms and 35% for J firms. The competitive condition for J firms will be

$$1.3 \; \pi J \geqq (\text{wJ}/\text{wC}) \cdot \pi J. \qquad (10-7)$$

We have assumed that the material costs for a J firm are smaller than those borne by a C firm. The reason is simple. Firms in country J may have more experience than firms in C. A J firm may have better design capabilities and may require lower quantities of material inputs than a C firm.

Suppose that J firms are competing with C firms when (10-6) or (10-7) are nearly equal. In these cases, the labor cost over output or labor/output ratio of J firms will be bigger than that of C firms. If the labor/output ratio is the measure of labor intensity, this means that J firms are more labor intensive than their competitors. This may seem paradoxical to those accustomed to thinking in factor proportions, but there is nothing strange about this result. If we suppose that $w\text{J}/w\text{C} = 5$ for the moment, the situation of near equality for (10-6) and (10-7) means that C firms need, respectively, 7.5 and 6.5 times the labor hours used in J firms to produce a unit of the product. In physical terms, C firms are operating with much more labor per unit than J firms.

10.5 Comparative Cost Criterion

In the above, we considered the case of two competing countries. Even when there are many countries, if we know the wage rate ratios of all of them, the basic logic remains the same, provided that transportation costs (including tariffs and transaction costs) are negligible and that the prices of the goods are the same in all countries. Indeed, the firms with the lowest costs will be able to compete on the world market. If a firm in country A is in such a situation, for any firms in country I we will have

$$w_\text{A}\, u_\text{A} + \langle \mathbf{b}_\text{A}, \mathbf{p} \rangle \leqq w_\text{I}\, u_\text{I} + \langle \mathbf{b}_\text{I}, \mathbf{p} \rangle. \qquad (10-8)$$

As we observed above, if the material input costs are comparable, it is the labor cost $w_\text{A} u_\text{A}$ that determines which country is competitive.

A remarkable point in the above argument is that competition over a product occurs among firms (or among production units of a multinational enterprise) and not among countries. According to Faccarello (2017) and Tabuchi (2017b), this is what Ricardo was thinking in his *Principles*. In this sense, the new theory of international values is more loyal to Ricardo's original theory of international trade than the textbook comparative advantage theory. The textbook interpretation is the traditional approach, originated by John Stuart Mill (1844, 1848) and inherited by Gottfried Haberler (1936) and Jacob Viner (1937) through Alfred Marshall (1879), but it should be abandoned now that a better and more loyal theory exists (Shiozawa 2017b, Tabuchi 2017b).

11 Competition with Large Wage Rate Handicaps

International competition over linked commodities is a struggle for existence. In this struggle, firms in high-wage-rate countries are handicapped by their wage rates. When a product is innovated, firms and production units in low-wage-rate countries, with more limited technical experience and work skills, may not be able at once to produce it either technologically or economically. When the product designs become standardized, however, technological catching up vis-à-vis high-wage-rate rivals may make their productivities high enough to functionally manufacture similar products with equal or lower unit costs.

As the product cycle theory (or *flying geese* theory) predicts, for a given new product, lower-wage developing countries may start by importing it from technologically advanced nations, but they may soon substitute such imports with domestic production and subsequently export the product itself (Akamatsu 1962, Vernon 1966). Accordingly, firms and sites making such a product in high-wage advanced nations may lose cost competitiveness and eventually decline.

Once domestic mass production of a certain product starts at the firms and sites in lower-wage-rate countries, they may increase their physical labor productivity (i.e., decrease labor input coefficient) and gain cost competitiveness through a combination of newer production technologies, economies of scale, and the experience curve effect. For instance, the famous Boston Consulting Group Report (1968) on the experience curve underlines that the total unit cost of producing and distributing a product tends to decline according to a certain constant percentage (e.g., 20%) with each doubling of a company's cumulative output. If this applies to the less experienced firms in low-wage-rate countries, they will certainly increase their productivity as time passes. Such evolution poses a great threat to firms in high-wage-rate countries because, in order to survive in this linked competition, they must increase their productivity at the same rate as their competitors if the wage rate ratio remains constant.

Hence, there is a sort of race between two aspects, each moving at its pace. One is the catching up pace of the physical labor productivity ratio of the two countries. This can be expressed as

$(\Delta u_J/u_J)$ / $(\Delta u_C/u_C)$. The other is the pace of wage catching up, expressed as $(\Delta w_C/w_C)$ / $(\Delta w_J/w_J)$. If the pace of productivity ratio catching up is faster than the pace of wage catching up, i.e., if

$$\frac{\Delta u_J}{u_J} - \frac{\Delta u_C}{u_C} > \frac{\Delta w_C}{w_C} - \frac{\Delta w_J}{w_J} \qquad (11-1)$$

with a constant margin, it is invetable that we will have the relation[50]

$$u_J\, w_J > u_C\, w_C \qquad (11-2)$$

after certain lapse of time. In other words, if trend (11-1) continues, low-wage country C will secure a cost advantage over high-wage country J. This process corresponds exactly to what the flying geese theory suggests.[51]

Yet, in the real history of international competition, it is not always the case that advanced nations lose competitiveness in older products or mature industries one after the other. For instance, Japan still exports considerable amounts of steel products, an apparently old sector, whereas some of its newer industries, such as that of semiconductors, have lost competitiveness since the 1990s. Thus, high-wage nations may lose competitiveness in certain mature industries (e.g., textiles in Japan), whereas they may retain comparative advantages in other mature industries (e.g., steel in Japan). We need to integrate the static theory of comparative advantages with the newer theories of product cycles in order to explain the abovementioned international trade dynamics. We believe that the new theory of international values provides a valid basis, since it contains the logic of how values and competitive techniques change when new production techniques are adopted.

In order to explain the reality of global competition, let us assume that the question regarding the viability of a single-product firm fundamentally depends on two constants: (1) the speed at which production costs are reduced and (2) the possible lower bound in reducing production costs. If the firm manufactures many products, it might abandon some of them, but it still has to remain competitive enough in producing other goods. The same logic works in the steel and textile industries, although the destiny of the two industries is completely different.

Imagine a situation in which near equality (10-1) holds. Our fundamental premise in this section is the constancy of wage rates (expressed by an international currency). This assumption is removed in the next section. First, let us suppose that the material cost difference $\langle \mathbf{b}_C, \mathbf{p}\rangle - \langle \mathbf{b}_J, \mathbf{p}\rangle$ is negligible with regard to unit labor costs. In order for a firm in high-wage-rate country J to survive, inequality (10-3) must be retained. If we use $\pi_J(t)$ and $\pi_C(t)$ to indicate the labor productivity (inverse of the labor input coefficient) of countries J and C at time t, this is equivalent to

[50]Expression (11-1) really means that $\{\log[\, w_J(t)\, u_J(t) / w_C(t)\, u_C(t)\,]\}' > 0$ for a time vaiable t. If this is always greater than constant δ, it is easy to see by integrating on the interval $[0, T]$ that $\log\, [w_J(T)\, u_J(T)\,] - \log\, [w_J(T)\, u_J(T)] > \delta T + \{\log(w_J(0)\, w_J(0) - \log\, (w_J(0)\, u_J(t)\} > 0$ for sufficiently large T.

[51]Here we assume a situation of pure labor input economy. When the material costs are the same in both countries, the argument applies with no modifications, but it needs to be modified appropriately if the material costs are different. See the pertinent part of Sect. 10.

$$\pi_J(t) \geqq (w_J(t)/w_C(t)) \cdot \pi_C(t). \qquad (11-3)$$

It is evident that the labor productivity of J firms must rise faster than that of C firms if the wage rate ratio remains constant. Can this race continue for a long time? The answer depends on the lower bound in (2). Indeed, does labor productivity grow without limits?

Production is a complex process and there are many ways to drive productivity up. Introducing a limit switch might eliminate machine supervising time, a small change in process may reduce handling time, and machining speed may be increased. However, if the upper bound for labor productivity exists, these efforts may become ineffective in the long run. The survival game cannot continue forever in the present form. Even though, if we have enough "grace period," the firm can adopt a new strategy by abandoning the current product and switch to a new product.

Hence, even a rough estimate of this upper bound might prove extremely useful. If we know it, we can guess how long the game will last. Two variables need to be kept in mind here. One is the density of effective work or value-adding time ratio. If we call $v(S)$ the value-adding time of site S within the total work hours measured in person-hours and $h(S)$ the total work hours of site S measured in person-hours, the value-adding time ratio is given by $v(S) / h(S)$. Another variable is the speed of value-adding. If we call $x(S)$ the volume of production of site S measured in units of the product, the speed of value-adding is given by $x(S) / v(S)$. All three variables should be analyzed over the same period of time, for example, a day or a week. Then, productivity $\pi(S)$ of site S is given by the formula

$$\pi(S) = x(S)/h(S) = x(S)/v(S) \cdot v(S)/h(S)$$

Value-adding is the design information transmission process from the worker/ equipment complex to the materials to be worked, as argued in Fujimoto (1999, Chap. 1). Taiichi Ohno, the innovator of the Toyota Production System, emphasized the importance of the density of value-adding time and adopted it as the leading index in his continuous improvement efforts (Ohno 1978).

In recent years, the value-adding time ratio $x(S) / v(S)$, or density of design information transmission, has been less than 10% in Japan's best-performing production sites. In other words, over 90% of the daily working time (e.g., 8 hours) is non-value-adding time, such as walking time, waiting time, equipment down time, and setup time. This number may sound very low, but the flip side is that there are ample opportunities for drastic productivity improvements. That is, by eliminating the abovementioned non-value-adding time (waste time or *muda*, in a broad sense) through continuous improvements of value flows at its sites, thereby increasing the value-adding time ratio x(S) / v(S) from, say, 5% to 15%, a firm can increase its physical productivity x(S) / h(S) by 3 times, other things being equal. In other words, even without increasing its speed of value-adding x(S) / v(S), which is often given by its production technology, significant increases in labor productivity are often possible even without radical process innovations.

Suppose that the upper bound of labor productivity can be increased three times yet for firms J. Suppose also that firms in countries J and C increase their labor productivity by 10% annually. It is evident that J firms will reach said upper bound much earlier than C firms. Even in this case, it is possible for firms in J to survive more than 11 years.

On the other hand, if massive inbound/outbound flows of workforce continue in the industrial regions of lower-wage country C, which makes its workers' average years of service shorter and average turnover ratio higher, the sites in country C may reach the upper bound of their productivities earlier than those in country J. This is likely to happen in those industries whose productivity improvements depend on coordination among multi-skilled workers, whose training and capability building require longer-term employment. To the extent that J's labor market is more stable than that of C, the upper bound of productivities at country J's coordination-rich production sites may be significantly higher than that at country C's sites, which may enhance the chances of survival of the former (Chapter "A Design-Information-Flow View of Industries, Firms, and Sites").

Besides the aspects illustrated above, we have to take into account changes in international wage handicap affecting higher-wage countries. We have assumed here that the wage rates of the two countries analyzed remain constant, but, as explained in the next section, if all industries in country C improve their labor productivity, country C's wage rate will also go up. The multiplier w_J / w_C will decrease and firms in country J will be able to slow down their productivity growth without losing cost competitiveness. As seen above, another possible scenario is that the pace of country C's productivity growth decelerates much earlier than that of country J when it approaches a certain level.

All these scenarios depend on the efforts of workers, production engineers, and managers at the production sites, as well as on the types of product technologies and architectures. According to our framework of design-based comparative advantage (Chapter "A Design-Information-Flow View of Industries, Firms, and Sites"), when product design and production are coordination intensive (i.e., integral product-process architecture), workers and engineers at relatively coordination-rich sites may achieve a higher upper bound of productivity and a higher pace of productivity increase than their rivals. In addition, if they possess *dynamic* or *evolutionary capabilities* to raise productivity, they may indeed survive for quite some time (Shiozawa 1990; Teece and Pisano 1994; Fujimoto 1999).

Another possible solution is to abandon the production of the product in question and switch to a new product or products. Investments in research and development are indispensable to pursue this strategy, which is also a way to prolong the firm's life. However, after a period, the wage rate discrepancy may decrease to nearly 1, and competition between country J and country C will enter a new phase.

When the material cost difference $\langle \mathbf{b}_C, \mathbf{p} \rangle - \langle \mathbf{b}_J, \mathbf{p} \rangle$ is substantial, a different feature of competition may arise. Among many input goods, there may exist some item whose processing is technically difficult and requires larger input coefficients. As long as C firms are affected by this kind of technical obstacle, J firms can survive regardless of the wage rate ratio.

In conclusion, comparative cost advantage is never determined by country characteristics, such as factor proportions. It is a most dynamic and complex element that requires a close investigation of the production process. The major variables that play a part in determining cost advantage are the wage rate ratio and the productivity ratio between two competing countries. In the next section, we will examine how the wage rate of a country moves.

12 Wage Rate Movements

In order to avoid unnecessary complications in our description, we now return to the assumption that no transportation cost is required. The situation in which wage rates stay constant is clear. Suppose that a regular international value $\mathbf{v} = (\mathbf{w}, \mathbf{p})$ is supported by a spanning tree T of production techniques. Let S be the set of production techniques that are competitive based on value $\mathbf{v}$. Adopting the same symbols used in Sect. 6, we can say that the international value stays constant when the economy with production volume vector $\mathbf{s} = (s_h)$ satisfies the following three conditions where $C(h)$ is the production capacity of the industry h and $\mathbf{d}$ is the world final demand[52]:

1.
$$s_h > 0 \text{ only when } h \in S \quad \text{and} \quad s_h \leqq C(h) \; \forall h \in S.$$

2.
$$J\mathbf{s} \leqq \mathbf{q}.$$

3.
$$A\mathbf{s} = \mathbf{d}.$$

In plain words, these conditions mean that, by using competitive production techniques, the economy can produce the world final demand $\mathbf{d}$ as the net product of each country's labor capacity. The international value is uniquely determined and remains constant as long as these condititons are satisfied. This is a direct consequence of Theorem 6-1, because by definition $T \subset S$.

In sharp contrast to neoclassical economics, a mere change in final demand does not change the value. Changes in demand composition are adjusted through corresponding changes in production volumes, but our focus here is on how wage rates and prices change. We should distinguish between two cases: (1) changes in the production technique of a product that is produced in one country, which we call monopolistic product, and (2) changes in the production technique of a product produced competitively in more than two countries, which, following Graham, we call *link commodity*.

---

[52]Compare these with the four conditions in Sect. 7. Conditions (1) and (3) are here replaced by a single condition (3). It is understood that T is in a general position. The world final demand $\mathbf{d}$ is not rquired to be on the production possibility frontier.

The direct effects of changes in the production technique of a monopolistic product are rather simple. Suppose that a production technique improvement is achieved. If we disregard the case of quality improvement, this means that the cost of production is lowered, which is the general form of productivity improvement. Firm managers have two options. Firstly, they can lower the product price. As long as the assumption that the markup rate is fixed remains valid, this is the only possible course of action. Yet, another solution exists, i.e., to raise the wage rate of their workers. In a more realistic case, firm mangers will rationally allocate the fruits of the productivity improvement. For example, they may decide to limit the price reduction to one third of the cost-down effect and distribute the remaining two thirds as a bonus or wage increase. Allocation percentages may depend on circumstances outside of firm. If price competition is severe, a larger portion will be assigned to price reduction. It is worth remembering that a number of firms normally compete in a country that exports a monopolistic product. Therefore, if the wage rates of other firms in the same industry are higher, the portion distributed as bonus or wage rate increase will be greater. Indeed, increasing the wage rate above the industry average actually constitutes rational judgment, because rewarding workers at the production site (genba), who have contributed to the improvement, is always a reasonable and wise managerial approach. In recent years, efforts to maximize shareholders' value have been rampant, but providing fair recompense to those who have helped make the firm more competitive is crucial, otherwise the company will lose its competency in the long run.

While the direct effects of productivity improvement are relatively simple, its indirect effects might be rather complex. Indeed, if the product is an input good, its pricing down affects the production cost of all the products that use it. The cost reduction is not proportional, because it depends on the composition of the input costs. If the product price drops substantially, this may bring about a disproportional change in cost. Firms have in general different coefficients for different inputs, even among firms producing the same product. Therefore, in an extreme situation, cost reductions may cause some firms to lose competitiveness with regard to certain products.

Another indirect effect of productivity improvement is a change in real wage rates. This may not be substantial when productivity is improved for only one or few products. But, if many products improve their productivity, the effect of these technical changes will be considerable. The real wage rate will grow in two ways, i.e., through pricing down or through wage rate improvement. This process is quite complicated, and it is difficult to predict what will actually happen. We might find answers in the results of empirical and historical research based on interviews and statistics.

The technical improvement of link commodities is more difficult to investigate, as many ramifications are possible. For the sake of simplicity, suppose that two firms in two different countries are competing. If firm A succeeds in innovating its production process and firm B cannot keep up, firm B will stop producing the product in question. This means that the link commodity will disappear. If, through this change, the set of competitive production techniques ceases to be spanning, the international value becomes unstable (Sect. 8). The relative wage rates may change between the countries. The whole picture of this change is very complex and is not described here, but the following discussion will analyze some typical cases.

Let us call country A and country B the two countries where firm A and firm B are located respectively. First, let us examine what happens in country A. If country A is below the full employment level, there is no immediate change in the wage rate. The price of the product is reduced but no specific variation occurs, except for any indirect effects of pricing down when the product is an input good. If country A has achieved almost full employment, the fact that firm B has abandoned production may induce an increase in country A's wage rate. The reason for this wage rate increase is different from the case of wage rates going up after productivity growth. Suppose that firms A and B originally shared the market fifty-fifty. The dropping out of firm B means that demand is now doubled. Firm A faces a difficult situation, because it has to increase its production volume but it cannot attract enough workers. If firm A employs a considerable share of the workforce, this induces a general labor shortage, and other firms in country A are obliged to compete with each other and will eventually raise their wage rates to attract more workers. As a result, the general level of wage rates in country A will increase, thus triggering a change in relative wage rates at the international level.

In contrast to the situation in country A, country B is affected by a decrease in employment. If this decrease is not compensated for by the arrival of new competitive industries, unemployment inevitably occurs or increases. This does not simply cause a reduction in the nominal wage rate (measured by the local currency). It is commonly observed that the wage rate remains constant even when the unemployment rate is high. However, the real wage rate may fall due to fluctuations in the exchange rate between the local currency and international currencies. We argued in Fujimoto and Shiozawa (2011–2012 Sect. 6) that the exchange rate in the floating system is determined so as to redress the balance of trade. This may not be true, because Japan experienced more than 40 years of trade surplus (1964–2010). Exchange rates seem to be influenced more by the balance of payments than by the balance of trade. The balance of payments includes services, income, transfers, and investments and does not closely reflect the state of the balance of trade (and service). Exchange rates among currencies in the floating system are extremely volatile and easily influenced by speculation. Their market often predicts the future correctly, but it overreacts to any minor news and unsupported expectations. It is, in a sense, self-fulfilling. If many speculators believe that the exchange rate will rise (or drop), it will indeed move in the expected direction so as to fulfill their expectations.

Note that the new theory of international values does not use the balance of trade (and services) and the balance of payments as determinants of exchange rates. International values and relative wage rates are determined independently of international payments. However, the foreign exchange market may react to changes in the competitiveness of a nation. Whatever their causes may be, major shifts in the exchange rate imply changes in the wage rate measured in an international currency. If the exchange rate moves and then remains at the new level, this will be the change in relative wage rates. Such exchange rate adjustments are not incorporated into the new theory, and, in this regard, we may say that the new theory of international values is not well harmonized with the existence of foreign exchange markets (see Shiozawa 2017a Sect. 6).

Another interesting case of technical change affecting link commodities is when two or more countries improve their production techniques by emulation. If each country improves its production technique independently, the pricing down rate will vary from country to country. But emulation may lead to a situation in which all concerned firms succeed in achieving the same pricing down rate. The product will continue to be a link commodity, while its price will drop at the same pace in all concerned countries. We will look more closely at the effects of emulation in the next section.

Exchange rate changes have a similar effect on firms in a country where the currency is evaluated compared to foreign currencies. Those firms face the situation that their product price is raised in terms of foreign currencies even though it remains constant with regard to the home currency. Relatively speaking, this is equivalent to a situation in which the competitors' product price was cut down owing to their product cost down. The only difference is that the effect is felt by all the firms in the country. If many of them have the dynamic capability of bringing down costs, they may succeed in nullifying the effects of evaluation. The product prices in foreign currencies may stay as competitive as before the evaluation and the currency may be reevaluated again. This causes a dynamic loop between the exchange rates and the general efforts of firms to cut down costs. This is an example of a micro-macro loop (Fujimoto-Shiozawa 2011–2012 Sect. 8). Krugman (1979 p 262) and Dosi et al. (1990 p 22) compared it to Red Queen's race. In the case of Japan, something similar happened between the Nixon shock and Deng Xiaoping's reform and opening-up. As the yen appreciated, Japanese firms continued their productivity race and caused the yen to rise further. All export-oriented enterprises were obliged to keep cutting down their costs in order to maintain their position in international competition.

13 Overall Patterns of Specialization

It is often understood that the Ricardian trade theory assumes complete specialization. In Sect. 3, we showed that this is a misleading tradition originated at the time of John Stuart Mill, and the new theory of international values does not assume complete specialization. As emphasized in Sect. 3 above, complete specialization is impossible when the number of commodities is greater than the number of countries. However, if there are no transportation costs (including transaction costs and tariffs), the trade economy is in general highly specialized. Indeed, if M is the number of countries and N the number of commodities, the number of competitive production techniques is in general $N + M-1$. The number of goods commonly produced in more than two countries (*link commodity* or *common commodity* in Graham's terminology) depends on the spanning tree, but does not exceed $M-1$ in general. This seems highly unrealistic, as we observe that a large number of commodities, probably in the order of hundreds of thousands of goods, are commonly produced in more than two countries.

This state of affairs can be explained within the framework of the new theory of international values. There are three possible reasons or mechanisms that cause this multiplicity of commonly produced goods: (1) transportation costs, (2) fine differentiation, and (3) emulation effects. Let us briefly explain each of these mechanisms.

13.1 Transportation Costs

We have already addressed the question of transportation costs when discussing the second meaning of the near equality in Sect. 10. As we noted there, positive transportation costs may drastically change the features of specialization.

Transportation costs is here an umbrella term which comprises transaction costs and tariffs to cross borders. Although they have greatly decreased in recent years, transportation costs are still an important factor that separates the market of a country from that of another country. Suppose, for the sake of simplicity, that transportation costs amount to 20% (i.e., 1/5) of the value of the commodity. Then, condition (10-1) that a product is commonly competitive can be relaxed into

$$5/6\left(w_C\,u_C + \langle \mathbf{b}_C, \mathbf{p}\rangle\right) < w_J\,u_J + \langle \mathbf{b}_J, \mathbf{p}\rangle < 6/5\left(w_C\,u_C + \langle \mathbf{b}_C, \mathbf{p}\rangle\right) \qquad (12-1)$$

If we focus on the cost of country J, it can move from 5/6 up to 6/5 of the cost of country C. The total range of permissible cost rate over the cost of country C is $(6/5)^2 = 1.44$. In this ample range of production costs, the pattern of specialization does not change and the two countries can produce the same product in common.

The proportion of transaction costs over product price changes from commodity to commodity and, in some cases, it is prohibitively high. For example, most face-to-face services are difficult to transport to another country, but ICT has drastically changed the cost of transportation. Now many services, such as claim centers (customer centers), text editing, programming, and other services, are outsourced from the USA to India. In spite of these exceptions, most services are still carried out domestically and major differences in service prices are observed across countries. In high-wage-rate countries, service prices and prices of other non-traded commodities are often several times higher than in low-wage-rate countries. This can be explained if we assume that the disparity in physical labor productivity of non-tradable commodities is relatively small compared with that of tradable commodities. The famous Balassa-Samuelson hypothesis (in its original form) assumes differences in productivity growth rates, but it seems better to assume simply that the relative disparity in productivity is smaller in non-tradable commodities and in services in particular.

Tradability is a fuzzy concept. There is a wide range of transition from highly tradable goods to practically non-tradable goods. As for highly tradable goods, international price competition is severe, whereas no such competition exists in non-tradable goods. The law of one price does not apply to the latter. This explains why there are so many commodities that are produced in two or more countries.

13.2 Fine Product Differentiation

In international input-output tables, we observe positive entries almost everywhere. Among developed countries in particular, trade within the same industry is generally bilateral. This is not the real problem of the new theory of international values but rather a question of classification and tabulation.

The new theory assumes that the number of commodities is enormous, possibly exceeding 10 million. Let us suppose for the moment that they are all different goods. The United Nations' Harmonized Commodity Description and Coding Systems (2017 revision) features about 5300 items appearing as headings and sub-headings. If we divide 10 million by 5300, each item includes on average more than 1800 different commodities. A possible situation is that country A exports many products to country B and country B exports completely different products to country A. In this case, trade statistics tables show bidirectional intra-industry trade within the same commodity category, but, in reality, this may simply be the result of aggregation, which is inevitable in statistics.

The real problem lies in the concept of *the same commodity*. In a modern economy, most commodities are characterized by a brand name. If we understand *the same* to mean the same product of the same brand, the case of bidirectional trade between countries may be rarer. But the case of Latin American IBM users once prefered machines imported from Japan to those made in the USA teaches us that commodities of the same brand are not the same to all buyers if they are produced in a different country or factory.

Trade theories traditionally assume that a commodity is always the same regardless of where it is produced—and the new theory of international values in no transportation case belongs to this category. But Staffan B. Linder (1961) proposes to regard goods produced in different countries as different commodities,[53] thus implying that nations trade with one another in similar but differentiated goods. Dosi et al. (1990 p 165) contrasted their technology gap model with Linder's treatment. The former is essentially a supply-side model, whereas the latter is a demand-based theory. We can make a similar comparison. The new theory is a supply-side theory, in the sense that the international value is essentially determined by the distribution of production techniques. However, Dosi et al. point out that the two theories are theoretically compatible. The new theory provides a finer interpretation of how wage

[53]This is not exactly the same as Linder's hypothesis.

rates and prices are determined, whereas Linder explores how demand for a similar commodity is distributed among different producers. We can even say that the two approaches are not only compatible but in fact complementary. The new theory mainly investigates how the international value is determined. When this is done, it is the procurers who decide how much to buy and, consequently, how much the product sells. Then, the producers produce as much as their product sells (Sraffa principle).

If we accept Linder's proposal, we can explain why similar (or practically the same) products are exported and imported between two countries. We can cite the idea of *love of variety* to explain why bidirectional trade within a finely classified category is widely observed, especially between developed countries. The richer people become, the stronger their love of variety. Yet, it is worth noting that love of variety is not a phenomenon which can be assumed only at the individual level. A group of individuals may display love of variety, while each individual has a *strong preference*, in the sense that each individual chooses a single variant from among a group of products.[54]

Linder's proposal is useful but cannot be accepted in full. In the new theory of international values, the near equality of type (10-1) is essential for the definition of international values.

13.3 Emulation Effect

When we announced Theorems 6-1 and 6-2 (or Theorems 5-1 and 5-3), we assumed the input coefficients to be in a general position. The concept of *general position* has a precise definition, but here it is enough to imagine that the coefficients are chosen by pure chance. In this case, it is normally difficult for the set of competitive production techniques to contain more production techniques than $M + N-1$ if the equality is interpreted in a strict way. However, there is an economic mechanism that causes near equality (10-1) to hold more often than when the input coefficients are chosen by pure chance.

As we observed in Sect. 10, when a firm runs the risk of losing competitiveness, its employees will try to improve productivity in order to save their firm and their jobs. So, relation (10-1) is found more often than it would if it were left to pure chance. This is the result of rivalry, or mutual emulation, which increases the number of goods produced in common.

[54]The Dixit-Stiglitz utility function is given by $U(x_1, \ldots, x_n) = (x_1^{\sigma} + \ldots + x_n^{\sigma})^{1/\sigma}$ with $\sigma < 1$. Each individual displaying a strong preference has a similar utility function but with $\sigma > 1$. See chapter "Product Variety for Effective Demand Creation" of this book, Sect. 2.1.1.

14 Some Notes on the Demand Side

So far, we have mentioned demand only in passing, but this does not mean that we believe it to be of small importance. This chapter is based on the new theory of international values, which forms a part of the classical theory of value (Shiozawa 2016). The former naturally inherits the major characteristics of the latter. One is the basic separation between values and quantities (Pasinetti 1993 p 19). In sharp contrast to the neoclassical theory of value, which assumes the simultaneous determination of value and quantity, the classical theory of value does not regard values as determined by a balance of demand and supply. Instead, it claims that prices are determined mainly by production costs, and production volume is adjusted to the demand for each product (Sraffa's principle or principle of effective demand at the firm level; see also Shiozawa et al. in press).

The main objective of this chapter was to investigate how firms in advanced countries fare in international price competition while being handicapped by big differences in wage rates. In this investigation, demand has played only a subordinate role, because demand is only passively determined when values (i.e., wage rates and prices) are determined. However, demand plays a decisive role in determining the production volume of a firm or a plant. Production volume is directly related to various factors, such as working hours, number of employees, gross profit, and profit rate.[55] Therefore, demand has a critical meaning for firms and workers, since the survival of a plant may depend on whether it can secure enough demand for its products.

Of course, product prices influence product demand.[56] Any differentiated product can get a certain amount of demand, whatever its price, but demand may be extremely low if the price is high compared to product quality. Even when the price is competitive, i.e., comparable to the prices of other competing products, the firm or plant may not achieve sufficient demand volumes. Sales efforts are always vital for the subsistence of a firm or plant.

Even when a firm or plant succeeds in keeping its product costs competitive, it may still have to face a serious problem. Suppose that a firm or production unit in a high-wage-rate country is forced to decrease its product price to deal with competition from firms in low-wage-rate countries. Also suppose that the product price is kept competitive owing to pricing down through increased productivity. Finally suppose that the firm or unit manages to keep demand nearly constant. If the cut in cost is achieved through a reduction in material input coefficients, there is no problem for the firm or plant, although this may trigger a reduction in demand for firms producing input goods. However, if the cost drops thanks to a reduction in physical labor productivity, the working hours needed to manufacture the product will go down because, by assumption, demand for the product remains stable.

[55]See footnote 33.

[56]Product price is only one of the determinants of product demand. Many other factors, like weather, trade cycle, fads, and various events, contribute to changes in demand flows.

In international competition for link commodities, firms or plants in high-wage-rate countries face the following dilemma: increase productivity or maintain current employment levels. This conundrum cannot be solved by considering only the production of a single product. To keep employment levels constant, managers are required to explore new potential demand for the product. If this is impossible, the firm will necessarily have to develop a new product or products. This is one of the reasons why firms and plants are increasingly becoming multiproduct entities.

It is worth noting that there is complete parallelism among plants, firms, industries, and countries with regard to employment maintenance (Fujimoto 2012a). Full employment is one of the primary targets of a country's economic policies. Let us imagine a high-wage-rate country competing on the international market. To retain competitiveness, it needs to boost labor productivity. If overall demand for the country's products remains constant, total working hours and employment levels will most likely drop.

These arguments are rather banal when referring to policy debate at the national level, but they are seldom heard when observing the firm-level situation. The reason for this is easily identified, since in standard economic analyses, the firm-level objective is solely the pursuit of profit. However, as we emphasized in Sect. 10 above, Fujimoto (2012a), Fujimoto (2018a, b, and c) and many other papers published in Japanese by Fujimoto, firms and plants (or production sites) are going concerns. A firm or a plant as going concern is a combination of different stakeholders, including workers. Maintaining employment is an imperative not only to avoid resistance by workers but also because of pressures from the local community. Productivity enhancement and demand creation are two efforts in this direction that are widely observed in Japanese firms and plants (see also Ikuine 2018a, b and other chapters in Fujimoto and Ikuine 2018).

The main requirement for a firm (or plant of a multinational enterprise) is to ensure its subsistence. Hence, the criterion of *competitiveness* must include a subsistence condition. If this is the case, it is also evident that competitiveness must be supported by a *dynamic capability* to pursue the combined imperatives of productivity enhancement and demand creation (Shiozawa 1990; Teece and Pisano 1994; Fujimoto 2012a). Such capability building competition is fought at a deep level among firms and plants, although on the surface competition manifests itself in terms of price, quality, and new products.

15 Conclusions

At the beginning of this chapter, we claimed that international competition among firms and among production sites is a game with wage rates as handicaps. Although it is instinctively trivial and evident, this image of international competition is quite different from the traditional comparative advantage argument. Supporting the new view requires a new theory. Sections 2, 3, 4, 5, and 6 presented an introduction to the new theory of international values. The second part of the chapter, Sects. 7, 8, and 9, offered some preliminary remarks on how to use it. In the third part (Sects. 10, 11, and 12), we proved that our claim about the nature of international competition is justified on the basis of our new approach. In the last part of our discussion (Sects. 13 and 14), we offered some remarks regarding the wider scope of application of the new theory.

References

Akamatsu K (1962) A historical pattern of economic growth in developing countries. Dev Econ 1:3–25

Althusser L (1965) Sur la dialectique matérialiste, Dans Lous Althusser, Pour Marx. François Maspéro, Paris. Repropduit en 1996 avec un avant-propos d' Etinenne Balibar, La Decouverte, Paris. English edition: Althusser L (1969) For Marx (trans: Brewster B). Verso, London

Armington PS (1969) A theory of demand for products distinguished by place of production. IMF Staff Pap 16:159–176

Arrow KJ (1962) The economic implications of learning by doing. Rev Econ Stud 29:155–173

Berge C (1971) Principles of combinatorics. Academic Press, New York

Blonigen BA, Wilson WW (1999) Explaining Armington: what determines substitutability between home and foreign goods? Can J Econ/Rev Can d'Econ 32(1):1–21

Boston Consulting Group (1968) Perspectives on experience. Boston Consulting Group Inc., Boston

Chipman JS (1965) A survey of the theory of international trade: part I the classical theory. Econometrica 33(3):477–519

David PA (2001) Path dependence, its critics and the quest for 'historical economics'. In: Garrouste P, Ioannides S (eds) Evolution and path dependence in economic ideas: past and present. Edward Elgar, Cheltenham, pp 15–40

Deardorff AV (2005) Ricardian comparative advantage with intermediate inputs. N A J Econ Fin 16 (1):11–34

Dorbusch R et al (1977) Comparative advantage, trade, and payments in a Ricardian model with a continuum of goods. Am Econ Rev 67(5):823–839

Dosi G et al (1990) The economics of technical change and international trade. New York University Press, New York

Faccarello G (2017) A calm investigation into Mr Rciardo's principles of international trade. In: Senga et al (eds) Ricardo and international trade. Routledge, Abingdon, pp 85–119

Freeman C (1997) The economics of technical change. In: Archibugi D, Michie J (eds) (1998) Trade, growth and technical change. Cambridge University Press, Cambridge

Fujimoto T (1999) The evolution of a manufacturing system at Toyota. Oxford University Press, New York

Fujimoto T (2007) Architecture-based comparative advantage: a design information view of manufacturing. Evol Inst Econ Rev 4(1):55–112

Fujimoto T (2011) Keizai Kyoshitsu (Economics classroom). The Nikkei, March 29, 2011

Fujimoto T (2012a) Evolution of firms and industries. Evol Inst Econ Rev 9(1):1–10

Fujimoto T (2012b) Kyōsōryoku Kōchiku no tame no Genka Keisan Shiron: Sekkei Jōhō Tensharon ni motozuku Zenbu Chokusetsu Genka Keisan no Kanōsei (A preliminary note on harmonizing manufacturing (Monozukuri) management and cost accounting: a possibility of full-and-direct costing based on design-information theory of manufacturing). MMRC Discussion Paper Series no. 410

Fujimoto T (2017) An architectural analysis of green vehicles – possibilities of technological, architectural and firm diversity. Int J Automot Technol Manag 17(2):123–150

Fujimoto T (2018a) A design-information-flow view of industries, firms, and sites. In: Fujimoto T, Ikuine F (eds) Industrial competitiveness and design evolution, Evolutionary economics and social complexity science, vol 12. Springer, Tokyo, pp 5–41

Fujimoto T (2018b) Capability building and demand creation in "Genba-Oriented Firms". In: Fujimoto T, Ikuine F (eds) Industrial competitiveness and design evolution, Evolutionary economics and social complexity science, vol 12. Springer, Tokyo, pp 123–154
Fujimoto T (2018c) Evolution of organizational capabilities in manufacturing-the case of the Toyota Motor Corporation. In: Fujimoto T, Ikuine F (eds) Industrial competitiveness and design evolution, Evolutionary economics and social complexity science, vol 12. Springer, Tokyo, pp 191–221
Fujimoto T, Ikuine F (2018) Evolution of industries and firms — Capability building and demand creation. Spinger, Singapore
Fujimoto T, Shiozawa Y (2011–12) Inter and intra company competition in the age of global competition: a micro and macro interpretation of Ricardian trade theory. Evol Inst Econ Rev 8 (1):1–37, 8(2):193–231
Graham FD (1923) The theory of international values re-examined. Q J Econ 38(1):54–86
Graham FD (1932) The theory of international values re-examined. Q J Econ 46(4):581–616
Graham FD (1948) The theory of international values. Princeton University Press, Princeton
Haberler G (1936) The theory of international trade. William Hodge, Edinburgh. Originally published in German, 1933
Ikuine F (2018a) Creating new demand: the impact of competition, the formation of submarkets and platforms. In: Fujimoto T, Ikuine F (eds) Industrial competitiveness and design evolution, Evolutionary economics and social complexity science, vol 12. Springer, Tokyo, pp 317–331
Ikuine F (2018b) Decline in demand creation: the development productivity dilemma and its consequences. In: Fujimoto T, Ikuine F (eds) Industrial competitiveness and design evolution, Evolutionary economics and social complexity science, vol 12. Springer, Tokyo, pp 333–347
Jones RW (1961) Comparative advantage and the theory of tariffs: a multi-country, multi-commodity model. Rev Econ Stud 28(3):161–175
Jones RW (2000) Globalization and the theory of input trade. MIT Press, Cambridge, MA
Jones RW, Kierzkowski H (1990) The role of services in production and international trade: a theoretical framework. In: Jones RW, Krueger AO (eds) The political economy of international trade. Basil Blackwell, Oxford
Jones R, Kierzkowski H (2001) A framework for fragmentation. In: Arndt S, Kierzkowski H (ed) Fragmentation: new production patterns in the world economy and international trade. Oxford University Press, Oxford, Chapter 2, pp 17–34
Kinoshita E (2003) Waga Kōseki: Kokusai Keizaron Tankyū no Tabi (My wake: a journey in pursuit of international economics). Tohoku Daigaku Shuppankai, Sendai
Krugman P (1979) A model of innovation, technology transfer and the world distribution of income. J Polit Econ 87(2):253–266
Krugman P (1991) History versus expectations. Q J Econ 106(2):651–667
Krugman P, Obstfeld M (2009) International economics: theory and policy, 8th edn. Pearson as Addison-Wesley, New York
Linder SB (1961) An essay on trade and transformation. Almqvist & Wiksells, Stockholm
Malerba F et al (2016) Innovation and the evolution of industries: history-friendly models. Cambridge University Press, Cambridge
Maneschi A (2004) The true meaning of David Ricardo's four magic numbers. J Int Econ 62 (2):433–443
Marshall A (1879) The pure theory of foreign trade. Privately published and circulated by Sidgwick
McKenzie L (1953) Specialization and efficiency in world production. Rev Econ Stud 21 (3):165–180
Mill JS (1844) Essays on some unsettled questions of political economy. John W. Parker, London. See in particular: Essay I of the laws of interchange between nations, and the distribution of the gains of commerce among the countries of the commercial world
Mill JS (1848) Principles of political economy with some of their applications to social philosophy. John W. Parker, London. See in particular: III.XVIII of international values
Ogawa T (2017) Analysis of production-efficient patterns of specialization allowing intermediates inputs: the meaning of Shiozawa's model from the viewpoint of modern economics. In: Shiozawa et al (eds) A new construction of Ricardian theory of international values. Springer Nature, Singapore, pp 123–148

Ohno T (1978) Toyota seisan hōshiki (The Toyota production system). Daiyamondo-sha, Tokyo
Ohlin B (1933) Interregional and international trade. Harvard University Press, Cambridge, MA
Oka T (2017) The relation between value and demand in the new theory of international values. In: Shiozawa et al (eds) A new construction of Ricardian theory of international values. Springer Nature, Singapore, pp 99–121
Pasinetti L (1993) Structural economic dynamics: a theory of the economic consequences of human learning. Cambridge University Press, Cambridge
Petri PA (1980) A Ricardian model of market sharing. J Int Econ 10(2):201–211
Pullen J (2006) Did Ricardo really have a law of comparative advantage: a comparison of Ricardo's Version and the modern version. Hist Econ Rev 44(1):59–75
Ricardo D (1951[1817–21]) On the principles of political economy and taxation. Edited by P. Sraffa, Volume 1 of The works and correspondence of David Ricardo. Cambridge University Press, Cambridge. https://books.google.co.jp/books/about/On_the_Principles_of_Political_Economy_a.html?id=cUBKAAAAYAAJ&redir_esc=y
Ruffin R (2002) David Ricardo's discovery of comparative advantage. Hist Polit Econ 34 (4):727–748
Sato H (2017) An overview of research into international values in Japan. In: Shiozawa et al (eds) A new construction of Ricardian theory of international values. Springer Nature, Singapore, pp 281–303
Shiozawa Y (1990) Shakai no Gijutsuteki Nōryoku: Seiteki Gainen to Dōteki Gainen (Technical capability of the society: static and dynamic concepts). In: Nakaoka T (ed) Gijutsu Keisei no Kokusai Hikaku (International comparison of technology formation: social capability for industrialization). Chikuma Shobo, Tokyo, pp 333–361
Shiozawa Y (2007) A new construction of Ricardian Trade Theory – a many-country, many-commodity case with intermediate goods and choice of production techniques. Evol Inst Econ Rev 3(2):141–187
Shiozawa Y (2014) Rikādo Bōeki Mondai no Saishū Kaiketsu (A final solution of Ricardo problem on international values). Iwanami Shoten, Tokyo
Shiozawa Y (2015) International trade theory and exotic algebra. Evol Inst Econ Rev 12 (1):177–212
Shiozawa Y (2016) The revival of the classical theory of values. In: Yokokawa et al (eds) The rejuvenation of political economy. Routledge, Abingdon UK, pp 151–172
Shiozawa Y (2017a) The new theory of international values: an overview. In: Shiozawa et al (eds) A new construction of Ricardian theory of international values. Springer Nature, Singapore, pp 3–73
Shiozawa Y (2017b) An origin of the Neoclassical Revolution: Mill's "Reversion" and its consequences. In: Shiozawa et al (eds) A new construction of Ricardian theory of international values. Springer Nature, Singapore, pp 191–241
Shiozawa Y (2017c) On Ricardo's two rectification problems. In: Senga et al (eds) Ricardo and international trade. Routledge, Abingdon, pp 195–215
Shiozawa Y, Morioka M, Taniguchi K (in press) Microfoundations of evolutionary economics. Springer, Singapore
Tabuchi T (2017a) Comparative advantage in the light of the old value theories. In: Shiozawa et al (eds) A new construction of Ricardian theory of international values. Springer Nature, Singapore, pp 265–280
Tabuchi T (2017b) Yukizawa's interpretation of Ricardo's 'Theory of Comparative Costs'. In: Senga et al (eds) Ricardo and international trade. Routledge, Abingdon, pp 48–59
Tatsumoto H (2018) Evolution of business ecosystems. In: Fujimoto T, Ikuine F (eds) Industrial competitiveness and design evolution, Evolutionary economics and social complexity science, vol 12. Springer, Tokyo, pp 155–188
Teece DJ, Pisano G (1994) The dynamic capability of firms: an introduction. Ind Corp Chang 3 (3):537–556
Vernon R (1966) International investment and international trade in the product cycle. Q J Econ 80 (2):190–207
Viner J (1937) Studies in the theory of international trade. Harper, New York

Product Variety for Effective Demand Creation

Yoshinori Shiozawa

Abstract Product variety is a crucial dimension for successful management. While numerous papers deal with this subject, few provide insights into estimating optimal product variety in a concrete situation. Their results are much too sensitive to the assumptions, and their conclusions are often contradictory. This chapter presents a new method to estimate optimal variety for a firm, which depends on the notion of estimated coverage function, defined and explained in the text in detail. All the complex issues related to optimal variety estimation revolve around the estimation of this function, allowing us to discern which issues are crucial and which are not. The formula obtained here is simple and intuitive, and no similar formula has ever appeared in the literature before. It provides various hints regarding optimal variety estimation by means of costs and sales expectations. The results can be extended to the case of oligopolistic competition, and our investigation is concluded by drawing a comparison between a monopoly case and an oligopoly case.

1 Introduction

Product variety in an economy is a crucial variable when considering economic development, growth, and demand satiation (Thirsk 1978; Pasinetti 1981; Saviotti 1996; Witt 2001; Aoki and Yoshikawa 2002; Matsumae 2005; Saviotti and Pyka 2008; de Vries 2008; Ciarli and Lorentz 2010). Based on the fact that the number of products circulating in human society has increased from several hundreds some fourteen thousand years ago to millions of commodities available today, Kauffman

---

The main part of this chapter comes from a paper originally published in 2012 under the title "Estimating Optimal Product Variety of Firms," Evolutionary and Institutional Economics Review 9(1):11–35. The title has been modified in accordance with the theme of this book, and a few corrections and changes have been made.

Y. Shiozawa (✉)
Osaka City University, Osaka, Japan
e-mail: y@shiozawa.net

T. Fujimoto, F. Ikuine (eds.), *Industrial Competitiveness and Design Evolution*, Evolutionary Economics and Social Complexity Science 12,
https://doi.org/10.1007/978-4-431-55145-4_3

(2000, Chap. 9) underlines that economic growth is closely related to variety growth for what concerns both goods and services.[1] Achieving optimal product variety is one of the most strategic decision-making processes for firms.

The ultimate goal of this chapter is to build a general theory of product variety applicable to both specific product groups, such as beers and automobiles, and to the economy as a whole (Saviotti 1996, p 97). Nevertheless, this is a rather daunting task, and here we confine our attention to a very specific problem, i.e., estimating optimal variety for monopolistic firms (and, partly, oligopolistic firms). Certain aspects, such as variety creation, product architecture, and variety implementation, are not addressed here, and insights into these topics may be found elsewhere (e.g., Ramadas 2003).

The question of determining optimal product variety is an actual problem that all producers of consumer goods face (Ho and Tang 1998; Sorenson 2000; Krishnan and Ulrich 2001). Logically, optimal variety should be decided before embarking on any product development project. Although it is evident that calculating the number of items to produce is necessary, this does not mean that an accurate estimation is easy to obtain. On the contrary, it is extremely difficult, both in theory and in practice, to correctly estimate the optimal number of items and optimal profile of product specifications (these two expressions are defined later).

After the seminal paper by Hotelling (1929) on oligopolistic competition, Lancaster (1966) introduced the idea of characteristics space. There have been countless discussions on this topic (survey papers include Waterson 1989; Lancaster 1990; Böckem 1994; Manez and Waterson 2001; and Brenner 2001; while Murata 2009 and Caminal and Granero 2012 are two examples of recent research papers). However, most of them appear to be simplified narratives, and many resort to sophisticated and sometimes complicated calculus, while their models have little relevance to the real world.

The investigation that follows focuses on a possible estimation of product variety made by the staff of a firm on behalf of its top management. Yet, our concern is not what the consequences of this estimation might be for the economy as a whole (except Sect. 5.5). Therefore, our analysis is not embedded in a general equilibrium framework. We are rather doubtful about the possibility of determining socially optimal product variety. Instead, an attempt is made to make the discussion as concrete as possible by providing detailed background assumptions. As the basics of our reasoning are clear, we hope that others might extend the present formulation to more complex cases.

The chapter is organized as follows. Section 2 forms the core of our exploration, and Sect. 2.2 presents a new method for optimal estimation that relies on a function estimating maximal sales for a given number of products. Once this function is given, optimal product variety is easy to calculate by means of Eq. (3). To the best of

[1]The number of products available today may be even greater. In our opinion, more than one hundred billion commodities were being produced in Japan at the beginning of the twenty-first century.

our knowledge, no similar formula has appeared in the literature before. Our new formula sheds light on how optimal variety changes in relation to changes in fixed costs, variable unit costs, product lifecycle, and other characteristics. All the difficulties in estimating optimal product variety are confined to the estimation of the expected coverage function.

Before the formula is introduced, all the necessary assumptions (or explanations of the situation in which the new method is applicable) are provided in Sect. 2.1. Although some explicit arguments about the differences between the Dixit-Stiglitz utility and the strong preference hypothesis are presented as background discussion, no particular assumptions on the Lancaster characteristics space are made. The space can be of any dimension, with any properties taken as axes. Therefore, once the sales function is estimated, our results are valid for both vertical and horizontal differentiation, in the sense used by Manez and Waterson (2001). Section 2.3 offers some remarks about the formula, while Sect. 2.4 considers when the decision-making should be formulated as an integer problem.

Section 3 explains how our approach can be extended to broader situations by providing a suitable interpretation. Section 4 introduces some additional comments to ensure that the formulation is correctly applied. Section 5 addresses the matter of oligopolistic firms competing over the same group of products. The main assumption in this section is that each firm keeps its market share. While this hypothesis simulating oligopolistic competition may be too strong, it allows the total number of products to be compared in both the monopoly and oligopoly cases. In the literature on product variety and multiproduct firms, many contradictory observations have been made about whether oligopolistic competition leads to an excess or shortage of products compared to the socially optimal number. While the present chapter says nothing about social welfare problems, based on the assumption of the market shares that firms presume, a proposition is obtained indicating that a greater number of product varieties would be produced in the oligopolistic market than in the monopolistic market. Section 6 concludes our analysis.

2 Calculation of Optimal Variety

2.1 Background Assumptions

Before introducing the mathematical formulation, the situation assumed in this analysis must be explained. Lancaster's four possible assumptions are taken as a frame of reference, as they allow variety within a product group to persist in a market economy (Lancaster 1990, p 190):

1. Each individual consumer seeks variety in his or her own consumption.
2. Different consumers want different variants because tastes vary.
3. Individual firms can increase profits by producing a variety of models.

4. Firms can increase profits by differentiating their products from those of their competitors.

2.1.1 Strong Preference Hypothesis

In relation to Lancaster's first two assumptions, the present study more specifically assumes that consumers display *strong preference*. Each individual chooses a single variant from among a group of products based on a range of relative prices and level of income.

This assumption is in sharp contrast to the well-known Dixit-Stiglitz utility function (Dixit and Stiglitz 1977), which is sometimes referred to as *love of variety* behavior (see, e.g., Helpman and Krugman 1985). A Dixit-Stiglitz utility function is convex to the origin (i.e., the set of all the points that are preferred or indifferent to a point is convex). A consumer with this utility chooses to buy all kinds of varieties at a given time, and each variety has the same relative utility for any unit of money consumed.

Strong preference is the extreme opposite of the Dixit-Stiglitz preference (i.e., the preference represented by a Dixit-Stiglitz utility function). Each consumer chooses only one among many variants within the same product group. In the case of automobiles, these variants may be brands, models, and specifications. While the indifference set of the Dixit-Stiglitz preference is convex to the origin, strong preference gives an indifference curve which is strongly concave to the origin. Concrete examples of utility functions with strong preference are easily given:

$$U(x_1, x_2, \ldots, x_n) = \left(x_1{}^{\xi} + x_2{}^{\xi} + \ldots + x_n{}^{\xi}\right)^{1/\xi},$$

where ξ is any positive number greater than 1. These functions have the same form as Dixit-Stiglitz utility functions, and the only difference is that index ξ is smaller than 1 in the Dixit-Stiglitz preference case, whereas it is bigger than 1 in the strong preference case. Indeed, strong preference utility functions are nothing but a L^p-norm or p-norm. As such, a set of points defined by

$$U(|x_1|, |x_2|, \ldots, |x_n|) \leqq C$$

is convex (in other words, the indifference curve is strongly concave to the origin).

In the Dixit-Stiglitz preference case, it is normally assumed that individuals have the same utility functions. A representative individual chooses all the possible variants from among the product group, and the rest of the people act similarly. In the strong preference case, each individual has his or her own parameter (a_1, a_2, ..., a_n), and the function of this individual may be given by

$$U(x_1, x_2, \ldots, x_n) = \left((x_1/a_1)^{\xi} + (x_2/a_2)^{\xi} + \ldots + (x_n/a_n)^{\xi}\right)^{1/\xi}.$$

As opposed to Lancaster's characteristics space, a point of which specifies a product, these parameters belong to the personal characteristics space, a point of which specifies the character of an individual. If the price vector $(p_1, p_2, \ldots, p_n)$ for the group of products is given, then almost all individuals (measured according to the ordinary Lebesgue measure) have only one product as their favorite, among the set of vectors which satisfy a budget constraint. Individuals with the Dixit-Stiglitz preference can be represented by a person (since they are all similar), and the maximal utility point is given as an inner solution. In the strong preference case, individuals are all different, and each has his or her favorite specifications. Products chosen by each individual are corner solutions.[2]

In the real economy, the Dixit-Stiglitz preference may apply when choosing a restaurant. On different occasions, a person may prefer to eat Japanese, Chinese, French, and so on. Conversely, strong preference may apply when choosing a car, PC, home, and others. But there are many groups of products which do not belong to either of these typical extreme categories.

The strong preference assumption is closely related to the concept of the same group of products, which has not yet been defined. Any set of products may display a certain degree of similarity (see Watanabe's Ugly Duckling theorem, Watanabe 1969). In this sense, any set of products can be taken as a starting point for the present discussion. It is convenient to consider that a set of products forms the same group of products when the strong preference property holds.

2.1.2 Simple Cost Assumption

In relation to Lancaster's third and fourth assumption, two more concrete assumptions are made here. One is the simple cost assumption, and the other is Sraffa's principle of production or effective demand principle at the firm or product level.

The simple cost assumption means that a firm incurs two kinds of costs in the production of a new commodity. The first kind is fixed costs, which cover all expenses related to development, designing, setting up of a specific line for the product, and training.[3] The other is proportional costs, which cover the wages of production workers, all inputs (materials, parts, fuel, and utilities), and depreciation of machines and installations. While they are important, cost interactions among different products are disregarded. For example, the development cost of a product generally decreases through learning by doing, as the firm accumulates experience in developing similar products, but we ignore such interactions (for these aspects, see Sorenson 2000).

[2] Indeed, one of the reasons why strong preference is not adopted by the majority of papers dealing with the general equilibrium framework is that corner solutions are difficult to analyze within this framework.

[3] Clark et al. (1987 p 734) explains that product development goes through four stages: concept generation, product planning, product engineering, and production engineering.

Firms are price setters. It is simply assumed that they set their product prices according to the full cost principle. That is, firms have a customarily determined markup ratio m, and the price of a product is set at $(1 + m)\ c$, where c is the proportional unit cost. Markup ratios can be influenced by the competitive conditions of the market for a given group of products, but in this chapter, markup ratios are assumed to be fixed at a certain level.

Price competition does indeed occur in the market, but it revolves around efforts to reduce (proportional) production costs, and no firm tries to cut down its margins or markup ratios. Further elements of background competition are sales promotion and other marketing efforts. It is assumed that marketing expenses are proportional to the sales volumes of each product and that they are already taken into account when markup ratios are determined.

It is also assumed that all products have a predetermined product cycle time. For example, a certain car model is restyled in 4-, 5- or 6-year cycles. Cycle time may differ across countries and industries and may vary depending on changes in economic conditions. Yet, here we assume a fixed cycle time. The production period is much shorter than the product cycle. For example, a car can be produced in a day, whereas its product cycle time is the order of years.

2.1.3 Sraffa's Principle of Production

A specific but important assumption in this chapter is that firms manufacture each product in the exact quantities that they expect to sell. This is Sraffa's principle (Shiozawa 1990, Chap. 5). There is no need to enter into the details of this quantity adjustment. Roughly speaking, based on the actual sales of a particular product in several previous periods, a production line manager decides how many pieces of each variant should be manufactured in each coming period. This type of behavior is in accordance with the observation that Sraffa made in his seminal paper (Sraffa 1926, p 543). As he pointed it out, the main difficulty for a firm wanting to increase its production is not a potential rise in marginal costs but the impossibility of increasing its sales volumes. In a sense, Sraffa's principle is the effective demand principle at the firm level, which determines its production volumes.

In the literature, it is often assumed that price determines demand. The demand curve is a fictitious concept that takes on an actual meaning only in specific situations.[4] In this study, by contrast, prices and quantities are assumed to be

[4]At a particular point in time, i.e., at the time of product design, development managers may consider a demand curve with selling price as the independent variable. If the expected sales volumes are too small, the engineers try to deduce the appropriate unit cost in order to obtain satisfactory expected sales volumes. This procedure is widely known as target costing (Monden and Hamada 1991; Kato 1993). Estimating optimal product variety is closely related to this procedure. Yet, the problem is that the target costing procedure comes after the number of variants to be developed is decided. So, when estimating optimal product variety, managers should assume normal development performance (based on their past experience), including cost planning. Sraffa's

primarily separated and to move independently. Competition mainly takes place in sales efforts. Once the specifications of a product and its production costs are determined, its price is fixed for the whole product cycle time. It may eventually change if sales expectations prove to have been too optimistic or if its proportional production costs change considerably and unexpectedly. The assumption that markup ratios and prices are fixed does not mean that competition is disregarded. On the contrary, we assume here that fierce competition takes place in relation to product quantities. Sales efforts are vital even for a monopolist because, without them, good results cannot be expected.

2.1.4 Once the Estimation Is Made

Once it is decided, after the estimation process, that the variety of products should be modified, the firm should change its product lineup. So, in logical order, the estimation of optimal variety should come before the development and design of each product. But, at that time, the firm possesses no detailed data on sales volumes and costs for the product to be developed, and all information to be used in the calculations relies on expectations. Firms often have extensive experience in production, sales, and development, and those in charge of the estimation rely on past experience to calculate potential future sales and costs. Both are highly speculative, but, generally speaking, cost estimates may be more reliable than sales volumes estimates, because development and production take place within the firm, whereas sales very much depend on external conditions.[5]

2.1.5 Expected Coverage Function

The crucial assumption of this analysis is that a firm is able to obtain the estimated coverage function $\psi(N)$. We assume that managers try to estimate optimal product variety when a product domain is given. In this estimation, expected sales volumes per period of production are important. The estimated coverage function is defined by means of the expected sales volumes. Sales volume may change from period to period, but it is assumed that average volumes can be estimated. For a given N, which is the number of variants, let $D(N)$ be the maximal average sales volume (in physical terms and not in value) per period of production. The detailed meaning of maximal sales volume is given in Sect. 4. Let D be the maximal of $D(N)$ when N is increased infinitely. This sales volume is named *potential sales volume* and denoted

---

principle does not concern this phase of development, since it deals with how daily production volumes are determined after prices are set.

[5]Clark and Fujimoto (1991) distinguish between fundamental and peripheral varieties, but no such explicit distinction is made in the present chapter. Readers may assume product variety to be either, although this makes a big difference for production and development management.

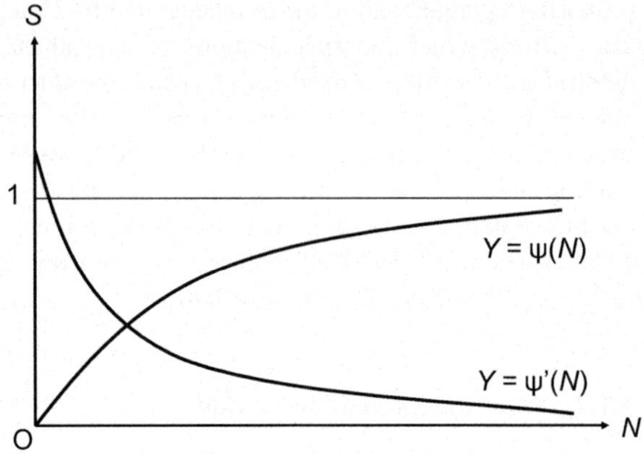

Fig. 1 Expected coverage function and its derivative

as D throughout this study. The *estimated coverage function* $\psi(N)$ is defined as the ratio $D(N)/D$. The expected sales function is defined as $\psi(N)\ D$. Thus, the value $\psi(N)$ can be interpreted as the retention ratio of the firm when consumers cannot find a product with their ideal specifications in said firm's product line.

The assumption made here is that the potential sales volume D is limited. This assumption, i.e., that D is bounded, is eventually removed in Sect. 5.

The expected coverage function ψ is assumed to have the following properties:

(i) $\psi(N) > 0$,
(ii) $\psi(0) = 0$,
(iii) $\psi'(N) > 0$,
(iv) $\psi''(\mathrm{N}) < 0$,
(v) $\lim_{\mathrm{N} \to \infty} \psi(\mathrm{N}) = 1$.

The typical shape of the expected coverage function is shown in Fig. 1. Some explanations are needed. The expected coverage function $\psi(N)$ is only defined for natural numbers N. But we adopt the usual convention that this function can be treated as continuous and differentiable (piecewisely in Sect. 5), in order to simplify the description and analysis. The integer problem is explained in Sect. 2.4.

Assumptions (i) and (ii) are self-evident. Assumption (iii) states that the expected coverage function is monotonically increasing. This is plausible, as an increase in variety boosts the chances to satisfy more people. Assumption (iv) states that the expected coverage function is concave. This property holds when an increase in sales volume through additional variety is not greater than that seen when the variety is smaller. Assumption (v) seems very strong, but it follows from the fact that D is defined as the potential expected sales volume when the number of variants is increased infinitely.

From assumptions (i) to (v), the following properties of the derivative function $\psi'(N)$ are also obtained:

(vi) $\psi'(N) > 0$,
(vii) $\psi''(N) < 0$,
(viii) $\lim_{N \to \infty} \psi'(N) = 0$.

Condition (viii) is derived by reduction to absurdity. Suppose that

$$\lim_{N \to \infty} \psi'(N) > \varepsilon$$

for some positive ε. Then

$$\psi(M) - \psi(0) = \int_0^M \psi'(x)\mathrm{d}x > \varepsilon M.$$

This generates a contradiction, since ψ is assumed to be bounded.

2.1.6 Some Other Specific Assumptions

This section assumes that the fixed cost is the same for all products. This common fixed cost is F. The proportional unit cost is also assumed to be the same, and it is indicated as v. In Sect. 3, these assumptions are removed, and an interpretation is given so that the equal cost assumption can be extended to more general cases.

2.2 *Maximization of Expected Profit*

We now move on to the estimation of expected profit. Let T be the product cycle time, with all products being synchronized (this is a very unrealistic assumption). Let S be the sales volume for all periods throughout the product cycle, which is equal to $\psi(N)\,D\,T$. Then, the expected sales profits for the firm are $p\,S$, where p is the product unit price. The total cost for all products throughout the product cycle, from the development of each variety to final production at the end of the product cycle, will be $F\,N + v\,S$. The for all products can be rewritten as

$$m\,v\,\psi(N)D\,T - F\,N.$$

Now, the estimation of optimal variety can be reduced to a simple maximization problem. Indeed, the problem is to maximize

$$m\,v\,\psi(N)\,D\,T - F\,N \qquad (1)$$

where N moves among natural numbers. When N can simply be assumed to be real, the problem is reduced to simple calculus, i.e., we take the derivative of the expression and equate it to 0.

The derivative of the maxim and of problem (1) takes the form

$$m v D T \psi'(N) - F. \tag{2}$$

Then, all results can be stated based on a single parameter $E = (m\ v\ D\ T) / F$. Because of the conditions of $\psi'(N)$, two possible cases are identified.

Case I: When E is less than or equal to $1/\psi'(0)$, then there is no positive solution. The firm's optimal variety is 0, which means that it is better to leave the present market situation unchanged.

Case II: When E is greater than $1/\psi'(0)$, then there is a single solution N^* for which the firm's total profit will be maximized.[6] More explicitly, N^* satisfies the equation

$$\psi'(N^*) = 1/E = F/(m v D T). \tag{3}$$

We confine our discussion to case II. What can be said about the behavior and order of N^*? A simple estimate from above can be given by E, i.e.,

$$N^* < E. \tag{4}$$

When we describe a graph of $S = \psi(N)$, this can be easily seen (Fig. 2). A slightly finer estimate $N^\dagger$ from above is given by the equation

$$\psi(N^\dagger) = (1/E)N^\dagger.$$

When $N = N^\dagger$,

$$m v \psi(N^\dagger) D T - F N^\dagger = E\{\psi(N^\dagger) - (1/E) N^\dagger\} = 0.$$

So, the estimate $N^* < N^\dagger$ simply says that, at the optimal variety level, the total profit of the firm is positive.

Further useful information is given by Eq. (3). As $\psi'(N)$ decreases (conditions (iv) or (vii)), it is easy to see that E and N^* always move in the same direction. This means that the bigger parameter E is, the greater the estimated optimal variety N^* will be, although the amount of changes is not proportional. As a very rough estimate, E may be seen as a surrogate of variable N^*, in the sense that the movement of N^* can be known based on that of E. The next section examines how the parameter moves according to changes in various situations.

[6] When N is not an integer, it is necessary to find an integer solution. This problem is discussed in Sect. 2.4.

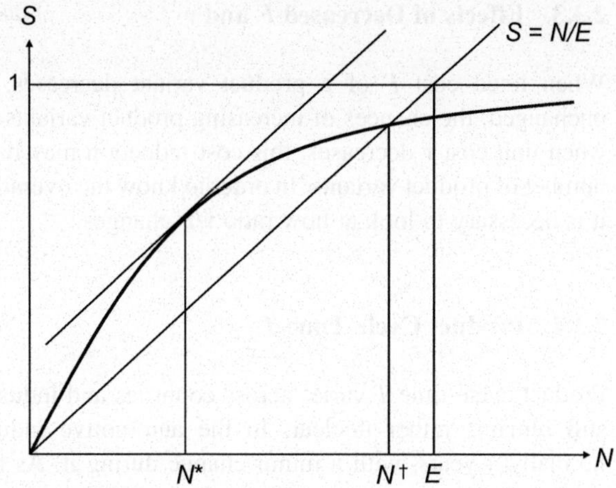

Fig. 2 N^* is majorated by E

2.3 *Observations*

Parameter $E = (m\,v\,D\,T)\,/\,F$ is the product of several multipliers and a divisor. In the above discussion, these factors were assumed to be given and fixed. But, of course, factors *m* and *T* are strategic variables that firms can change when necessary. Factors *v* and *F* are costs, the reduction of which is an eternal target for any firm. As product price is determined by $(1 + m)\,v$, if *m* and *v* change substantially, potential demand in a given period may change accordingly. Some typical situations are examined below.

2.3.1 Break-Even Point *F*/(*m v*)

Let $B = F\,/(m\,v)$. If the amount of product units sold is greater than *B*, the firm can recuperate its fixed cost investments, i.e., costs incurred for development, designing, setting up of the production line, dies and molds, training, and others. When the firm sells a unit of the product, it can gain an amount equal to *m v* without accounting the fixed cost. If the firm has a level of demand equal to *B*, then it can recuperate all the costs incurred in the initial phase of product development.

2.3.2 Total Potential Sales Volume *D T*

Arithmetic product *D T* is the estimated maximal possible sales volume for a given product that can be expected during the whole product cycle. In order for the firm to make a profit, *D T* must exceed $B = F/(m\,v)$. Thus, inequality

$$D\,T/B = m\,v\,D\,T\,/\,F > 1$$

is the necessary and sufficient condition for the firm to produce at least one item in the product group at a profit.

2.3.3 Effects of Decreased F and v

When fixed cost F of a product variant decreases, while other factors remain unchanged, the chances of increasing product variants improve. On the other hand, when unit cost v decreases, this cost reduction may have an opposite effect on the number of product variants. In order to know the overall effect of said cost reduction, it is necessary to look at how ratio v/F changes.

2.3.4 Product Cycle Time T

Product cycle time T varies across countries and industries, and what determines it still remains rather unclear. In the automotive industry, Japan's cycle time is normally 4 years, with a minor change during it. As for Europe, the cycle time is customarily 5 years, whereas the majority of US automobile producers set their cycle time at 6 years.

At the beginning of the twenty-first century, GM and Toyota had comparable total demand D. If we assume that GM's cycle time T is 1.5 times longer than that of Toyota and that the ratio $m\, v\, /F$ is comparable for both companies, then it is plausible for GM to have more models than Toyota. However, as this is not the case, it may be concluded that ratio $m\, v\, /F$ for Toyota is smaller than 2/3 of that for GM.

2.4 Discrete or Integer Problem

As stated in Sect. 2.2, to estimate optimal variety in a finer way, we need to obtain an integer solution for N^*. In principle, this is not difficult, as it is sufficient to take a difference operator instead of a differential operator.

We provide the following definitions:

$$\Delta\psi(N) = \psi(N+1) - \psi(N)$$

and

$$\Delta^2\psi(N) = \{\psi(N+2) - 2\psi(N+1) + \psi(N)\}/2.$$

Then, functions $\psi(N)$, $\Delta\psi(N)$, and $\Delta^2\psi(N)$ satisfy the five parallel conditions indicated in Sect. 2.1. Moreover, $\Delta\psi(N)$ and $\Delta^2\psi(N)$ satisfy the three parallel conditions below:

(vi′) $\Delta\psi(N) > 0$.
(vii′) $\Delta^2\psi(N) < 0$.
(viii′) $\lim_{N \to \infty} \Delta\psi(N) = 0$.

The only difference from the discreet problem case is that it is not necessary to obtain an exact equation

$$\Delta\psi(N) = 1/E = F/(m v D T) \tag{5}$$

for an integer N. Therefore, it is wiser to calculate optimal variety through the following dynamic programming:

Step 1 Set $N = 0$.
Step 2 If $\Delta\psi(N) > E$, then let $N = N + 1$.
Step 3 Continue step 2 until $\Delta\psi(N) < E$. Then let $N^* = N$ and stop.

This program stops when conditions (vii′) and (viii′) hold for $\Delta\psi(N)$.

3 Extension by Reinterpretation

At the end of Sect. 2.1 rather strict assumptions were made. These assumptions can be relaxed by reinterpretation.

3.1 Varying Unit Costs

If fixed costs are the same for all prospective products while unit costs vary, the next convention is useful. Indeed, let unit costs for product J be v_J and fixed costs be F, which is the same for all products. Choose any one product to be the standard and indicate its unit cost simply as v_S. Then, the number of variants N is counted as usual, by means of the number of development projects to be started. As for potential sales volume D and expected sales volume $\psi(N)\,D$, each expected sales volume must be summed up after being multiplied by weight v_J / v_S. Thus, the estimated coverage function $\psi(N)$ is defined for each integer N. The key parameter E is calculated as $(m\, v_S D\, T)\, /F$ with unit cost v_S for the standard product.

3.2 Varying Fixed Costs

A sort of conjugate calculation is possible when fixed costs vary and unit costs remain the same. This time, the number of products must be weighted by the factor F_J / F_S, where F_J is the fixed cost for variant J and F_S is the fixed cost for an arbitrarily chosen standard product. In this case, the expected sales function is not necessarily defined for integers but may be defined for some positive numbers. Parameter E is also given as $(m\, v\, D\, T_S) / F$, with fixed cost F_S for the standard product.

The shorthand assumption that the expected coverage function is defined on all nonnegative numbers becomes a useful convention when we want to extend the optimal variety estimation to the case of varying fixed costs.

3.3 Varying Fixed Costs and Unit Costs

Taking one product as the standard, the expected coverage function can be constructed like in the cases presented in Sects. 3.1 and 3.2. However, the operation becomes much more complicated, and it is difficult to obtain an intuitive meaning for D and N. Nonetheless, the principle remains the same, but parameter E loses its meaning.

In this complex case, in which both fixed costs and unit costs vary from product to product, it is wiser to make a similar calculation to that of the dynamic programming for the integer problem.

4 Exact Meaning of Maximal Sales Volume

The analysis above in search of optimal solutions may have a normative meaning only when the maximal sales volume is interpreted in an exact sense. It is appropriate to provide a detailed definition of the concept of *maximal sales volume*.

First, it should be pointed out that the term *optimal variety* is in fact ambiguous, as it has two different meanings.[7] One is the number of variants that a firm plans to produce. This was the meaning used in Sects. 2 and 3. The other meaning concerns the profiles of the firm's products, with specifications for each of them. For a given number of variants N, a firm can opt for different profiles, and selecting the best one is a crucial task. As for a single-product firm, the choice of specifications for its only product is sometimes called positioning. In the case of a multiproduct firm, selecting the best profile is also vital, since it is a strategic decision. This is why it is necessary to distinguish between the number of variants and the profile of variants.

To clarify the concept of maximal sales volume, the above distinction must be taken into account. Indeed, the maximal sales volume is defined for a given number of products N, but it should be calculated considering all the profiles of N variants. A more concrete example may clarify the matter.

Take a two-dimensional Lancaster space, and assume that the consumers are distributed according to a normal distribution given by the density function

$$\alpha \cdot \exp\{(x+y)^2/2 + (x-y)^2/4\},$$

where α is a coefficient so that the integral over the entire $\mathbf{R}^2$ space is 1. A graphical description of this distribution is given in Fig. 3.

Assume that the consumers buy one of the products in the group when the Euclidean distance in this Lancaster space is smaller than δ. Said distance δ may be called reservation distance, just like reservation price (Waterson 1989). Parameter

[7] Meade (1974) adopted the terminology *number and variety of products*, where *number* stands for what we call *variants* of products and *variety* stands for what we call *profiles*.

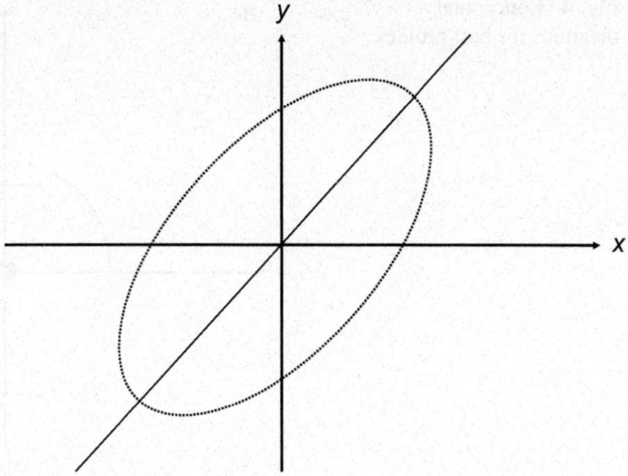

Fig. 3 A distribution of consumers in $\mathbf{R}^2$

δ varies depending on the price level of the group of products, the consumer's income level, and whether the prospective purchase is the first or second. A consumer looking for a second car must have a smaller δ than other consumers who make their first purchase. This kind of detail is abstracted in the following.

When $N = 1$, the best specification for the firm is (0, 0) (Fig. 4, Panel A). When $N = 2$, the best profile is the one displayed in Fig. 4, Panel B. When $N = 3$, the profile is different again (Panel C).[8] The most important aspect here is that the arrangement of the best profiles changes when N changes. The choice of the best profile for N cannot be attained by simply adding a new product with new specifications. To obtain the best profile, all the products must be rearranged with new specifications. In other words, best profiles are not achieved through sequential optimization, by merely introducing new products with good specifications.

The need to redesign all their products anew is a heavy burden for firms seeking to expand their product range. Starting the development of several products at the same time requires extra work and skills, which often involves recruiting new development staff. Moving from an incumbent model to a new model, with a new name and completely different specifications, increases the risk of losing a portion of former customers. Therefore, it is reasonable for a firm to introduce a new product while keeping its older products as they were before.[9] However, the meaning of maximal sales volume should be understood as the expected maximal demand when all product designs are newly developed and suitably arranged while considering overall effects to attract as many potential consumers as possible.

[8]The panels in Fig. 4 are conceptual ones. They illustrate how best profiles (arrangements of specifications) change when N changes.

[9]An evolutionary perspective may be useful as a practical solution for a firm (Sorenson 2000). However, this chapter confines itself to analyzing the decision-making process when all these complications are negligible.

Fig. 4 Conceptual diagrams for best profiles

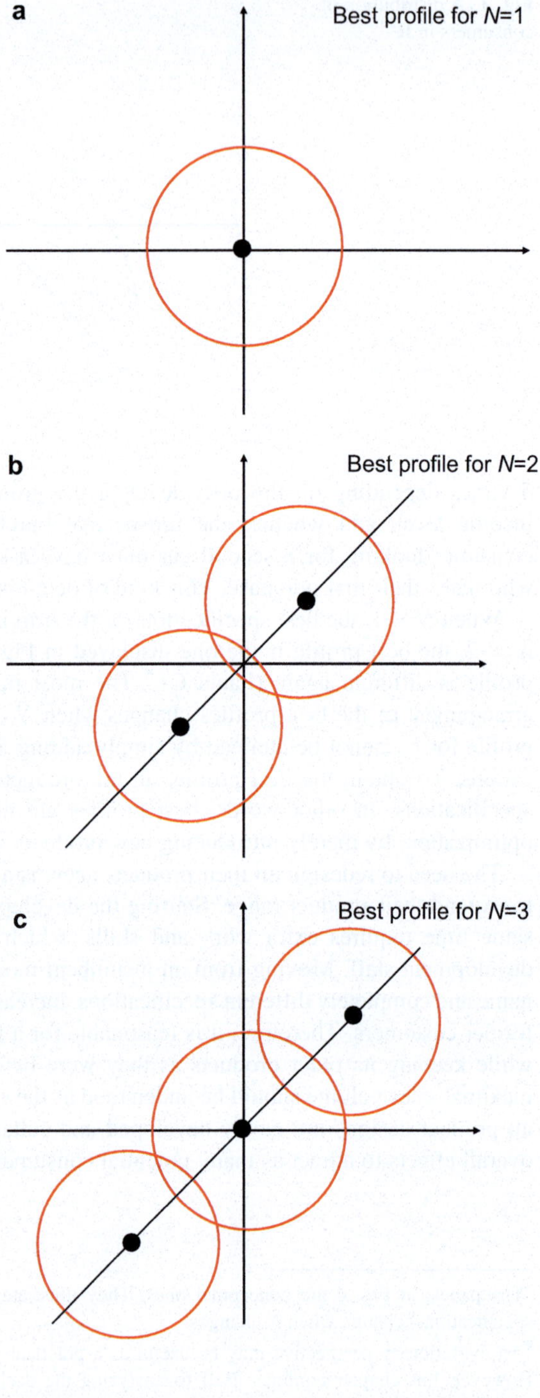

5 Oligopoly Market with Multiproduct Firms

When there are more than two firms competing in the market for the same group of products, the analyses in the previous sections are invalid. Even when competition is between two firms manufacturing a single product, it is known that the Nash equilibrium does not necessarily exist (see Hotelling's analysis of locational competition).[10] If a firm adopts a given product specifications profile, the other firms will be influenced by these decisions. Yet, in this section, we suppose that each firm chooses the best profile for itself independently, without taking its competitors' choices into account, and all complications arising from the question of profile choice are neglected.

5.1 *Market Shares of Firms*

Suppose that firms are attributed shares of the market. Let them be

$$s_1, s_2, \ldots, s_K > 0 \quad \text{and} \quad s_1 + s_2 + \ldots + s_K = 1,$$

when there are K firms in the market. A draconian assumption made here is that each firm uses this share when estimating optimal variety for itself. As a result, each firm J in the market perceives that its expected maximal sales volume is

$$s_J \psi(N) D.$$

We also assume that all firms have the same fixed cost F, the same unit cost v, the same product cycle T, and the same markup ratio m. Then, each firm uses this estimation and determines that its optimal product variety N_J^* is given by the formula

$$s_J \psi'(N_J{}^*) = 1/E = mvDT/F. \tag{6}$$

Formula (6) is not convenient for calculating how many products (i.e., how many variants) will be produced. When the market admits a single monopolist, is the total number of variants for all firms less than, equal to, or greater than the optimal variety? To examine this, let us take the multiplicative inverse (reciprocal) of formula (6)

$$1/\psi'(N_J) = s_J F/(mvTD). \tag{7}$$

The function $S = 1/\psi'(N)$ is defined on the semi-positive axis $[0, \infty)$. Take the inverse function of this function and write it as

$$N = V(S). \tag{8}$$

This function V is defined on an interval $[S_0, \infty)$ where $S_0 = 1/\psi'(0)$.

[10]See Hotelling (1929) and Lancaster (1990 p 198). Eaton and Lipsey (1975) show that there can be no equilibrium.

Determining whether the total number of variants for all firms is bigger than the optimal variety, for a monopolist depends roughly on whether function $N = V(S)$ is concave or not. Details are discussed later.

Let us take a short break and see how concavity is related to the later examination. Suppose an arbitrary function $N = f(S)$ is defined on the semi-positive axis $[0, \infty)$. Assume also $f(0) = 0$, for the moment. For such a function, the following two propositions hold:

1. If f is convex, then for any positive x and for any set of shares $s_1, s_2, \ldots, s_K$,

$$f(x) > f(s_1 x) + f(s_2 x) + \cdots + f(s_K x). \tag{9}$$

2. On the contrary, if f is concave, then

$$f(x) < f(s_1 x) + f(s_2 x) + \cdots + f(s_K x). \tag{10}$$

These two propositions are easily proven. Here, we prove proposition (2). If $f(0) = 0$ and f is concave, for any s in $(0, 1)$,

$$f(s x) > 0 + sf(x)$$

by definition of the concave function. Then, summing up these inequalities, we obtain the following (Fig. 5):

$$f(x) = (s_1 + s_2 + \ldots + s_K)f(x) < f(s_1x) + f(s_2x) + \cdots + f(s_Kx).$$

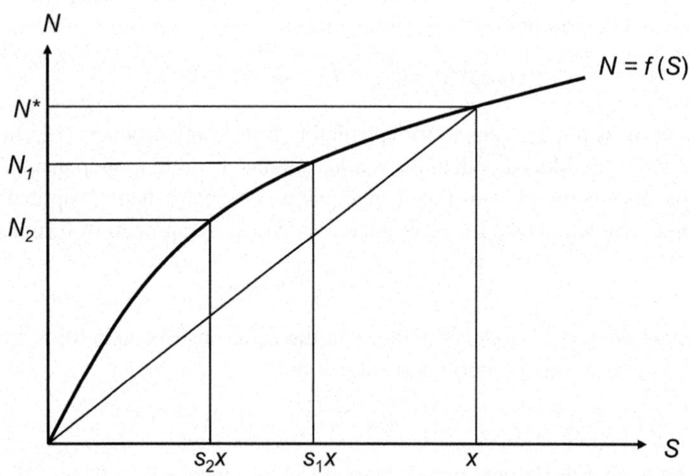

$N_1 > s_1 N^*$ and $N_2 > s_2 N^*$. Then $N_1 + N_2 > N^*$, when $s_1 + s_2 = 1$.

Fig. 5 A concave function with $f(0) = 0$ and proportional division

The trouble with our case is that function $N = V(S)$ is not necessarily defined on the whole semi-positive axis. As the general case is complicated and difficult to see, in the following sections, we will examine some concrete functions for both the concave and the convex cases. Implications for the oligopoly market case will be discussed after the examination of these concrete functions.

5.2 *Concave Case*

We will now examine some typical cases of functions V. Let

$$\psi(N) = 1 - 1/(1 + N/A). \tag{11}$$

Here A is any positive constant. This is the most typical case, and $\psi(N)$ satisfies conditions (i) to (v) in Sect. 2.1. An easy calculation leads to the following:

$$\psi'(N) = (1/A) \cdot 1/(1 + N/A)^2 \quad 1/\psi'(N) = A(1 + N/A)^2.$$

Then, we see that

$$V(S) = \sqrt{AS} - A \qquad \forall S \geqq A.$$

Here are two other simple examples. Functions ψ and V are given as Eqs. (12) and (13):

$$\psi(N) = 1 - e^{-N/A}, \quad V(S) = A \log S/A \quad \forall S \geqq A, \tag{12}$$

$$\psi(N) = 1 - (1 + N/A)^{-\alpha}, \quad V(S) = A((\alpha S/A)^{1/(1+\alpha)} - 1) \quad \forall S \geqq A. \tag{13}$$

Index α in Eq. (13) is any positive constant. The example given as Eq. (11) is a special case of the example given as Eq. (13) when $\alpha = 1$. In these examples, the increasing order of the coverage function is moderate, in the sense that saturation proceeds by a power of N. In contrast to moderate examples, the expected coverage function of the example given as Eq. (12) saturates rapidly when N exceeds A.

Note that, in these examples, function V is always concave, but it is only defined on the interval $[A, \infty)$. An extended function V can be defined along the entire semi-positive axis by the convention $V(x) = 0$ for $0 \leqq x < A$. Then, even if the original function V defined on $[A, \infty)$ is concave, the extended function is not concave (see Fig. 6). However, we claim that, in most cases, estimation (10) is valid. Indeed, let E be $m\, v\, D\, T/\, F$ as before. Then $V(E)$ gives the optimal variety for the fictitious monopolistic firm. Let L be the coordinate where a straight line connecting the origin and point $(E, V(E))$ intersects curve $N = V(S)$. Constant L is bigger than A, but not very different from A. Moreover, when $N = V(E)$ is large, the slope of the straight line becomes lower and L approaches A.

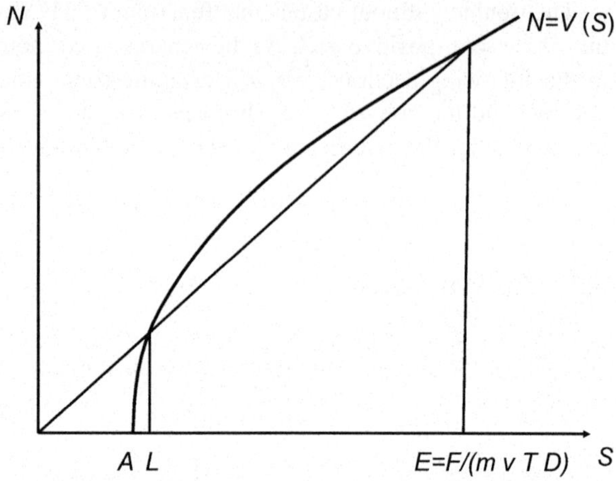

Fig. 6 Concave function *V*(*S*) with restricted definition domain

Now, let us go back to a market of *K* oligopolists. As firms are oligopolists, they produce at least one product, and their market shares are sufficient to ensure that their optimal variety $V(s_J E)$ is at least bigger than or equal to 1. It is very probable that

$$s_J E > L$$

for any firm *J*. In this case,

$$V(s_J E) > sJV(E).$$

Summing up these inequalities for all firms, we obtain

$$V(s_J E) > (s_1 + s_2 + \cdots + s_K)V(E). \tag{14}$$

This is the same as claiming

$$N_1 + N_2 + \cdots + N_K > N, \tag{14a}$$

where N_J is the optimal variety for firm *J* and *N* is the optimal variety for the fictitious monopolist.

Thus, without any great loss of generality, we can claim that, if oligopolists produce their optimal variety and if function *V* is concave on its domain of definition, then the total number of variants for all oligopolists will exceed the monopolist optimal variety.

5.3 Convex Case

Are functions *V* defined above always concave on their domain of definition? The answer is not necessarily. First, let us take an example as given in equation

$$\psi(N) = (N/A)^{\alpha} \tag{15}$$

where $1 > \alpha > 0$ and $A > 0$. By means of a simple calculation,

$$1/\psi'(N) = (A/\alpha) \cdot (N/A)^{1-\alpha}.$$

Thus, function V is given by

$$V(S) = A^{-1/(1-\alpha)} \cdot (\alpha S)^{1/(1-\alpha)}. \tag{16}$$

This function is defined along the entire semi-positive axis and is convex for $1/(1-\alpha) > 1$.

Considering a function $\psi(N)$ like (15), since $V(S)$ is convex, following a similar argument to that made for the concave case, we may have an estimation such as

$$N_1 + N_2 + \cdots + N_K < N. \tag{14b}$$

In this case, the function of Eq. (16) is defined along the entire semi-positive axis. So, there are no complications, just like in the general concave case. Can we thus say that both cases (14a) and (14b) are dependent of the shape of functions $\psi(N)$? The answer is no.

Note that the function of Eq. (15) is not bounded. Hence, this function does not satisfy property (v) of the expected coverage functions, since property (v) of function ψ is equivalent to the expected sales volume being bounded. As we explained previously, it is only a conceptual content which matters when we set $\lim_{N->\infty}\psi(N) = 1$. If the expected sales volume remains bounded, it is always possible to set in this way through a simple change of units. But when the expected sales volume is not bounded, an expression like the coverage function loses its meaning. However, if demand for a group of products can be stimulated without bounds by increasing the variety of products, function $\psi(N)$ can be defined by fixing D at any convenient amount.

The convexity of function $V(S)$ is closely related to the unbounded nature of the expected sales volume. Indeed, if we suppose $V(S)$ to be convex on an interval which extends to infinity, then, for some positive β, we can take

$$V(S) > \beta \cdot N \qquad \forall S > S_0.$$

By definition $V(S) = 1/\psi'(N)$, when $N = V(S)$. The above inequality is equivalent to

$$1/\psi'(N) > \beta \cdot N \text{ or } \psi'(N) > (1/\beta) \cdot (1/N)$$

for all $N > \mathbf{N}_0$. Taking an integration of the latter inequality, we obtain

$$\psi(U) - \psi(N_0) = \int_{N_0}^{U} \psi'(N)\,\mathrm{d}N$$
$$= (1/\beta) \int_{N_0}^{U} 1/N\,\mathrm{d}N = \log U - \log N_0.$$

When U increases infinitely, then $\log U$ increases infinitely. This means function $\psi(N)$ is unbounded.

It is worth mentioning that even if function $\psi(N)$ is bounded (or smaller than 1), function $V(S)$ may have a small part where it is convex. To construct such a function, it is sufficient to connect continuously (in both $\psi(N)$ and $\psi'(N)$) two or three parts from bounded and unbounded ψ. But the convex part cannot be very large, for reasons similar to those of convex $V(S)$.

Furthermore, the fact that $\psi(N)$ is unbounded does not guarantee that $V(S)$ will be convex. An example like Eq. (17) exists:

$$\psi(N) = B\log\{1 + (N/A)\} + C\{1 - 1/(1 + N/A)^{\sigma}\} \tag{17}$$

for A, B, C, and $\sigma > 0$.

When $\sigma = 1$, function $V(N)$ is explicitly expressed as follows:

$$V(S) = 1/2\left\{BS - 2A + \sqrt{B^2S^2 + 4ACS}\right\}.$$

For other σ values, the reciprocal of $\psi'(N)$ is not easily solved as a function of S. But, it is possible to see that $1/\psi'(N)$ has a positive second order derivative in N and that function $V(S)$ is concave where it is defined.

It is interesting to note that the increasing orders of such functions are very slow. In fact, if function $\psi(N)$ satisfies conditions (i) to (iv) but not (v) and if the corresponding function $V(S)$ is concave where it is defined, then $\psi(N)$ can be majorated by a logarithmic function $B' \log(1 + N/A)$ for a suitable positive B'.

5.4 *Observations and Discussion*

The product variety literature is characterized by many contradictory assertions. For example, as Lancaster (1990, p 192) points out, Chamberlin (1933) concludes that monopolistic competition will always lead to more products than the socially optimal level, whereas Dixit and Stiglitz (1977) claims that the market will always have too few products.[11] Although we do not consider socially optimal variety, the observations in Sects. 5.3 and 5.4 may help shed light on this. To the extent that our assumptions are valid, it is highly probable that the inequality of (14a), rather than that of (14b), will hold.

If our construction is somehow linked with those contradictory assertions, two observations are worth mentioning. One is that relations like (14a) and (14b) are too sensitive to the specific assumptions introduced on an ad hoc basis. Another aspect is

[11] See also Itoh (1983). Caminal and Granero (2012) compares the cases of multiproduct firms and single-product firms using a *spokes model* and concludes that multiproduct firms have a competitive disadvantage vis-à-vis single-product firms, which contradicts the observations made in this analysis.

that these contradictory results are due to the conflicting characteristics of the Dixit-Stiglitz preference, on the one hand, and strong preference, on the other hand.

5.5 Clustering and Overcapacities

As total variety in the market for a group of products is important, it might be interesting to investigate how variants are distributed in the case of an oligopoly market.

Suppose that the market comprises one giant player and a few smaller firms. The giant may choose the optimal profile for its products, and these will be distributed rather widely across the Lancaster space. Smaller firms, in particular when their optimal variety is a single product, will tend to cluster around the volume zone since, if they ignore or find it difficult to guess the other firms' behavior, they will choose the position where a single variant will attract as many consumers as possible. Not many such positions exist, so that the natural consequence will be clustering. Clustering is widely observed and indicates that implicit coordination or collusion among firms is often rather difficult.[12]

The fact that inequality (14a) is likely to hold indicates that, in an oligopoly, firms normally operate in a situation of overcapacity, even though each firm estimates its share and expected sales reliably. This proposition is easily explained by Fig. 7. The total number of variants for market $N^{\#}$ exceeds the optimal variety N^{*} of the

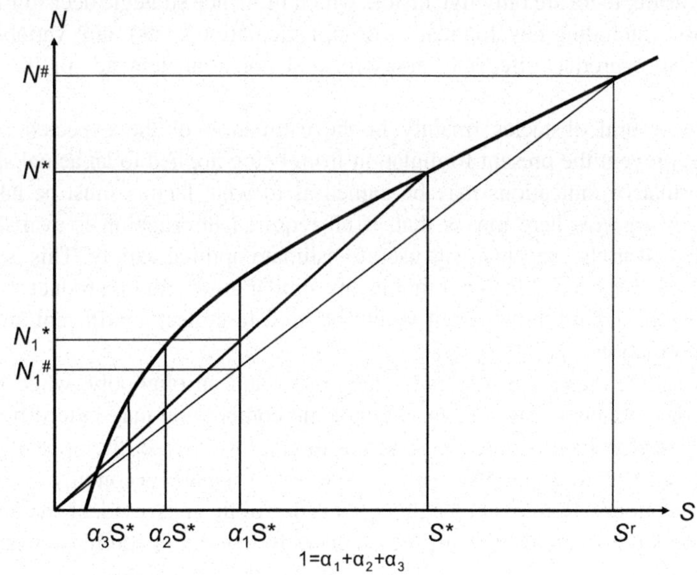

Fig. 7 Overcapacities are almost inevitable

[12]Another possible strategy for small firms is to find a niche that is not covered by the products of the big firm.

monopolistic estimation. So, the total amount of sales for firms may exceed expected sales S^*. But increasing variety to a level higher than N^* will not increase actual sales proportionally. Thus, realized sales $N_1^{\#}$ for firm 1, for example, will be lower than expected sales N_1^*. The same reasoning applies to other firms too. When it sets its production capacities, a firm aims at making sure that said capacities are sufficient to produce what it expects to sell. Yet, the above situation indicates that an oligopoly market will be affected by overcapacity in spite of firms forecasting their sales proceeds reliably. This may explain why it is normal for a market economy to have overcapacities and to be a *pressure economy* (in the sense used by Kornai 1980).

6 Conclusions

The Dixit-Stiglitz formulation is most frequently used in the analysis of product variety. However, this formulation only focuses on each individual's *love of variety* and does not give useful insights into the matter of product variety arising from individual differences. Conversely, the present study favors the strong preference hypothesis and provides a general framework for the analysis of multiproduct firms operating in markets where individuals have different tastes.

The merit of this work is that it presents an explicit formula that may help guess the direction of change in a firm's optimal variety. Putting the expected coverage function aside, there are only five factors which influence strategic decision-making: fixed costs (including development and implementation costs), unit variable costs, markup ratio, product lifecycle, and expected potential demand for the product group.

While several obstacles (mainly in the estimation of the expected coverage function) prevent the present formulation from being applied to an actual situation, some normative indications may be gained as to what factors must be taken into account. *Normative* here means that, if all required information is available, one must proceed in the way that was used to estimate optimal variety. This is a virtue that we have not been able to not find in any similar argument on product variety. In real decision-making, however, it would be hard to collect a sufficient amount of exact information.

As for the comparison between a monopoly and an oligopoly, while the total number of products in the case of oligopolistic competition may indeed be smaller than the number of products found in the monopolistic case, the opposite case is much more likely to occur. This partly explains why a market economy tends to be a pressure economy. Remarkably, although it is one of the most evident characteristics of free market economies, this aspect is generally ignored in equilibrium analysis.

Acknowledgments Professor Takahiro Fujimoto of Tokyo University is gratefully acknowledged for sparking the author's interest in this topic. Without his encouragement, the author would not have dared start any investigation into product variety.

References

Aoki M, Yoshikawa H (2002) Demand saturation-creation and economic growth. J Econ Behav Organ 48(2):127–154
Böckem S (1994) A generalized model of horizontal product differentiation. J Ind Econ 42 (3):287–298
Brenner S (2001) Determinants of product differentiation: a survey. Discussion Paper. Humbolt University, Berlin
Caminal R, Granero LM (2012) Multi-product firms and product variety. Economica 79 (314):303–328
Chamberlin EH (1933) The theory of monopolistic competition. Harvard University Press, Cambridge, MA
Ciarli T, Lorentz A (2010) Product variety and changes in consumption patterns: the effects of structural change on growth. Working Paper, Max Planck Institute of Economics
Clark KB, Fujimoto T (1991) Product development performance. Harvard Business School Press, Boston
Clark KB et al (1987) Product development in the world auto industry. Brook Pap Econ Act 3:729–781
de Vries J (2008) The industrious revolution /consumer behavior and the household economy, 1650 to the present. Cambridge University Press, Cambridge
Dixit AK, Stiglitz JE (1977) Monopolistic competition and optimum products variety. Am Econ Rev 67:297–308
Eaton BC, Lipsey RG (1975) The principle of minimum differentiation reconsidered: some new developments in the theory of spatial competition. Rev Econ Stud 42(1):27–49
Helpman E, Krugman P (1985) Market structure and foreign trade. The MIT Press, Cambridge, MA
Ho TH, Tang CS (eds) (1998) Product variety management: research advances. Kluwer Academic Publishers, Boston
Hotelling H (1929) Stability in competition. Econ J 39:41–57. Reprinted in: Stigler GS, Boulding KE (eds) (1952) Readings in price theory. Richard D. Irwin, Chicago
Itoh M (1983) Monopoly, products differentiation and economic welfare. J Econ Theory 31:88–104
Kato Y (1993) Target costing support systems: lessons from leading Japanese companies. Manag Account Res 4(1):33–47
Kauffman S (2000) Investigations. Oxford University Press, Oxford
Kornai J (1980) Economics of shortage (2 volumes). North-Holland, Amsterdam
Krishnan V, Ulrich KT (2001) Product development decisions: a review of the literature. Manag Sci 47(1):1–21
Lancaster KJ (1966) A new approach to consumer theory. J Polit Econ 74:132–157
Lancaster KJ (1990) The economics of product variety: a survey. Mark Sci 9(3):189–206
Manez JA, Waterson M (2001.) Multiproduct firms and product differentiation: A survey. Warwick Economic Research Papers. http://go.warwick.ac.uk.wrap
Matsumae T (2005) A study on the consistency between empirical studies and growth models with demand satiation and structural change. Evol Inst Econ Rev 1(2):197–220
Meade JE (1974) The optimal balance between economies of scale and variety of products: An illustrative model. Economica, New Series 41(164):359–367
Monden Y, Hamada K (1991) Target costing and kaizen costing in Japanese automobile companies. J Manag Account Res 3(Fall):16–34
Murata Y (2009) On the number and the composition of varieties. Econ J 119:1065–1087
Pasinetti L (1981) Structural change and economic growth, a theoretical essay on the dynamics of the wealth of nations. Cambridge University Press, Cambridge
Ramadas K (2003) Managing product variety: an integrative view and research directions. Prod Oper Manag 12:79–101. https://doi.org/10.1111/j.1937-5956.2003.tb00199.x
Saviotti PP (1996) Technological evolution, variety and the economy. Edward Elgar, Cheltenham

Saviotti PP, Pyka A (2008) Product variety, competition and economic growth. J Evol Econ 18:323–347
Shiozawa Y (1990) Shijo no chitsujogaku (The science of the market order). Chikuma Shobo, Tokyo (in Japanese)
Sorenson O (2000) Letting the market work for you: an evolutionary perspective on product strategy. Strateg Manag J 21:577–592
Sraffa P (1926) The Laws of returns under competitive conditions. Econ J 36(144):535–550
Thirsk J (1978) Economic policy and projects. Clarendon Press, Oxford
Watanabe S (1969) Knowing and guessing: A quantitative study of inference and information. Wiley, New York, pp 376–377
Waterson M (1989) Models of product differentiation. Bull Econ Res 41:1–24
Witt U (ed) (2001) Escaping satiation. Springer, Berlin

Capability Building and Demand Creation in "Genba-Oriented Firms"

Takahiro Fujimoto

Abstract This chapter analyzes the behavioral patterns of *genba-oriented firms*, which pursue both target profit (markup) rates and stable employment, by using a simple evolutionary economic model with Ricardian cost function and design-based product differentiation, called the PXNW model, as opposed to the standard neo-classical profit-maximizing model of firms. Here, *genba* refers to a manufacturing site where value-carrying design information flows to the customers.

After illustrating the basic logic of the PXNW model, a process analysis is performed to explain the sequence of events and activities leading from a steady state to another steady state with lower prices or higher wages. Finally, we present several actual cases of Japanese local factories or small and medium manufacturing firms (SMEs) in various sectors whose behavior is consistent with the patterns of firm activities predicted by the PXNW model.

The process analysis shows that, when the production prices are stable, a genba-oriented firm can move from one steady state, achieving both goals of target markup ratio and stable employment, to another one with higher wages or lower prices, provided that it can simultaneously improve physical labor productivities (often through process innovations) and create effective demand (often through product innovations).

In brief, certain evolutionary economic models, such as the PXNW model, may better explain what was happening in the real economy with many genba-oriented firms during the Cold War and post-Cold War periods, particularly between the 1950s and the 2010s.

T. Fujimoto (✉)
Graduate School of Economics, The University of Tokyo, Bunkyo-Ku, Tokyo, Japan
e-mail: fujimoto@e.u-tokyo.ac.jp

T. Fujimoto, F. Ikuine (eds.), *Industrial Competitiveness and Design Evolution*, Evolutionary Economics and Social Complexity Science 12,
https://doi.org/10.1007/978-4-431-55145-4_4

1 Demand Creation for Employment: A New Perspective

1.1 Introduction

1.1.1 Purpose of This Chapter

This chapter discusses manufacturing firms' efforts for effective demand creation. The firms that we have in mind here are not necessarily the profit-maximizing firms, which appear in today's standard economic textbooks but rather the so-called manufacturing site-oriented firms or genba-oriented firms, which we have observed in our field studies of actual manufacturing sites in the period of global competition. This chapter argues that, when they face intense international cost competition, such firms tend to pursue (i) *continuous capability building* for physical productivity improvements that lower their products' supply curves and (ii) *effective demand creation* that expands the products' demand curves. The first part of this book primarily examines the capability building of such firms and sites, while the second part mostly investigates their effective demand creation, including stepping up sales efforts, increasing product variety, finding new market segments and applications, as well as developing new demand-creating products.

Within the history of economic theories, the economic model adopted here is based on classical economics emphasizing production (*plutology* in Sir J. Hicks' terminology; Hicks 1976) rather than neoclassical economics focusing on exchange (*catallactics*).[1] We also argue that effective demand creation to achieve full employment may be the task of not only J.M. Keynes's government at the macroeconomic level (Keynes 1936) but also of numerous genba-oriented firms at the microscopic level of the national economy.

1.1.2 The Basic Logic Behind the Behavior of Genba-Oriented Firms

To begin with, we summarize the basic logic of this chapter as follows:

(i) We can construct an economic model for genba-oriented firms as firms that pursue target markup ratios (r^*) and target numbers of employees (N^*) at the same time, rather than simply maximizing profits. Since a genba is part of a firm

[1]While there is no clear consensus on the price (value) theory of classical economics, this book adopts the notion that normal price p is determined by unit production cost C and markup ratio r, following the full-cost principle, or $p = (1 + r)C$. The following remark by David Ricardo explains this price theory: "The value of a thing is the 'cost of production' including profits" (Ricardo 1817). Shiozawa (2017) provides a clear discussion of this matter: "There is no unified understanding of what it [the classical theory of value] is." "I have chosen Ricardo as the representative of classical economics." "My definition of Ricardo's theory of value, and consequently, of the classical theory of value, is the cost of production theory." This book adopts Shiozawa's view, in that it does not use the famous "labor theory of value" but regards the broader concept of unit production cost as the basis of the classical theory of price (value).

and of a community at the same time, and since one of the most important objectives of a community is likely to be stable employment, we assume that a genba-oriented firm is also a community-oriented firm that aims to achieve stable employment ($N = N^*$) in addition to stable profit ratios ($r = r^*$).

(ii) When genba-oriented firms or sites in high-wage advanced nations face intense cost competition vis-à-vis those in low-wage emerging nations, the former are forced to increase physical productivities or lower input coefficients, in order to make up for international wage gaps and thereby survive unit cost competition. Unless their sales quantities grow fast enough to absorb the redundant workforce created by the abovementioned productivity improvements, the high-wage firms and sites will have to reduce the number of their employees to achieve the target profit ratios ($r = r^*$), which, however, means missing the other target, i.e., stable employment ($N < N^*$).

(iii) In this competitive situation, the necessary condition for a genba-oriented firm to simultaneously achieve its target markup ratio (r^*) and employment level (N^*), while maintaining real wage rates, is to lower its unit cost by improving physical productivities and to create effective demand quantities at the same time. Graphically, this means simultaneously shifting the supply curves and corresponding cost curves downward and the demand curve outward.

(iv) In the context of innovation theories, the above conditions are often accompanied by the combination of process innovations for productivity improvements and product innovations for effective demand creation. We may call these *grassroots innovations* or innovations to stabilize employment at the level of sites, communities, and firms.

In this chapter, we will show a simple Ricardian (i.e., classical or plutology) economic model that can approximate the typical behavior of genba-oriented firms. We will also present some actual cases of such genba-oriented firms observed during the post-Cold War period.

1.2 A Model of Demand Creation for Employment

1.2.1 Design and Downward Sloping Demand Curves

Since products having different designs are mutually differentiated on the market, each firm or site experiences a situation akin to that of monopolistic competition, in which competing products are similar but heterogeneous and each product is characterized by a downward sloping demand curve (Chamberlin 1933).

What is different from the conventional monopolistic competition model, however, is that a firm with a new product is neither a simple price taker nor a simple price setter. The process in this model is more like a dialog between the firm and the future market through design, a new factor. That is, the firm introduces a new design to the market, forecasts market responses in the form of predicted demand curves, and determines a new product's price and quantity plan. In other words, design is an additional independent variable that the firm can manipulate. We will explain this model later in this book.

To better understand the trend of the downward sloping demand curve, let us assume that a customer's reservation price (i.e., the price the customer is willing to pay) for a product with given design information is determined by the said product's functions and the customer's tastes and that product functions are realized when the customer-user operates the product's structure in certain use environments. That is, when consumer *i* decides whether to buy product *j* (e.g., a passenger car) at a given time, consumer *i*'s reservation price p^r_{ij} for product *j* can be expressed as follows:

$$p^r_{ij} = f_{ij}\left(\tilde{F}_{ij}\right) = f_{ij}\left(S_j, \tilde{O}_{ij}, \tilde{E}_{ij}\right)$$

where the product's structure is S_i, the user's predicted pattern of operation is $\tilde{O}_{ij}$, the predicted use environment is $\tilde{E}_{ij}$, the product's expected function is $\tilde{F}_{ij}$, and the customer's taste, which converts product functions into a reservation price, is f_{ij}.

It is obvious that the reservation prices (p^r_{ij}) of the same product structure (Sj) will be different depending on individual customers, due to the distribution of their tastes (f_{ij}), as well as predicted operation pattern ($\tilde{O}_{ij}$) and predicted use environments ($\tilde{E}_{ij}$). If we aggregate all potential customers' individual demand functions connecting selling price p^s_j with demand quantity 1 or 0 (1 if $p^r_{ij} \geq p^s_j$; 0 if $p^r_{ij} < p^s_j$), we can create a market demand function for product *j* that is downward sloping (Fig. 1).

On the supply curve side, if we follow the classical assumption that a product's price is determined not by demand-supply equilibrium but by unit production cost plus markup regardless of quantity ($p^s_j = (1 + r)C_j$), the individual supply curve for product *j* can be drawn as a flat line, and the intersection of a product's *individual flat*

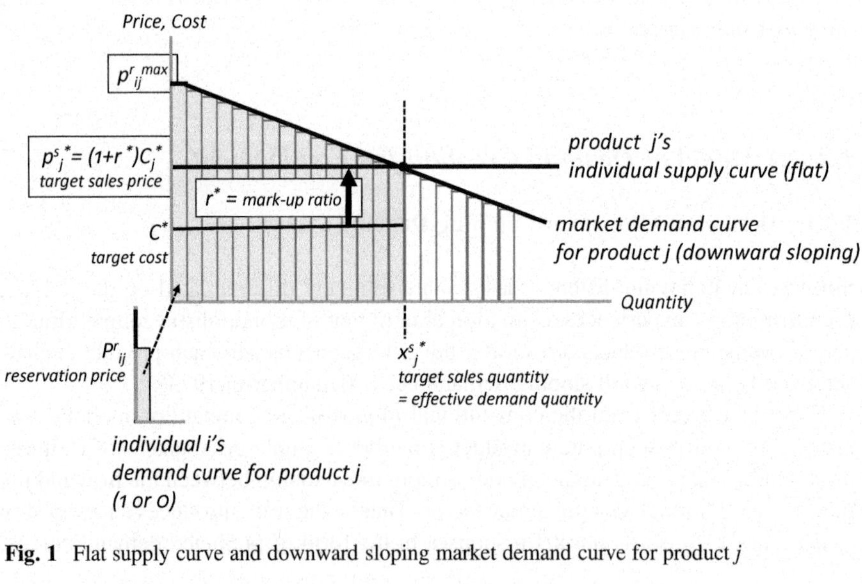

Fig. 1 Flat supply curve and downward sloping market demand curve for product *j*

supply curve and *downward sloping market demand curve* indicates the *effective demand quantity* for product *j*.

Thus, our model starts from the traditional assumption of normal price as production cost plus normal profit, but it introduces an additional factor, i.e., design. As a result, products are now differentiated by design, so each product design will have a downward sloping demand curve representing a summary of how the market evaluates this particular product. Because effective demand quantity is determined at the intersection of the supply and demand curves, design quality will influence demand quantity by shifting the demand curve.

In any case, in the present production economy model—unlike the standard neoclassical model, where prices and quantities are simultaneously determined by demand-supply equilibrium—prices and quantities are not determined simultaneously, partly because the supply curve is flat and partly because product design affects the demand curve and quantity.

1.2.2 Demand Creation and Innovation as New Design

In this context, *design information* is information about the causal relations between product *j*'s structures and predicted functions when customer i uses it, or $(S_j, \tilde{F}_{ij})$. So, if a firm can successfully improve the quality of its product designs, potential customers are likely to revise their reservation prices p^r_{ij}, so that the market demand curve is likely to be shifted upward.

As shown in Fig. 2, given the supply curve or given the planned unit production cost (C_j^*), markup ratio (r^*), and supply price ($p^{s*}_j = (1 + r^*)C_j^*$) of product *j*, the upward shift of the demand curve through effective design will boost effective demand quantity (x^s_j) and, consequently, employment levels at the site manufacturing the said product. Conversely, given the target production quantity (x^{s*}_j) required

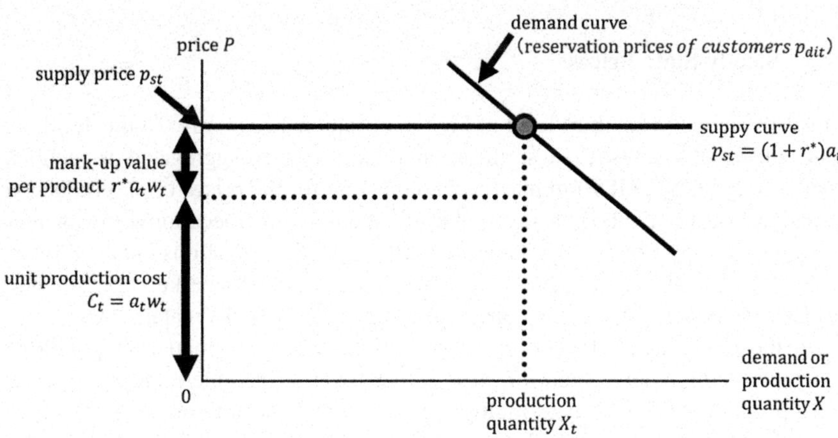

Fig. 2 A product's demand and supply curves (PX quadrant)

for the site to sustain current employment levels, higher design quality, thanks to a shift of the demand curve, will drive up the supply price (p^s_j), thereby leading to higher site-specific wage rates and/or profit rates, at least temporarily.

This is the power of effective design, which should be incorporated into the analysis of the modern production economy. With design as an independent variable that can shift the market demand curve, it may be possible for a firm's product, and site, to achieve target employment levels, corresponding target sales quantity (x^{s*}_j), target selling price (p^{s*}_j), markup ratio (r^*), and target unit cost (C^*_j) at the same time as a steady state. We will discuss this model in detail later in this book.

Joseph Schumpeter defined *innovation* as a new combination of materials and forces that generates economic development (Schumpeter 1934). Within the production economy framework adopted in this book, we can reinterpret the concept of innovation as new combination of an artifact's functions and structures creating economic value, which, according to our definition of design, is nothing but *new design*. The concept of design was not known to economists when Schumpeter published his first version of *The Theory of Economic Development*, but it seems obvious that his famous examples of new combinations (e.g., new good, new production method) involve new designs and artifacts.

We can thus regard the abovementioned upward shift of the demand curve in Fig. 1 as *innovation as new design* for generating additional effective demand at the firm level. While Schumpeter focused on discontinuous innovations for major economic development at the national economy level, we also identify incremental innovations causing small shifts of the demand curve for individual products, which can have a significant economic impact when many products and sites implement them at the same time.

Furthermore, if the aforementioned target sales quantity (x^{s*}_j) is set for securing employment at the production site and effective new design is used to achieve it, we may regard this *as innovation for employment*. Many such grassroots innovations are observed in small community-based and site-oriented firms in Japan, for instance.

1.2.3 Steady-State Prices

Let us now move to the third level of Fig. 1, the national economy. At this level, the value added of domestic sites, firms, and industries is aggregated to national products (e.g., GNP, GDP) in macroeconomics. As for the prices of all products in the national economy, the general equilibrium theory in microeconomics mathematically proves—by solving simultaneous price-quantity equations—that a set of prices able to achieve demand-supply equilibriums in all markets exists under perfect competition. This is the consensus on the side of Hicks' catallactics.

On the other hand, there is no clear consensus on the matter in plutology. Nonetheless, this book's point of view is that the most complete theory of prices developed thus far to analyze multiple products and markets at the level of the national economy is Piero Sraffa's *Production of Commodities by Means of Commodities* (Sraffa 1960).

Sraffa's concise but mathematically precise book adopts the aforementioned Ricardian (or classical) view that price is determined by unit production cost and markup ratio ($p = (1 + r)C$). Yet, unlike Ricardo, who reduced unit production cost to direct labor cost ($C = aw$), where a is the labor input coefficient and w is the average hourly wage, Sraffa considered not only labor cost but what today's cost accounting calls direct material costs and factory overheads (e.g., production tools and equipment), each of which can also be calculated as input coefficient times input price. That is, $C = (a_0w + \sum a_jp_j)$. So, the corresponding selling price is $p_j = (1 + r)(a_0w + \sum a_jp_j)$, where a_0 is the labor input coefficient and a_j is the input coefficient of product j. Note here that there is a circular relation in which inputs become outputs and outputs become inputs (p_j is used twice in this equation).

Sraffa (1960) proved that we can solve these simultaneous equations in the cases of non-growth economies and growing economies, as discussed later. An important aspect is that, in Sraffa's price system, the prices, including the wage rate (w, $_jp_j$), are determined by input coefficients (a_0, a_j), which are the inverse of physical factor productivities, and markup ratio (r), which is related to distribution of income between capital and workers. That is, in Sraffa's system, given the markup ratio, prices and wages are determined by the productive performance of manufacturing sites. Thus, we adopt Sraffa's model of prices as the basis for the production economy framework put forward in this book.

Moreover, this pricing system of the modern production economy does not rely on the classical labor theory of value or subsistence theory of wages, which is not realistic in most industrialized nations today. Sraffa's prices are based only on productivities (input coefficients) and profit (markup) ratios regardless of subsistence wage, however, defined. The prices of modern plutology are not equilibrium prices either, unlike in catallactics, but rather *steady-state prices* or the prices that sustain a stable pattern of flows or circulations of merchandise and value added.

To sum up, the modern production economy adopts the *steady-state price model*—in which all prices in the national economy are compatible with stable flows/circulations of materials, goods, and value added, given the sites' productivity and profit requirements—instead of the neoclassical general equilibrium model. We may regard this as a *flow-sustaining system of prices*. As such, the model follows the classical production-cost-plus-markup theory of value, but it departs from the now obsolete labor theory of value of the nineteenth century. In any case, the price (value) model adopted here is flow-oriented.

1.2.4 The PXNW Model: Assumptions

Based on the discussions concerning the demand curve and prices within the classical production economy model (plutology), let us now construct a simple model of a genba-oriented firm as a PXNW model (Fujimoto 2017), which combines a classical model of natural prices (i.e., flat supply curves) with Ricardian cost functions, product differentiation (i.e., downward sloping individual demand curves), and Keynes-type employment functions at the firm/site level. As such, the model includes the following assumptions:

(i) *Full-Cost Principle*: A firm is assumed to produce a single product for simplicity's sake. Its supply price (p_s) is determined by its unit cost (C) and a predetermined markup ratio (r^*), i.e., $p_s = (1 + r^*)\ C$. That is, the full-cost principle applies here (Hall and Hitch 1939).

(ii) *Ricardian Cost Function*: It is assumed that unit production cost C is ultimately represented only by unit labor cost, which is hourly wage rate (w) multiplied by unit labor cost (a). That is, $C = aw$, which is nothing but a Ricardian cost function (Ricardo 1817).

(iii) *Natural Price*: The above assumptions lead to the concept of natural price, which is assumed to be fixed, regardless of production quantity. That is, in the price-quantity space, the supply function is flat, or $p_s = (1 + r^*)\ C = (1 + r^*)\ aw$.

(iv) *Product Differentiation*: We also follow the modern assumption that products with similar designs are differentiated from one another, which results in a situation of product differentiation and oligopolistic competition (Chamberlin 1933). Hence, we have a downward sloping demand curve for each product design. Assuming a linear demand curve for simplicity's sake, we may describe it as p^r=A-Bx, where p^r is the reservation price of potential customers, A is the maximum reservation price, and B is a positive coefficient (B > 0).

(v) *JIT Production*: For the sake of simplicity, it is also assumed that, given fixed supply prices, the firm's demand forecasts for the next period are accurate and that its periodical production plans follow the said accurate demand forecasts. In other words, a just-in-time (JIT) production system is assumed here. For a given product, production quantity (x) for the next period is determined at the intersection of its demand and supply curves, which is nothing but effective demand quantities.

(vi) *Employment Function*: When the next period's production volume (x) is determined, the level of employees required (N) is predicted as a function of the production plan (x). This is similar to what J.M. Keynes called employment function, except that in this case employment is a function of effective demand (i.e., production) quantity rather than effective demand value (Keynes 1936). For simplicity's sake, a linear employment function is assumed here. That is, $N = ax$, where a is the aforementioned labor input coefficient.

(vii) *Labor Supply Curve*: We assume a situation with high levels of unemployment, so that the labor supply curve is flat at a given wage level (w).

1.2.5 The PX Quadrant (Product Demand and Supply Curves)

This model assumes a flat supply curve with a constant supply price for period t, p_{st}, regardless of periodical production volumes x_t for products with identical design. With the assumption of product differentiation and monopolistic competition, the

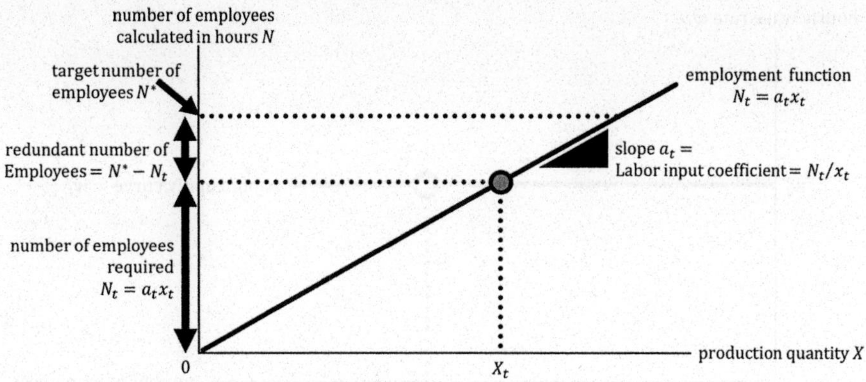

Fig. 3 Employment function (XN quadrant)

product in question also has a downward sloping demand curve, with the different reservation prices of different customers i at period t in descending order, or p_{dit} (Fig. 2). The supply price p_{st} follows the full-cost principle. That is, when a_t is the labor input coefficient (person-hours per unit), w_t is the average hourly wage rate, and r^* is the target markup ratio permitted by the capital market, our Ricardian production cost per unit, C_t, is $C_t = a_t \cdot w_t$, and the supply price is $p_{st} = (1 + r^*) C_t = (1 + r^*)a_t \cdot w_t$. On the other hand, production quantity at period t, or x_t, is equal to periodical effective demand quantity, which is shown at the intersection of the flat supply curve and the downward sloping demand curve. To sum up, the firm in question produces as many items as the effective demand quantity, x_t, given the normal supply price p_{st} and the customers' reservation prices p_{dit}.

1.2.6 The XN Quadrant (Employment Function)

We use the Ricardian linear labor input coefficient, $a_t = N_t/x_t$, where x_t is production (i.e., effective demand) quantity at period t and N_t is the required labor input (assuming 8 h of daily work), as the basis for the linear employment function $N_t = a_t \cdot x_t$ (Fig. 3).

1.2.7 The NW Quadrant (Labor Supply Curve)

Here we assume a flat labor supply curve with a fixed hourly wage rate, w_t, under a condition of insufficient aggregate demand at the macroeconomic level (Fig. 4). If aggregate demand grows rapidly and aggregate supply is insufficient, the labor supply curve slopes upward, but the situation here is that of advanced nations in the age of global competition, characterized by a gross national product that does not grow fast enough.

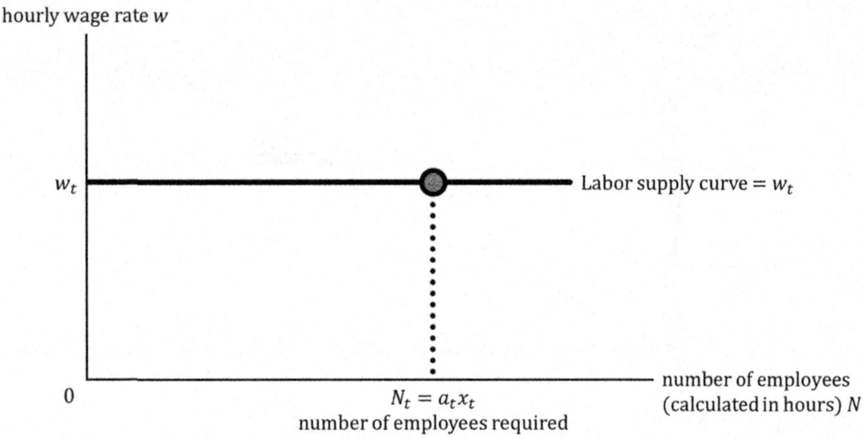

Fig. 4 Labor supply curve (NW quadrant)

When global cost competition becomes more intense, profit-maximizing firms in high-wage advanced nations, which can choose the location of their production sites internationally, may find it rational to close down their higher-wage domestic factories and move their production capacities to lower-wage emerging countries, such as China and India. Our genba-oriented firms, however, are more likely to maintain some of their domestic manufacturing sites by letting them improve physical productivities and thereby reduce unit production costs, given the hourly wage rate w_t.

1.2.8 The WP Quadrant (Cost Function)

Finally, we look at the unit production cost of the product in question at period t, or C_t. By adopting the classical Ricardian cost function, which assumes that all the items of production cost can be ultimately reduced to direct labor cost, the unit production cost can be expressed as labor input coefficient a_t (person-hours per unit) multiplied by average hourly wage rate w_t, or $C_t = a_t \cdot w_t$. In the WP quadrant, this is shown graphically as a half line with the gradient of a_t, which we may call the wage-cost curve (Fig. 5).

In a context of global competition, with significant pressure to reduce its normal supply price p_{st}, a genba-oriented firm in a high-cost country must reduce its unit cost C_t. As it is difficult to reduce the domestic hourly wage rate x_t, the said firm will have no choice but to significantly decrease its labor input coefficient a_t (i.e., to increase its physical labor productivity $1/a_t$). In fact, starting from the 1990s, many Japanese sites manufacturing tradable goods have made remarkable efforts to increase their productivities and thereby decrease unit costs vis-à-vis rival sites or firms in low-wage emerging counties.

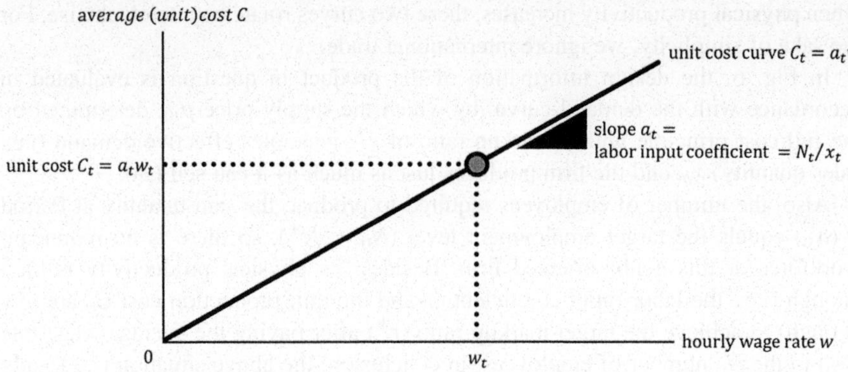

Fig. 5 Unit cost curve (WP quadrant)

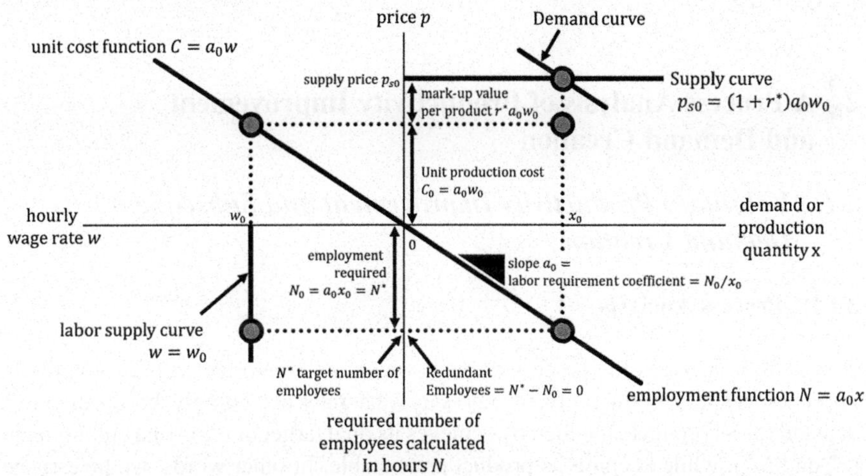

Fig. 6 PXNW diagram in Period 0 (target markup ratio r^* and target employment N^* achieved)

1.2.9 The Overall PXNW Model (Period 0)

The abovementioned four quadrants can be integrated into a PXNW diagram with four axes, i.e., price (P), quantity (X), employment (N), and wage (W) (Fig. 6).

This diagram shows how a genba-oriented firm maintains a steady state that stably achieves the target profit ratio (i.e., markup ratio $=r^*$) and the target number of employees (N^*) at the same time at Period 0. The demand and supply curves are in the first (PX) quadrant, the employment function is in the fourth (XN) quadrant, the labor supply curve is in the third (NW) quadrant, while the unit cost function is in the second (WP) quadrant. Note also that the employment function $N_t = a_t \cdot x_t$ and the unit cost function $C_t = a_t \cdot w_t$ share the same gradient a_t for any period t. Hence, when physical productivity increases, these two curves rotate counterclockwise. For the sake of simplicity, we ignore international trade.

In Fig. 6, the design information of the product in question is evaluated in accordance with the demand curve, by which the supply price p_{s0}, determined by the full-cost principle with a markup ratio of r^*, generates effective demand (i.e., sales quantity) x_0, and the firm produces just as much as it can sell (x_0).

Also, the number of employees required to produce the said quantity at Period 0 (N_0) equals the target employment level ($N_0 = N^*$), so there is no redundant workforce at this genba-oriented firm. Besides, its physical productivity is high enough (i.e., the labor input coefficient a_0 and the unit production cost C_0 are low enough) to achieve the target markup ratio (r^*) after paying the average wage rate (w_0) to the N_0 number of employees. In conclusion, the above situation is a steady state that satisfies (Simon 1947) the requirements of profit, survival, and employment of a genba-oriented firm.

2 A Process Analysis of Productivity Improvement and Demand Creation

2.1 *Voluntary Productivity Improvement and Subsequent Demand Creation*

2.1.1 Process Analysis

This section analyzes the process through which a genba-oriented firm voluntarily improves its site's productivity, increases its wage rates accordingly, boosts effective demand for its product, and thereby achieves its dual target of profit and employment (r^* and N^*), while keeping its product price stable. In other words, we investigate the case of a firm moving from a certain steady state to another improved steady state, with higher wages, unchanged profit rate, and stable employment.

Unlike equilibrium analysis, process analysis takes time to account, in that the sequence of events is important. Accordingly, we first investigate the processes initiated by firms and sites to voluntarily improve productivity through their manufacturing capability building, continuous improvements (kaizen), or process innovations (Fujimoto 1999). We can consider at least two cases, as shown below.

2.1.2 Returning to Steady State by Price Reduction

The first example is the case in which voluntary productivity improvements by the leading firm trigger price competition. This brings the above-normal profit of the leading firm back to a normal level, which A. Smith, K. Marx, and J. Schumpeter predicted in their economic models.

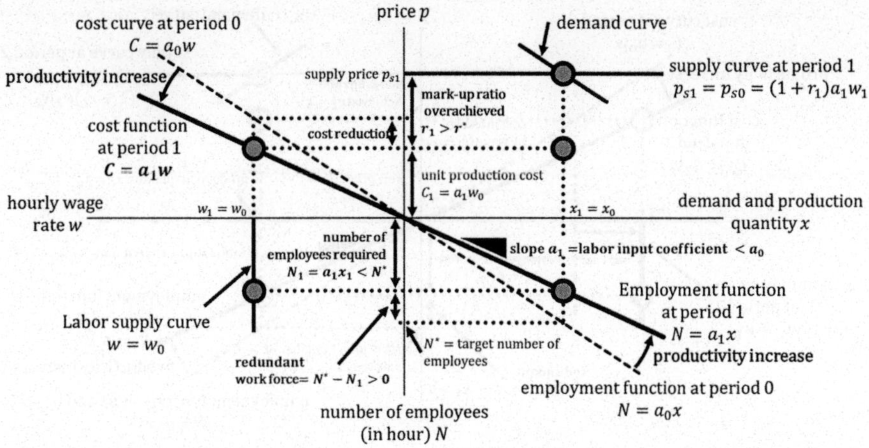

Fig. 7 PXNW diagram in Period 1 (markup ratio overachieved $r_1 > r^*$ and employment N^* underachieved)

Period 1a: Productivity improvements and corresponding unit cost reduction are initiated by the leading firm or by its manufacturing site in question ($a\downarrow$, $C\downarrow$). As a result, above-normal profits are generated by cost reduction, given the stable supply price ($r > r^*$), while redundant workforce is created through the same productivity increase ($a \cdot x = N < N^*$). This situation is described in Fig. 7, in which the counterclockwise rotation of both the employment function ($N_t = a_t \cdot x_t$) and the cost function ($C_t = a_t \cdot w_t$), caused by productivity increases, creates above-normal markup ratio and insufficient employment levels at the same time.

Period 2a: Above-normal profit triggers price-cost competition among both leaders and followers in capability building and innovations, which lower the product's supply price and bring the markup ratio back to the normal level ($p_s\downarrow$; $p_s = r^* \cdot C$).

Period 3a: The expansion of effective demand quantity by the genba-oriented firm, through an outward expansion of the demand curve in the PX quadrant of the PXNW diagram, will bring the firm's number of employees back to the target level ($x\uparrow$; $N = N^*$; see Fig. 9).

If these price reduction processes occur for most products of the national economy, the overall price levels of the whole economy will drop, and real wage rates (i.e., living standards of workers) will increase, given the flat monetary wage rate.

2.1.3 Returning to Steady State by Wage Increase

Alternatively, we can consider the case in which productivity improvements by the leading firm trigger upward wage renegotiation between the firm and the workers under the condition of stable product prices. What we have in mind here is a

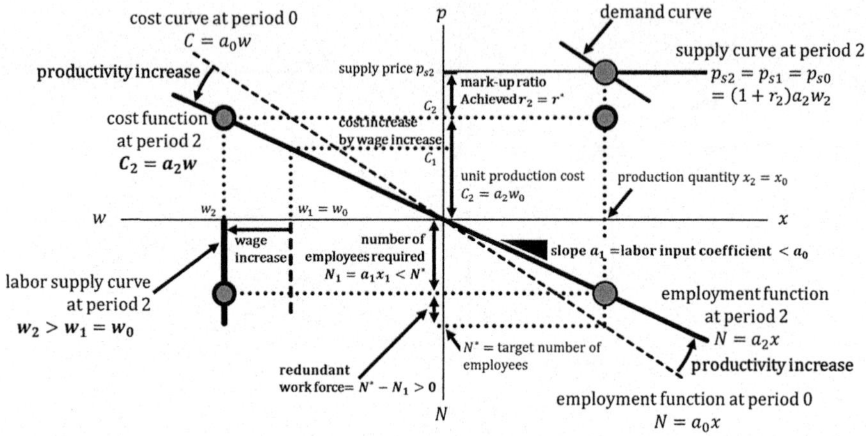

Fig. 8 PXNW diagram in Period 2 (markup ratio $r_2 = r^*$ through wage increase and employment N^* underachieved)

manufacturing firm in a relatively high-growth economy (e.g., Japan in the 1960s) that increases its productivities and wage rates simultaneously, without causing serious inflation or stagflation in trade goods sectors. This is what regulationist economists may call a "Fordist compromise" in the postwar era (Boyer and Saillard 2002). The process taking place between the two steady states may be described as follows:

Period 1b: As in the case of Period 1a (Fig. 7), productivity improvements and unit cost reductions are initiated by the leading firm ($a\downarrow$, $C\downarrow$). This results in above-normal profits through cost reduction, given the unchanged supply price ($r > r^*$). Redundant workforce is created through productivity increases ($a \cdot x = N < N^*$).

Period 2b: Above-normal profit triggers wage renegotiation between the firm and the workers in the productivity-leading firm, which results in above-average wages ($w\uparrow$; $w > w_0$). Hence, unit production cost goes up, and the product's markup ratio returns to the normal level ($a \cdot w = C\uparrow$; $p_s = r^* \cdot C$). In brief, wages increase but the price remains stable and competitive (Fig. 8).

Period 3b: The leading firm promises its workers that it will keep employment levels unchanged in exchange for the abovementioned restricted wage increase that maintains unit cost and prices stable. This compromise between the firm and the workers will lead the productivity-leading genba-oriented firm to expand effective demand quantity in order to keep the target employment level ($x\uparrow$; $N = N^*$), which implies an outward expansion of the demand curve (Fig. 9). The firm's efforts to secure stable employment will in turn boost the workers' motivation to cooperatively participate in kaizen activities for continuously improving the genba's productivity.

In the two cases above, both productivity improvements and effective demand creation are required for a genba-oriented firm to move from one steady state to

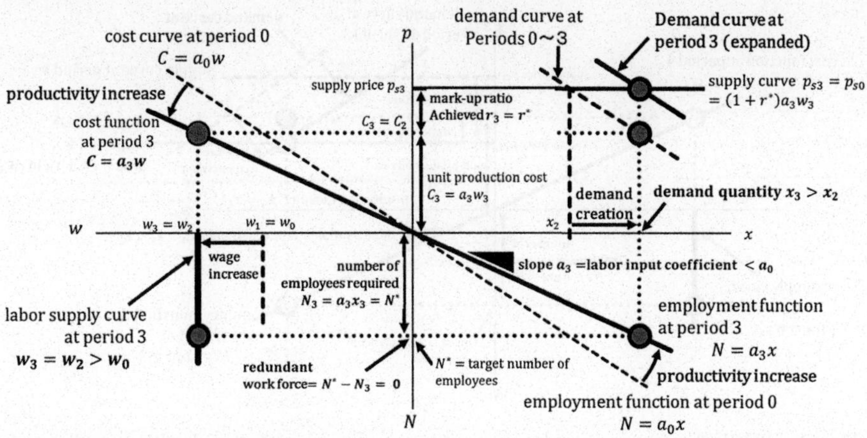

Fig. 9 PXNW diagram in period 3 (wage increase, markup ratio r^* and employment N^* achieved through demand creation)

another improved steady state in terms of real wage levels (i.e., workers' standards of living). With these findings in mind, let us move on to the next case, in which the process starts from price reductions under intense global competition.

2.2 *Productivity Increase Under Global Price Competition*

2.2.1 Price Reduction, Productivity, and Employment

In this section, we analyze the process through which, under intense competitive pressure to reduce prices, a genba-oriented firm moves from one steady state to another by increasing its site's productivity and expanding its product's effective demand at the same time, thereby achieving once again its dual target of profit and employment (r^* and N^*). As a real example, we may consider a firm manufacturing certain tradable goods (e.g., consumer appliances) in a high-wage advanced nation that faced global cost competition vis-à-vis rival firms and sites in low-wage emerging nations during the post-Cold War period. For the sake of graphical continuity, let us start from the steady state described in Fig. 9 in which target markup ratio (r^*) and target employment (N^*) are both achieved stably.

2.2.2 Price Reduction by Means of Global Competition (Period 4)

Figure 10 shows the situation of post-Cold War global competition, in which the product's supply price decreases sharply. In the case of monopolistic competition among differentiated products, this implies an inward shift of the given product

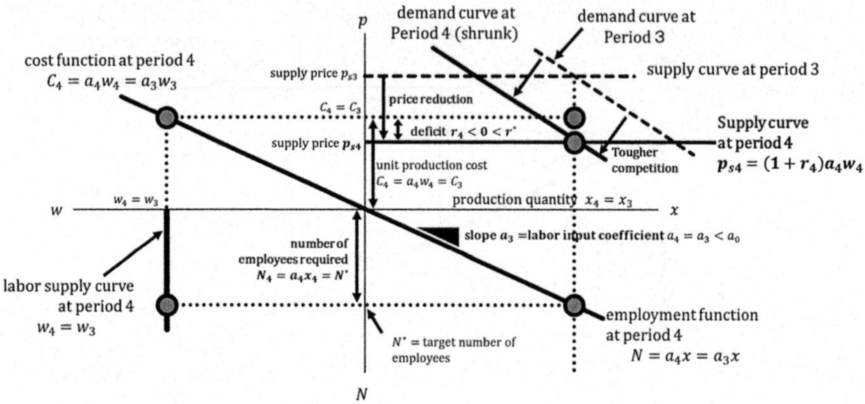

Fig. 10 PXNW diagram in Period 4 (price reduction through global competition, markup ratio negative, and employment N^* sustained)

demand curve. In this situation, which we may call Period 4, the product's effective demand quantity (x) and corresponding employment level ($N = a \cdot x$) of the genba-oriented firm in question cannot be maintained without a drop in price.

Whereas a profit-maximizing firm may reduce its production volume (x_4) and employment level (N_4), a genba-oriented firm, under pressure to contribute to the community, will try to maintain its level of production (x_4) and full employment ($N_4 = N^*$) by reducing its price ($p_{s4} < p_{sa}$). Accordingly, the markup ratio will become negative ($r_4 < 0 < r^*$), so it will be difficult for this company to maintain its domestic manufacturing site (genba) for a long time, unless significant productivity improvements are implemented.

2.2.3 Productivity Improvements as Countermeasures (Period 5)

During the "moratorium" period before they close down or declare bankruptcy, many genba-oriented firms and their domestic manufacturing sites will channel their efforts into improving their physical productivities. This was indeed the case in many small and medium enterprises and local factories of tradable goods industries in Japan between the 1990s and the 2010s (Fujimoto 2007, 2012, 2014). Some achieved threefold productivity improvements in 2 years or fivefold improvements in 5 years and, although many were still closed down, many others survived.

Figure 11 shows the situation of a firm that managed to survive. The rotation of the cost and employment functions through productivity enhancement ($a_5 < a_4$) brings the markup ratio back to the target level ($r_5 = r^*$). At the same time, however, the number of employees fails to reach the full employment target ($N_5 = N^*$). Hence, the genba-oriented firm faces the dilemma of profit (i.e., survival) versus employment.

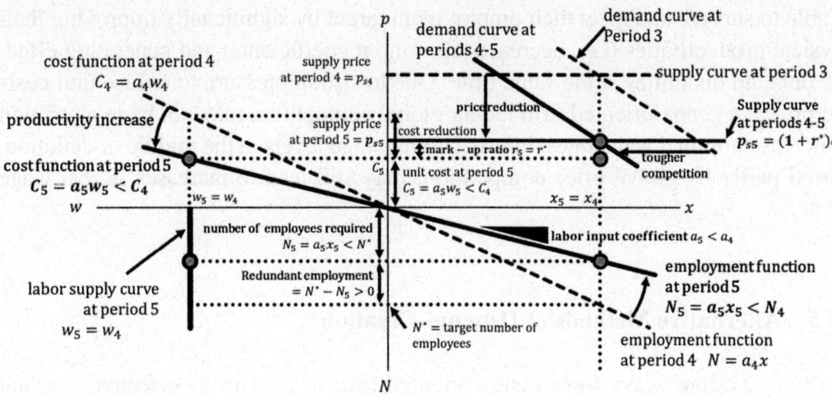

Fig. 11 PXNW diagram in Period 5 (target markup ratio r^* regained through productivity increase and employment N^* unachieved)

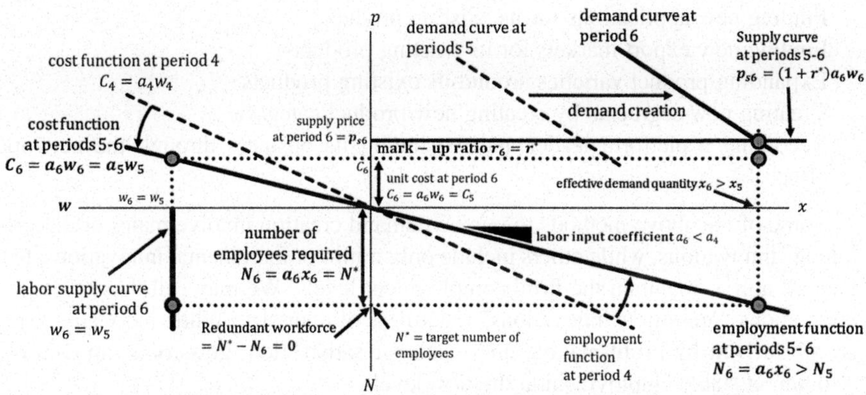

Fig. 12 PXNW diagram in Period 6 (both target markup ratio r^* and employment N^* achieved through demand creation)

2.2.4 Returning to Steady State by Demand Creation (Period 6)

In this predicament, in order to regain its steady state, i.e., to achieve its target markup ratio r^* and target employment level N^* at the same time, the genba-oriented firm and its manufacturing site have no choice but to increase productivity ($a_5 < a_4$) and expand the product's effective demand quantity ($x_6 > x_5$) simultaneously. In Fig. 12, the latter means shifting the demand curve (see the top right PX quadrant) significantly enough to absorb any redundant workforce at the manufacturing site (see the bottom left XW quadrant).

By doing the above, many genba-oriented firms in Japan, facing intense global price competition vis-à-vis sites and firms in low-wage emerging countries, will still be able to survive and meet their employment target by significantly improving their physical productivities (i.e., decrease labor input coefficients) and generating effective demand quantities at the same time. Due to strong pressure to reduce unit costs and prices, a genba-oriented firm facing global competition may not be in a position to increase nominal wage rates, but sustaining wage levels in the middle of deflation, caused partly by global price competition, may still lead to increases in real wage rates.

2.2.5 Alternative Methods of Demand Creation

There are various ways for a genba-oriented firm to expand its effective demand quantity:

- Strengthening marketing efforts for its existing products
- Lowering unit prices of its existing products
- Finding new applications for its existing products
- Finding new export markets for its existing products
- Expanding product varieties around its existing products
- Creating new segments by creating new product categories
- Becoming a platform leader and expanding its business through the network effect

Some of the above methods to pursue demand creation involve major or discontinuous innovations, while others include only minor or incremental innovations, but they all aim to maintain the firm's employment levels. We may call these types of innovations "grassroots innovations" regardless of their size. While innovations are usually driven by profit-making and customer satisfaction, grassroots innovations also aim at stable employment at the firm level.

3 Genba Cases: Survival with Employment

3.1 Cases of Genba-Oriented Firms in Post-Cold War Japan

We have so far described the behavioral patterns of genba-oriented firms mainly through process analysis, using a series of PXNW diagrams. Let us now look at some real cases of genba-based firms in Japan, a high-wage advanced nation that faced global price competition with low-wage emerging nations, such as China and India, during the post-Cold War era, i.e., between the 1990s and 2010s.

3.1.1 Textile Manufacturer ST in Ishikawa Prefecture[2]

ST is a textile manufacturer producing cloth for highly functional uniforms (over 70%), car seats, home interior goods, and sportswear. The firm has about 200 employees (of which 180 production workers) and reached approximately $40 M sales in 2017. It was established in 1954 as a spinout-weaving firm of textile wholesaler I, which still holds about 50% of its shares. Its three factories are all located in the same city, Hakui, in Ishikawa Prefecture, an area with a population of roughly 22 K. ST is one of the largest manufacturers in Hakui.

When the author visited the company in 2013, the Japanese economy—particularly its older sectors, such as that of textiles—was suffering due to both deep recession and strong appreciation of the yen. In order to compete effectively against textile manufacturers in lower-wage countries, such as China and Bangladesh, ST took two measures. First, it concentrated on difficult-to-weave products, such as cloth for high-technology uniforms and carbon fiber textiles. Indeed, it is the top manufacturer of uniform textiles in Japan (about 4% market share). Second, it made continuous efforts to develop human resources for multi-skilling and multi-machine operations so as to achieve higher physical productivities on the shop floor.

ST's factories have more than 400 automatic looms, where one operator handles as many as 60 machines with a reasonably high machine uptime ratio, which results in extremely high productivity compared to its rivals in lower-wage countries—high enough to overcome international wage gaps. However, this also means that the company may not need as many as 200 employees. Its president at that time, Mr. X, estimated that there would be, in fact, about 50 redundant workers when the factories achieved the abovementioned level of labor productivity.

As we were walking through the shop floor, I asked if those redundant workers would be dismissed to cut costs and increase profits. Mr. X immediately replied, "No, of course not. This is a rather big factory in this city. If we fire that many employees, I won't be able to walk down the main street from now on. We are scared of angry city people rather that angry stockholders, as we are a small non-listed company."

What Mr. X did instead was to visit the headquarters of one of his firm's parent companies, the large global textile manufacturer T, which has been a pioneer in carbon fiber products for years. He asked company T to allow ST to produce a carbon fiber textile for automobile interiors and other uses, a potentially growing business for both T and ST. Eventually, the new carbon fiber business was established at ST, and the orders were large enough to absorb the redundant workforce.

In this way, even after drastic productivity improvements in the manufacturing of cloth for uniforms, ST maintained a stable workforce of about 200 by securing the growing line of business of carbon fiber textiles.

---

[2]The author visited ST's site in Ishikawa Prefecture for research and data collection on August 30, 2013.

3.1.2 Ship Building Company OZ in Nagasaki Prefecture[3]

Company OZ is a medium-sized shipbuilding company located in Nagasaki Prefecture, with about $1B sales and 1300 employees. It has only one shipyard, with a single shipbuilding dock located on an island where coal mines were closed down in the 1970s. OZ was established in 1973, but its predecessor company started shipbuilding in Osaka in the 1930s. That is, the company moved its shipbuilding operations from Osaka to Nagasaki in the early 1970s by building a dock that was large enough to make big oil tankers, a seemingly promising business during Japan's high-growth era.

Soon after the shipyard started operations in 1973, however, the first oil crisis began and orders for such large ships stopped coming in almost entirely. So, from the very beginning, the company had to face heavy restructuring, and, by the end of the 1970s, it had reduced its workforce from 1800 to 800. In order for the company to survive, Mr. M., then OZ's president, shifted the focus of the business from large tankers of 300 K tons to smaller bulk carriers of 30 K to 80 K tons.

The shipbuilding dock was originally designed to build large oil tankers, but it was made slightly wider than necessary for the sake of future strategic flexibility. Taking advantage of this slack resource, OZ partitioned it into four sections, so that it could simultaneously build four smaller bulk carriers. Since a dock's production lead time is about 1 month, production capacity per year is usually about 10 ships, but OZ can build nearly 40 ships from a single dock.

OZ also adopted the Toyota Production System (TPS) for smoother flows of value-added and improved productivities. According to Mr. M, the introduction of TPS was somewhat unintended and emergent. In response to yet another shipbuilding recession, OZ was once forced to dispatch many of its employees to the Toyota Group companies to stabilize its regular workforce. Those employees then returned to OZ and brought the concept of TPS back to Nagasaki with them.

Thanks to the success of its new product strategy of focusing on smaller bulk carriers, the facility-based approach of building four ships in one dock, as well as operational excellence by adopting TPS for flow-oriented shipbuilding, OZ was recognized as one of the world's most productive shipyards, according to an international comparative survey conducted in 2001 by a major US university (an anonymous comparison of 10 Asian, 10 European, and 6 US shipyards).

Industry experts estimate that the productivity of superior Japanese shipyards is three to five times higher than that of China's shipyards. Partly because of this productivity advantage and partly because, in the case of shipyard welding workers, the international wage gap between China and Japan shrunk to about a third as of the mid-2010s, OZ has succeeded. After a long struggle in this apparently declining industry, it has achieved moderate sales growth, a very high profit ratio (e.g., much

[3]The author visited OZ's site in Nagasaki Prefecture for research and data collection on September 16, 2014. The author also interviewed Chairman M and Vice President H on February 27, 2015.

higher than that of Japan's better automakers), and rather stable employment (roughly 1000 employees since the end of the oil crisis in the 1970s).

When the author interviewed Mr. M in February 2015 and asked him about his management motto, the president made no mention of the company's high profit ratio but simply said "Its people." He then talked about the potato liquor distillery that was built by OZ just next to its shipyard in the 1980s, when the company had to face recession in the shipbuilding sector once again. Although it is a very small business, the fact that OZ started to make potato liquor—and to grow potatoes and tomatoes to secure jobs for those who could not leave the island when the shipbuilding business was stagnant—is regarded as highly symbolic and proof of its employment-conscious corporate culture.

To sum up, except in its early years during the oil crisis of the 1970s, OZ has pursued a reasonably stable employment policy for many years, with a product strategy focusing on bulk carriers and as high as possible physical productivities. Its profit ratios have been generally high and stable as well.

It is interesting to observe that, while large Korean shipbuilders have tended to concentrate on large tankers and higher-technology ships, such as LNG tankers and drill ships, there are profitable Japanese shipbuilders like OZ in the lower-technology segments, such as that of bulk carriers—a counterintuitive phenomenon in the post-Cold War competition. Indeed, OZ remains an internationally cost-competitive shipbuilder vis-à-vis its Chinese rivals, whose products are also differentiated downward from those made in Korea.

3.1.3 Photocopier Manufacturer SF in Mie Prefecture[4]

Company SF is a production subsidiary of the large and global copier/printer manufacturer F.[5] SF is located in Suzuka, Mie Prefecture, in the central part of Japan. It was established in 1982 as a component factory for manufacturer F.

SF produces key components of photocopiers and multifunctional office machines. In 2009, when the author visited its site, SF's sales revenues were 88B yen (roughly $800 M; 1.4B in 2005), with about 3% operating profit per sales ratio (4% in 2005). It employed about 1000 people (including about 200 temporary workers), with an additional 700 workers at on-site subcontractors.

Its parent company F is a global company that has many foreign production subsidiaries, including some in China (e.g., Shanghai and Shenzhen). Thus, in the 1990s and 2000s, SF faced tough cost competition against rival factories in low-wage emerging countries owned not only by other firms but also by company

[4]The author visited SF's site in Mie Prefecture for research and data collection on December 9, 2009, and January 25, 2016.

[5]In 2010, SF was merged into F's newly established national production subsidiary FM and became "FM Suzuka." Yet, since the author's initial visit to this site was in December 2009, we call it "SF" as a local production subsidiary.

F itself. By the year 2003, 70% of F's global production had already been concentrated in China, with plans to bring that figure up to 90% in the following years. Without drastic productivity improvements, rapid decreases in both production volumes and employment at domestic production sites, like SF, seemed inevitable.

In order to try and survive, SF adopted the flow-oriented Toyota Production System (TPS) in 2004 by inviting a first-rate consultant in this field, Mr. I, who had been the CEO of a Toyota Group parts supplier. The TPS improvement project (2004–2008) included 5S (sort, set in order, sweep, standardize, sustain), work standardization charts, frequent and punctual shipping, process layout changes (short straight lines with hand delivery), flow levelization (*heijunka*), small lot *kanban* (pull) delivery system between lines, automatic defect detection and response (*jidoka*), process control for building in quality, low-cost tools that utilize gravity (*karakuri*), production equipment improvements, product redesign for manufacturability, quality function deployment (QFD), front-loading by design reviews, and quality engineering (Taguchi Method).

Thanks to these activities, SF's overall physical productivity increased by 3.3 times and production lead times were reduced to one fifth between 2004 and 2008. In SF's 9 surface-mount technology (SMT) and reflow lines for printed circuit board (PCB), physical productivity increased by 5.6 times over the same 5 years, and work-in-process inventory was reduced by 80%. As a result, 1 worker was able to handle 2 SMT reflow lines, whereas 1 SMT line had on average 2.5 workers in F's Chinese factories. So, the SMT productivity gap between SF and the Chinese factories was five times. SF's production manager told the author that, as a result of productivity increases in the Japanese factory and wage rate increases in China, SF's production lines had caught up with the Chinese lines in unit production cost by 2009.

However, the capability-building competition between the Japanese and Chinese factories within F's global production network continues. SF has a dual role: it is both a commercial factory that has to survive by being cost-competitive and a mother factory that has to transfer its superb manufacturing capabilities and knowledge to its sister factories in emerging nations. As the productivity of the latter grows, partly due to SF's activities as a mother factory, SF itself has to keep improving its own productivities through further capability building. Yet, the productivity gap between SF and the factories in China may be sustained for a long time, to the extent that the turnover ratios of the Chinese factories may continue to be much higher than those of SF, which hamper workers' multi-skilling, a critical factor for productivity improvements in production operations of this type.

On the side of demand creation, F established a recycling/reuse business at its headquarters in the mid-1990s. Shortly thereafter, SF negotiated with F to secure this business and started to operate its disassembly/reassembly lines in 2005. F's photocopiers had been made available for use by customers at the over 10,000 outlets of X, a major chain of convenience stores across Japan. Since these machines were lower-end products, they were made in F's Chinese factories. Previously, when the machines' life was over, X would simply replace them. SF proposed changing this practice to a recycling business, mainly for environmental reasons. So, all 5-year-old photocopiers installed in X's stores are now collected and shipped (30 trucks per day) through F's central distribution centers to SF's recycling/reuse facility in Suzuka, which consists of a disassembly line and a reassembly line with synchronized takt time. About 150 rebuilt machines are shipped every day.

The incoming old photocopiers are first disassembled into components. About 60% of the parts are reused, while the remaining 40% are crushed and recycled after being classified into 36 material categories. The reuse ratio has been increasing over the years, and the copiers themselves have been redesigned to improve disassemblability. The reusable parts are taken out at the disassembly conveyor line, washed, inspected, and put into kitting boxes, which are moved immediately to the line side of the next reassembly line. About half of the parts used in the reassembled machines come from the disassembled ones, while the rest are new parts.

Disassembly, reassembly, and delivery lead times are very short at SF, and delivery lead times are more important than unit product cost to the customer (i.e., the convenience store chain), which aims to maximize utilization ratios of the photocopiers. Because of this, as far as its Japanese customers are concerned, the firm has achieved lead-time-based competitive advantage in this recycling/reuse business vis-à-vis Chinese rival factories, regardless of international wage differences. As SF's senior managers told us, "even if the Chinese factories' hourly wage were zero, we would still be competitive in this business model."

In brief, SF has survived the tough cost competition vis-à-vis rival firms and factories in China by drastically improving physical productivities and stabilizing its employment levels to a certain extent, thanks to the introduction of a new business model for machine recycling/reuse.

3.1.4 Printer Factory NF in Niigata Prefecture[6]

NF is a Japanese factory of FM, a domestic production subsidiary of the abovementioned global manufacturer F. FM was established in 2010 by consolidating F's five local manufacturing subsidiaries in Japan. As one of FM's factories, NF is located in the city of Kashiwazaki in Niigata Prefecture. It has about 270 regular employees and 100 temporary workers and produces larger printers and industrial printing machines.

The factory, established in 1974 by N, another Japanese electronics manufacturing firm that produced PC printers at the time, was bought by F in 2001. NF shifted its product mix to larger printers as it lost its competitive advantage in smaller

[6]The author visited NF's site in Niigata Prefecture for research and data collection on December 6, 2011.

printers for personal computers. About 90% of NF's employees are residents of Kashiwazaki, with 0.5% turnover ratio per month, and the firm aims to ensure stable operations, so that the local employees' children and grandchildren can continue to work at the plant. As such, NF emphasizes community-supporting activities. Moreover, it recovered quickly after two large earthquakes in 2004 and 2007 by collaborating intensely with the local community.

NF started to introduce the Toyota Production System (TPS) in 2000. Its basic manufacturing concept was to give first priority to flows of value added. It synchronized its workers' operations, streamlined its flows as short and flexible production lines, standardized operations, introduced the pull (kanban) production system, and developed the lines under stable control and steady evolution.

Nonetheless, NF's efforts for continuous kaizen were not drastic enough to secure its survival under intense global cost competition vis-à-vis rival factories and firms in lower-wage countries. Indeed, the manufacturing of smaller products was progressively transferred to F's overseas factories in emerging countries. To compensate for its shrinking operations, NF had to launch another drastic cost-reduction project for its main product, large printing machines.

The firm strived to improve physical productivity at its large printer assembly line by 3 times in 2 years, between 2010 and 2012, and managed to achieve a 2.8-time increase without any major capital investments. Through various kaizen activities targeting non-value-adding time of direct workers, including improvements of line layout and jigs, jidoka (automatic defects detection), and low-cost semiautomatic equipment for material handling, the line's value-adding time ratio increased from about 16% to 37% (2.3 times), and value-adding time itself was reduced (value-adding speed increased) by about 1.2 times during the same period. Note here that, other things being equal, physical productivity increases by $n \cdot m$ times when value-adding-time ratio increases by n times and value-adding speed increases by m times mathematically.

In 2016 NF had about 400 workers and aimed to keep steady employment levels. Roughly half of its employees were temporary workers, but NF was gradually moving them to regular contracts. As of 2011, partly due to the aforementioned productivity improvements, NF's unit cost—including production, transportation, and inventory—was already as low as that of its Chinese rivals, as far as domestic sales were concerned.

To sustain its employment levels, NF has been expanding in-house production of its own parts. Large printing machines are mostly exported to America and Europe, but the firm has also retained production of niche domestic products, such as printers for larger-size paper. For instance, it produces printers for A3 paper, 90% of which are sold domestically, with about 50% global market share.

3.1.5 Motorcycle Parts Supplier TB in Shizuoka Prefecture[7]

TB is a supplier of motorcycle parts (about 2/3) and automobile parts (about 1/3) located in the city of Hamamatsu, Shizuoka Prefecture. The firm was established in 2011 from the merger of two parts suppliers, T and B. It mainly produces engine components, shafts, valves, and cogs by means of metal cutting and other production processes. As of 2011, it had around 400 employees, working at 4 local factories in the same prefecture.

Company T was established in 1955 as a metal cog supplier for a local weaving machine manufacturer that subsequently closed down. In the same year, it started to sell motorcycle cogs to Y, a Japanese motorcycle manufacturer, and eventually became its subsidiary in the 1980s. Company B was also a subsidiary of Y. T tried to diversify into electric motor shafts but lost cost competition, so it shifted its focus to automobile parts, a more stable business environment than that of motorcycle parts. Although T was a subsidiary of Y, Y allowed and even promoted the expansion of T's business to other motorcycle makers and automobile suppliers.

With such diversification strategy, company T's number of employees grew from 50 in 1975 to 100 in the early 1980s—and nearly all of them had permanent contracts. Thanks to the luxury motorcycle boom in the United States, T's production and export of motorcycle engine parts grew rapidly. The revenues of companies T and B reached about 7 billion and 4 billion yen, respectively, in 2008, while T's employees became about 370 in the same year (270 regular employees and 100 temporary agency workers). Because of Japan's advantage in special purpose steel and lack of economies of scale in the luxury motorcycle business, T chose to export its parts rather than move production to the United States.

When the financial crisis struck in 2008, the US luxury motorcycle market virtually collapsed and the combined revenues of T and B halved in the course of 1 year (i.e., from 11B yen in 2008 to 5.5B yen in 2009), with over 10% operating loss to sales in 2009. Faced with this crisis, companies T and B made the following major changes to their business.

First, T sharply reduced employees at its domestic site from 370 in 2008 to 240 in 2011, mostly by terminating the contracts of its temporary agency workers. It should be noted here that the number of T's regular employees did not change much (about −10% in 3 years) and that almost 100% of the remaining employees had permanent contracts. That is, despite a 50% reduction in revenues and significant deficit, T's number of regular employees remained overall stable.

Second, in order to make up for the negative impact of the US market collapsing, T built its first overseas subsidiary, TI, in Indonesia, where both the economy and the motorcycle market were booming then. T's parent company Y had already started its local motorcycle assembly in Indonesia in the 1970s, and, after many ups and downs, Y's domestic production in this country reached 3.6 million units per year

[7]The author visited TB's site in Shizuoka Prefecture for research and data collection on June 11, 2011.

in 2011. Thus, TI's Indonesian production could enjoy economies of scale, and dividend revenue from TI's profits from Indonesia helped its parent company T improve its financial performance starting from 2010.

Third, as for diversification, T rapidly expanded its automobile parts business, which was less affected by the US financial crisis than that of motorcycles. T did most of its automobile parts business with a large Toyota Group supplier, A, with whom T had already had dealings in the 1980s. T started this initiative by buying A's used production lines, which were heavily depreciated, and by combining them with T's own used machine tools, a low-cost solution to expand production capacities. The materials are now supplied by company A, and T does subcontracted machining only to complete the parts. Additionally, with improved financial performance, T has begun to boost its production capacities by building new lines.

Fourth, thanks to the introduction of the Toyota Production System, T improved its physical productivity by drastically increasing the number of machine tools that one multi-skilled worker could operate. At the abovementioned used line for the automobile parts, one worker was in charge of two machining lines by repeatedly moving the works in process from one machine tool to another.

Fifth, in 2011, in order to strengthen their financial and other resources, companies T and B merged into one company, TB. The new company had revenues of about 7 billion yen and 390 employees in 2011.[8] In the same year, TB's profit became positive after 2 years of deficit. The number of TB's regular employees remained fairly stable, i.e., about 370 in 2016.

Hence, company TB overcame the 2008–2009 crisis through a combination of overseas production in Indonesia, business diversification into automobile parts, productivity improvements, and an M&A process. The number of its regular employees did not change much despite a 50% reduction in revenues and significant operating losses. The firm also had about 150 lower-tier suppliers, with between 1 and 30 employees each, but only one of them went bankrupt (several others disappeared due to lack of business successors).

3.2 *Summary of the Cases: Fit with the PXNW Model*

We have provided some examples of manufacturing firms and sites in Japan in the post-Cold War period, based on the author's direct observations and data collection. Although the above five cases refer to diverse industries and different periods, they share certain features that seem to be typical of genba-oriented firms: (i) significant efforts by the firms or sites (genba) to improve physical productivities for survival;

[8]In 2017, TB further merged with Y's supplier of casting parts and changed its name to Y Precision Parts Manufacturing, Y's 100% owned subsidiary, which now aims to produce highly functional precision parts, from casting to machining to assembly. In 2016, the new company had about 490 permanent employees and 70 temporary workers.

(ii) significant efforts by the firms and sites to generate additional effective demand by securing new orders or developing new businesses, as well as new products or variations, both domestically and internationally; (iii) relatively stable number of regular employees at the sites despite fluctuations in revenues and profits; and (iv) long-term survival, if not constant growth, of the firms and sites by securing minimum required profits in the long run.

Moreover, having visited over a thousand manufacturing sites in many countries since the 1980s, the author has repeatedly found patterns of actions and results similar to the abovementioned cases. Thus, we believe that genba-oriented firms are not only theoretically predicted but also empirically observed in certain industrial or business sectors. They are usually smaller manufacturing firms and sites strongly embedded in their home communities and regions.

Although global capitalism has been the main thrust of the world's economy since the 1990s, local firms and sites have made significant and simultaneous efforts for productivity improvements and demand creation in order to secure their survival and employment. We may call this pattern a stylized fact of genba-oriented firms facing intense international cost competition in some sectors of the global economy. When local actors meet global competition, the result may include grassroots process/product innovations or smaller-scale innovations implemented by these firms or sites, each of which is motivated by a collective will to survive by maintaining required profit levels and to contribute to the surrounding communities by securing reasonably stable levels of employment.

4 Conclusion: Demand Creation and Capability Building

4.1 Additional Process Analysis

In this chapter, we explored the concept of *genba(site)-oriented firm* and its economic model, based primarily on classical or Ricardian/Sraffian assumptions of stable markup ratios and flat supply curves, as well as Keynesian effective demand generation at the level of individual firms and sites. To analyze the behavior of genba-oriented firms, this chapter proposed a simple PXNW model that connects product prices (P), production quantities (X), employment levels (N), and hourly wage rates (W) through physical productivities (i.e., labor input coefficients) and downward sloping demand curves that shift depending upon the product's design qualities.

A prediction that follows from the above analysis is that, in order to improve the living standards of workers/consumers, a firm or site that produces a particular product needs to increase its physical labor productivity (i.e., decrease its labor input coefficient) and generate the product's effective demand quantity (i.e., expand the product's demand curve outward) at the same time. Let us examine this prediction through a process analysis of Periods 3–6 in the PXNW model explained earlier in this chapter (Figs. 9, 10, 11, and 12).

As this process diagram predicts, when strong pressure to reduce supply prices exists under intense global competition, the manufacturing firms in high-wage advanced nations may not have the option to increasing nominal wage rates, but they may still be able to drastically increase the physical productivities of their manufacturing sites through continuous capability building, thereby absorbing the impact of international price reductions. In this way, they can maintain their target markup (or profit) ratios and nominal wage rates.

When prices decrease due to global competition at the macroeconomic level, this may imply slight increases in real wage rates. In fact, between 2000 and 2010, the value-added productivities of Japan's manufacturing sector increased, while its nominal wage rates were mostly unchanged, and consumer prices dropped slightly. This was certainly a period of low growth, and Japan's overall unemployment rates were between 4% and 5%, but they did not soar even during the 2008–2009 financial crisis. Thus, employment levels remained relatively stable despite the difficult period.

These macroeconomic trends are generally consistent with the predicted behavior of genba-oriented manufacturing firms and sites shown in the process analysis of the PXNW model or the sequence diagrams shown in Fig. 13.

This process analysis, based on a simple classical model of one product market and one labor market, hypothetically describes how a genba-oriented firm moves from one steady state to another steady state with improved standards of living. Periods 0–6 correspond to those in the PXNW analysis in Figs. 6, 7, 8, 9, 10, 11, and 12.

Two types of cases are examined here. Case 1 regards spontaneous productivity increases during the high-growth phase of the Cold War, while Case 2 explores the responses by firms in high-wage countries to global price reductions during the post-Cold War period. Both cases start with a steady state and end with another steady state, with higher-wage rates relative to product prices. Specifically, Periods 0, 3, and 6 are steady states.

Let us focus on Case 2 or the process model for the post-Cold War global price competition. When external changes in technologies or prices destroy the initial steady state (Periods 3–4), a genba-oriented firm may take a series of actions to regain its target profit (i.e., markup) ratio (r^*) and employment level (N^*) by increasing its genba's physical labor productivity (a↓; Periods 4–5) and then expanding its product's effective demand quantity (x↑; Periods 5–6).

While the process analysis of Periods 0–3 is not discussed here, it should be noted that, in both cases, in order to regain the steady state, with target markup ratio (r^*) and employment level (N^*) achieved, the genba-oriented firm needs to increase productivity (a↓) and expand effective demand (x↑) at the same time or in a certain sequence.[9]

These two actions by genba-oriented firms, i.e., productivity improvement and demand creation, may involve grassroots process innovations and product innova-

[9]Fujita and Fujimoto (2017) provide some mathematical explanations for the assumption that (i) improving productivity and (ii) creating effective demand are two necessary conditions for genba-oriented firms to simultaneously achieve target profit ratios and employment levels while

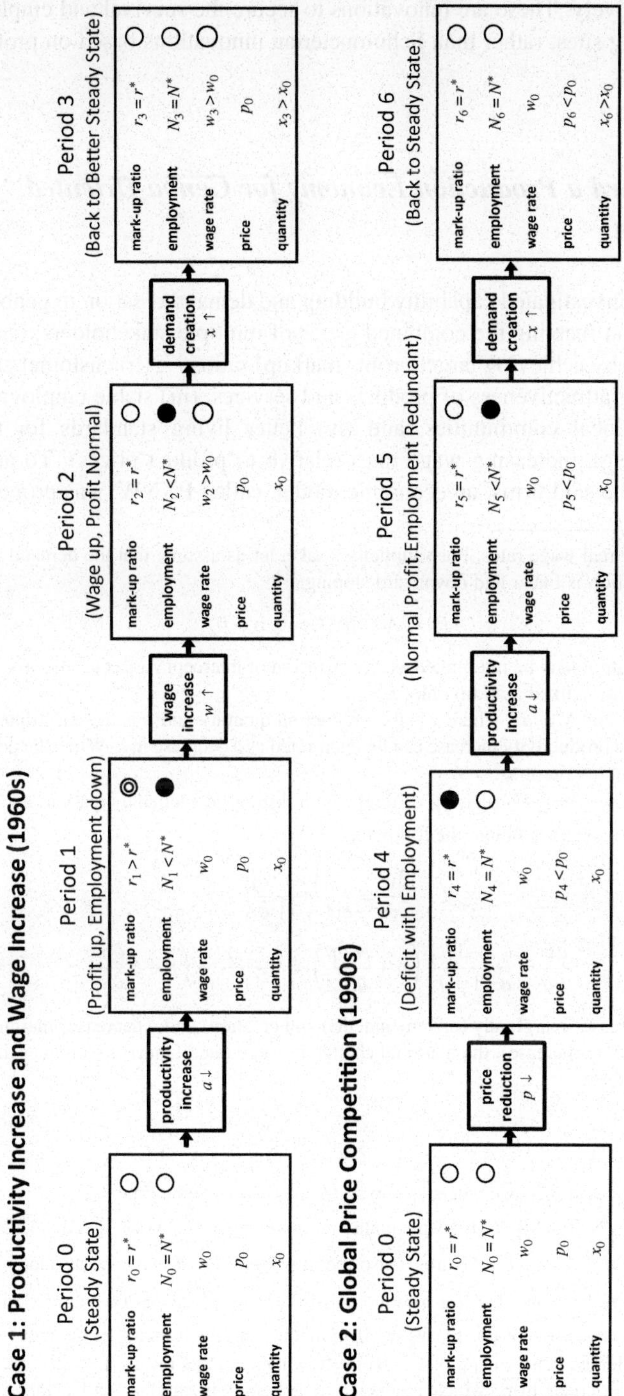

Fig. 13 Process analysis (from steady state to steady state)

tions, respectively. These are innovations to secure the survival and employment of manufacturing sites, rather than Schumpeterian innovations based on profit-seeking and growth.

4.2 Toward a Production Economy for Genba-Oriented Firms

This chapter investigated capability building and demand creation in genba-oriented firms, aimed at meeting the combined needs of multiple stakeholders: (i) the firms' own survival by achieving target profit (markup) ratios r^*, (ii) customer satisfaction by improving attractiveness of products and services, (iii) stable employment N^* to support the local communities, and (iv) better living standards for the firms' employees (e.g., increasing wage rates relative to product prices). To analyze the behavior of the said firms, an economic model, called PXNW, was proposed.

---

ensuring higher real wage rates. For simplicity's sake, let us assume that the demand curve of the product in question is linear and downward sloping:

$$P_d = A - Bx(\mathrm{A} > 0, \mathrm{B} > 0)$$

Demand creation may be interpreted as an expansion of intercept A. Let us also assume the full-cost principle with a fixed markup ratio, r:

$P_s = aw(1 + r)$is $N = ax$, where x is the production quantity and a is the Ricardian labor input coefficient. Here productivity increase can be interpreted as a decrease in a. When the nominal wage rate is w, we can rewrite it as:

$w = \frac{1}{a(1+r)}p(A,x) = \frac{1}{a(1+r)}p\left(A,\frac{N}{a}\right)$1 differentiation through A (demand creation) and a (physical productivity increase), we obtain the following:

$$\frac{\partial w}{\partial A} = \frac{1}{a(1+r)}\frac{\partial p}{\partial A}$$

$$\frac{\partial w}{\partial a} = \frac{-1}{a^2(1+r)}\left(p + \frac{N}{a}\frac{\partial p}{\partial x}\right) = \frac{-1}{a^2(1+r)}\left(p + x\frac{\partial p}{\partial x}\right)$$

When A increases marginally (incremental demand creation) and a decreases marginally (incremental productivity increase), the marginal change in wage rate w may be shown as the following total differentiation:

$$dw = \frac{\partial w}{\partial A}dA + \frac{\partial w}{\partial a}da$$

Then, wage w will increase when $\frac{\partial w}{\partial A} > 0$ and $\frac{\partial w}{\partial a} < 0$ at the same time.

Because $\frac{1}{a(1+r)} > 0$ and $\frac{\partial p}{\partial A} > 0$ (design quality improvement raises price, given the quantity x), we can say that $\frac{\partial w}{\partial A} > 0$ always holds true. Also, because $\frac{-1}{a^2(1+r)} < 0$, we can say that $\frac{\partial w}{\partial A} < 0$ when $p + x\frac{\partial p}{\partial x} > 0$, which means that, for a monopolistic firm that faces a downward sloping demand curve, its marginal revenue is positive ($\frac{dpx}{dx} > 0$) or that it is not maximizing its revenue (px) from this product. If the demand curve is $P_d = A - Bx$ as shown above, this means $A - 2Bx > 0$. To sum up, when a genba-oriented firm, which pursues a target markup ratio r^* and a target number of employees N^* simultaneously and produces a differentiated product, is not maximizing revenue, simultaneous productivity improvements ($a\downarrow$) and demand creation (a↑) always result in wage increases ($w\uparrow$).

Our process analysis of the PXNW model indicated that, when facing a situation where its selling prices go down ($p\downarrow$) or its wages go up ($w\uparrow$), a single-product genba-oriented firm can achieve its dual targets of markup ratio (r^*) and number of employees (N^*) if it increases its physical productivity ($a\downarrow$) and creates demand for its products ($x\uparrow$) at the same time. The case studies of several Japanese firms and sites (ST, OZ, SF, NF, TB) located in specific regions show that their behavior is generally consistent with what the PXNW model suggests.

Actions aimed at demand creation may range from simple sales efforts by the firms' owners, cultivating new customers, adding product variations, finding new applications for existing products, improving product designs, and so on. In any case, a policy implication of such demand-generating efforts is that Keynes' effective demand creation may be achieved not only through the governments' fiscal spending policies but also through these firms' efforts to secure their own employment and survival. Accordingly, certain industrial policies to promote such efforts by genba-oriented firms or community-oriented firms may be added to the governments' policy packages for full employment, particularly when many domestic firms appear to have full employment policies of their own.

From the viewpoint of innovation economics and management, these model analyses and case studies point to the concept of grassroots innovations, or innovations that are motivated by the manufacturing sites' will to survive and attain stable employment within the communities where they are located. When global price-cost competition is intense, for example, these firms may pursue (a) grassroots process innovations to increase productivities for survival and (b) grassroots product innovations to generate sufficient demand for stable employment.

Further research on genba-oriented firms and their behavior will prove valuable. Although our field-based empirical studies already suggest that there are many cases of local companies (mostly small and medium) that behave like genba-oriented firms, we will need more systematic data collection and statistical analyses. In order to see if genba-oriented firms are also found outside Japan, we may have to turn to international studies regarding community-based and employment-conscious firms.

From a theoretical point of view, we may regard our PXNW model of genba-oriented firms as an example of *production economy*, or what John Hicks called plutology (Hicks 1976), which is seen as an alternative economic approach vis-à-vis the standard neoclassical model of catallactics (i.e., theory of exchange). As is widely known, in the neoclassical theory of exchange, the firm model is that of a profit-maximizing firm, which is an essential part of the general equilibrium theory. The concept of genba-oriented firm, by contrast, is based on our empirical observation of many local firms struggling for survival in real competitive situations. They aim to control and improve flows of wealth or value-carrying design information and pursue stability of their value flows rather than profit maximization.

As such, the economic framework that can effectively describe and analyze the actual behavior of genba-oriented firms may be the production economy model (plutology or theory of the production of wealth), which adopts the concepts of flows of value added, steady state (reproduction of flows) rather than equilibrium, evolution rather than optimization, and detailed analysis of industries and industrial sites as the places where such value flows exist, thus including insights from classical economics (Ricardo 1817; Sraffa 1960). Whereas the standard neoclassical economic model is a mathematically sophisticated way to analyze efficient exchanges of wealth, the reality of industrial firms and sites, as well as their evolution, may be better captured by production economy or flow-oriented evolutionary economics for industrial analysis.

The PXNW model to study genba-oriented firms may be regarded as one of such attempts to explore the possibilities of evolutionary production economy. Continuous capability building and demand creation by community-oriented firms and sites are certainly the two key issues in such evolutionary analyses of industries, sites, and firms—and they are the two main topics of the present book as well.

References

Boyer R, Saillard Y (eds) (2002) Régulation theory: the state of the art. Routledge, London/New York

Chamberlin EH (1933) The theory of monopolistic competition: a re-orientation of the theory of value. Harvard University Press, Cambridge, MA

Fujimoto T (1999) The evolution of a manufacturing system at Toyota. Oxford University Press, Oxford

Fujimoto T (2007) Architecture-based comparative advantage: a design information view of manufacturing. Evolutionary and Institutional Economics Review 4:55–112

Fujimoto T (2012) An economic analysis of architecture and coordination: applying Ricardian comparative advantage to design costs and locations. Evolutionary and Institutional Economics Review 9(1):51–124

Fujimoto T (2014) Innovation management in Japan. In: Dodgson M et al (eds) The Oxford handbook of innovation management. Oxford University Press, Oxford

Fujimoto T (2017) Genba-oriented firms and product-process innovations: a preliminary analysis by the "PXNW model". Keizaigaku Ronshu, The Society of Economics, University of Tokyo 81 (3):2–19 (in Japanese)

Fujita Y, Fujimoto T (2017) Existence condition of "active Genba-oriented (site-oriented) firms": an endogenous model regarding productivity improvement and demand creation. Keizaigaku Ronshu, The Society of Economics, University of Tokyo 81(4):2–20 (in Japanese)

Hall RL, Hitch CJ (1939) Price theory and business behaviour. Oxf Econ Pap 2:12–45

Hicks JR (1976) 'Revolutions' in economics. In: Latsis S (ed) Method and appraisal in economics. Cambridge University Press, Cambridge, pp 207–218

Keynes JM (1936) The general theory of employment, interest and money. Macmillan, London

Ricardo D (1817) On the principles of political economy and taxation. John Murray, London

Schumpeter JA (1934) The theory of economic development: an inquiry into profits, capital, credit, interest, and the business cycle. Transaction Publishers, London

Shiozawa Y (2017) An origin of the neoclassical revolution: mill's 'revolution' and consequences. In: Shiozawa Y (ed) A new construction of Ricardian theory of international values. Springer, Berlin, pp 191–240

Simon HA (1947) Administrative behavior: a study of decision-making processes in administrative organization. The Macmillan Company, New York

Sraffa P (1960) Production of commodities by means of commodities: prelude to a critique of economic theory. Cambridge University Press, Cambridge

Evolution of Business Ecosystems

Hirofumi Tatsumoto

Abstract The aim of this chapter is to investigate the dynamics of business ecosystems, which are a special form of industrial cluster characterized by diversity of firms, complex relationships among them, and the presence of platform firms. Such characteristics come from network effects. Firms often set open standards to trigger network effects that are advantageous for their business models. Among them, platform firms, which strategically use the open standards and make the most of the network effects in relation to their business models, play a special role in business ecosystems, because their strategic behavior contributes not only to increasing their market presence but also to the expansion of the ecosystem by stimulating the entry of newcomer firms.

To predict the dynamics of business ecosystems, we should try to understand how product architecture influences firm collaboration within them and characterizes their industrial structures. Open-modular systems, such as PCs, tend to modify their industries and turn them into business ecosystems, whereas closed-integral systems, such as automobiles, yield an industrial form called cohesive clique. Owing to the linkages between architecture and industrial cluster, changes in architecture lead to actual changes in industrial clusters.

Two main factors have recently been affecting product architecture, i.e., digitalization and globalization. Digitalization works as a trigger for architectural change, and globalization acts as its amplifier, which means that products whose architectures have seemed stable for years might drastically change. A slight change in architecture often causes the industrial structure to quickly transform from a cohesive clique into a business ecosystem, which naturally causes changes in the profitability and competitiveness of firms. Faced with these architectural dynamics, each firm needs to thoroughly understand the connections between product architecture and industrial cluster.

H. Tatsumoto (✉)
Graduate School of Business Sciences, University of Tsukuba, Tsukuba, Japan
e-mail: tatsumoto@gssm.otsuka.tsukuba.ac.jp

T. Fujimoto, F. Ikuine (eds.), *Industrial Competitiveness and Design Evolution*,
Evolutionary Economics and Social Complexity Science 12,
https://doi.org/10.1007/978-4-431-55145-4_5

1 Business Ecosystems

1.1 What Is a Business Ecosystem?

The present book has so far analyzed the evolution of product designs and organizational capabilities, as well as their impact on product competitiveness and demand creation, assuming that such products are stand-alone and compete with other products. This chapter expands the research scope to analyze the evolution of industries in which certain products complement other products. The relationship among products is more complex in this case, since they interact not only by competing with but also by complementing one another.

When we find both competition and complementation in an industry or industrial cluster, we call the said industrial system a business ecosystem. Thus, a business ecosystem is an industrial cluster characterized by multiple products and by network effects among them. A product can be complementary to others, which means that its growth depends not only on itself but also on other products. For example, if sales of hardware increase, sales of software will increase too, as software is a complementary good to hardware. This mutual growth is the result of network effects among products.

A business ecosystem contains numerous network effects within it. Indeed, network effect is a significant strategic factor, since an additional user of a product has a positive effect on the value of that product to others. Thanks to the presence of network effects, the value of a product increases according to the number of users. The users benefit not only from the product itself but also from the number of other users, and, in some cases, they value the number of other users more than the product itself. The telephone is a classic example, as a telephone user enjoys greater benefits when he/she belongs to a large telephone network.

What makes the situation more complex is the fact that network effects can emerge from complementary goods. In the example above, when a user chooses a smartphone, he/she should consider not only the number of telephone users but also the variety of apps. Apps are typical complementary goods for smartphones, and their variety is often referred to as an indirect network effect, whereas the number of telephone users is referred to as a direct network effect. Both network effects affect user preferences and the business of firms.

Moreover, the evolution of a business ecosystem is different from that of an ordinary industrial cluster. One major difference is the presence of platform firms, which play leading roles in the evolution of the business ecosystem itself. Previous research has paid much attention to these firms and given them several names: keystone firms (Iansiti and Levien 2004a), platform leaders (Gawer and Cusumano 2002), platform providers, and technology enablers. The increasing number of studies on the subject indicates that there is considerable interest in their special role and overwhelming performance in business ecosystems.

Here, we will focus on the evolution of business ecosystems, with special attention to their dynamics, which are affected by architectural change. This chapter is organized as follows. The first section offers a brief overview of the concept of business ecosystem and illustrates its dynamics and growth model. The second section deals with architecture, which is the main subject of this book, and with firms' approaches to design problem-solving. The linkage between architecture and firms' behavior provides the theoretical basis for analyzing the impact of architectural change on business ecosystems. The third section first explores the recent developments of digitalization and globalization and then incorporates them into the analysis. Additionally, it discusses architectural change and its impact on business ecosystems by drawing a comparison between automobiles and personal computers. The fourth, and last, section presents some concluding remarks.

1.2 An Industry with a Layered Structure

The concept of business ecosystem dates back to the 1990s. Grove (1996) was the first person who identified the structural change occurring in the computer industry. In his book, Grove described this change as "vertical silo to horizontal layers." Before the mid-1990s, computer firms had their own operation chains based on the industrial structure of computers. IBM had its own chips, components, assembly manufacturers, and retailers, and its operation chain was dedicated only to the firm itself. Thus, operation chains were closed networks in those days. Yet, a major architectural change happened in computer systems in the mid-1990s, due to the shift from mainframes to personal computers. The architecture of computers became open, and manufacturers freely chose chips or components from suppliers to produce their goods. Computer retailers sold any brand of computer products. There were no longer closed relationships in the industry.

This architectural change of computer systems also caused a structural change in the computer industry. The operation chains became more open and firms flexibly connected or disconnected with other firms. Computer makers could easily modify their operation chains because computer systems now had open interfaces that enabled compatibility across many components. This open flexibility changed the industrial structure from vertical silo to layer by layer, with each layer standing for chipmakers, component providers, assembly manufacturers, or retailers. The layers had open relationship based on open standards. This open, layer-by-layer structure is the origin of the concept of business ecosystem.

Such new form of industrial structure originally emerged in personal computers and then prevailed in digital consumer goods, cell phones, and similar products. The Internet significantly accelerated this move, and all the industries related to the Internet came to form the same business ecosystem.

1.3 Analogy with Biology: Diversity, Symbiosis, and Keystone

The term *business ecosystem*, the definition that Iansiti and Levien introduced into the academic literature to describe this industrial structure, comes from biology (Iansiti and Levien 2004b), since analogies are drawn with biological ecosystems. Indeed, business ecosystems share some common features with biological ecosystems.

The first common feature is diversity. A biological ecosystem comprises many species that interact with one another in the natural environment. For example, the African savannah ecosystem includes a diverse community of species, such as tropical grasslands and various animals. Similarly, a business ecosystem is an industrial cluster comprising a diverse community of firms. Each firm plays its role by providing components, developing products, and delivering them to users.

Secondly, business ecosystems have complex relationships. Organisms in biological ecosystems interact both directly and indirectly. The food chain typically involves direct relationships, but there are also indirect relationships, like symbiosis. The case of sea anemones and clownfish is famous for its mutualistic relationship of symbiosis. Sea anemones provide clownfish with a safe home, and, in return, clownfish clean the anemone and offer nutrients in the form of waste. Firms in business ecosystems have relationships that are similar to those found in biological ecosystems. They interact directly with one another along the supply chain for parts/materials or services, but they also interact indirectly by means of compatible standards. When compatible standards are present, the sales of a product affect those of compatible or complementary products. Even when firms have no direct mutual relationships, i.e., buying or selling, they need to be aware of this indirect relationship. Indirect relationships in business ecosystems are referred to as network effects.

The third characteristic is the presence of special actors within the ecosystem. In biological ecosystems, some species have an outsized impact on the ecosystem. Paine, a famous ecologist studying biological diversity, conducted field research and found that removing a specific species from an environment can severely affect neighboring species (Paine 1966). He called these special species keystone species. Business ecosystems are much the same, in that they also comprise special firms. Iansiti and Levien called those special firms keystone firms (Iansiti and Levien 2004a). The keystone firms that they mentioned included many platform providers, such as Microsoft and Qualcomm, which are typically called platform firms—and this is the terminology that we will adopt in the remained of our analysis.

Many economists and management scholars have studied platform firms and their business strategies, because they often have dominant market shares and affect neighboring firms. National authorities investigate platform firms in relation to potential breaches of antitrust laws, economists want to theorize their business models in light of their economic behavior, and management scholars wish to understand the secret of their success.

In brief, business ecosystems are characterized by diversity of firms, complex relationships among them, and the presence of platform firms as special firms. These concepts, based on the analogy with biological ecosystems, are useful to describe this new form of industrial structure. Network effects, as an invisible engine, are the key aspect behind these characteristics of business ecosystems (Evans et al. 2006). Firms use network effects to deliver new value, to enlarge their market, or to gain competitive advantages.

When network effects are extended to more and more products or services, the industrial structure is likely to take on the form of a business ecosystem. As mentioned above, business ecosystems emerged first in the personal computer industry, followed by digital consumer goods and cell phones. The Internet strongly reinforced this shift. Today, business ecosystems are found in the machine tool and automotive industries too, which are usually regarded as legacy industries. Hence, this structural change has occurred not only in the new digital industry but also in traditional ones.

1.4 Open Standards as a Source of Network Effects

A business ecosystem is an industrial structure with network effects, which come from complementary relationships among products or services. In many cases, system products have complementary goods. For example, personal computers consist of hardware and software. Software and hardware have a complementary relationship, since for hardware makers software is a complementary good and vice versa.

System products and complementary relationships can easily be found in everyday life. DVD software and DVD players, smartphones and apps, and TV and TV programs are some typical cases. Even the automobile business, which is a very traditional industry, entails complementary relationships. Since system products have become common in our daily life, network effects are ubiquitous, and firms try to exploit them to boost their business.

Whether a business is new or old has nothing to do with the concept of business ecosystem. Complementary relationships are the key factor here, because they generate network effects. The things that one counts within a system determine the boundary of its business ecosystem, so the idea of business ecosystem is deeply rooted in how one sees related workings and recognizes them as a system. Through a broader lens, even a traditional industry can be regarded as a business ecosystem.

Let us look, for example, at the automobile business, which is a traditional industry. If we focus only on automakers and suppliers, it does not appear to have the industrial form of a business ecosystem, because there are only direct relationships and no complementary relationships along the supply chain. But, once we broaden the scope, we can indeed view the automobile industry as a business ecosystem. Within the whole transportation system, automobiles have a complementary relationship with the energy business. Refueling periodically is necessary to drive a vehicle, so that automobiles and gasoline are characterized by complementarity and network effects between them.

All the examples above share the common denominator of compatible standards among products. PCs and smartphones have APIs for developing software and apps, DVDs have various formats for developing content, and automobiles have clear gasoline quality standards to guarantee compatibility. Firms within each ecosystem share these compatible standards, which are called open standards because they are openly shared by the firms in the industry.

Open standards are a powerful source of network effects, since firms need them to collaborate with other firms. Collaboration is a key factor to address the increasing complexity of the system, due to the fact that no single firm can deal with the whole system alone. Open standards provide the pivotal basis for firm collaboration and eventually generate network effects.

Business ecosystems contain many open standards as a source of network effects, but these do not necessarily appear as formal standards. They can take on many forms, such as roadmaps for investments, open APIs, open-source software, or protocol documents. All of them work as open standards because firms openly share them and collaborate based on them, but it does not matter whether the standards are formal or not. Even when they do not entail penalties for protocol violation, they still work well as open standards. As a source of network effects, it is irrelevant whether the open standards are compulsory or voluntary, and firms often follow them by collaborating in an autonomous, decentralized manner.

Business ecosystems offer firms direct and indirect ways to interact with other firms. Supply chains are a direct means of interaction, which is typical in many industries. The use of open standards is an indirect form of interaction that enables firms to collaborate in an autonomous, decentralized manner and is specific to the case of business ecosystems. The notable fact is that open standards generate the network effects that characterize the dynamics of business ecosystems.

1.5 Emergence of Platform Firms

Network effects characterize the behavior of firms in a business ecosystem. Firms have different ways of exploiting network effects depending on their business models, and their strategic behavior is mainly divided into three types: designing of network effects, matching of two markets, and bundling a set of goods.

These strategic actions put in place by firms can contribute to their competitiveness. Although exploiting network effects is common to all the firms in a business ecosystem, platform firms, which play a special role, are in a position to make the most of network effects and usually pursue all three actions above.

The first action is the designing of network effects. As they strongly impact on business growth and competitiveness, designing desirable network effects is a basic strategy. As discussed above, open standards are a powerful source of network effects. Consequently, firms often set open standards to trigger network effects that are advantageous for their business models. There are three approaches to setting open standards, which can be de facto, de jure, or consensus standards.

De facto and de jure standards are more traditional forms. De facto standards are automatically defined through market share dominance, while de jure standards are established by public committees, such as ANSI and ISO/IEC. As a new approach, consensus standards are set by firms gathering together in consortia, forums, or alliances. They are thus similar to de jure standards but differ in the fact that firms come together according to their own will, so multiple consortia might be created to pursue similar targets. We will discuss the differences among these three approaches to standards setting later on in this chapter. Platform firms are likely to organize the consortia tasked with setting open standards. Their purpose is not only to establish a basis for collaboration but also to try and make desirable network effects emerge in the ecosystem.

The second strategic action, the matching of two markets, is another typical firm approach aimed at exploiting network effects. For example, a firm develops a navigation app for car drivers and collects information on where they go or what they are likely to do. Based on this information, the firm is able to suggest a restaurant or shopping mall where the drivers might want to stop. In this case, the navigation firm acts not only by giving information on the goal but also by matching two markets, the navigation service and the other services that the car drivers might want to receive according to their preferences. Matching works well if the two products are linked by a network effect, since consumption of one good leads to increased consumption of the other good. The network effect between two products means that the user of a product is likely to benefit from the other product when he/she jointly uses them.

Scholars have studied matching as the theory of two-sided markets or pricing (Rochet and Tirole 2003). Thanks to network effects, matching between two markets contributes to the expansion of the total market, ultimately supporting the growth of the business ecosystem as a whole. Of course, this also leads to an increasing presence of firms that provide matching services, which are often referred to as platform firms. So platform firms play a special role in business ecosystems.

Firms often use matching as a critical part of their business models, especially in Internet services. Google, Facebook, and Amazon are the technology giants that provide matching services. These platform firms work as matching agents between two or more markets by offering open interfaces for third parties to use their services. The third parties use the open interfaces, collect information about the users, and reinforce the network effect by matching. Real businesses also use matching of two groups. For example, in the area of the Internet of Things, or IoT, there are many IoT devices that generate data about the usage of a product/service. Matching in real business connects these usage data with other products/services. As seen above in the case of a car navigation app, matching connects the navigation service with the restaurant or shopping service.

The third way to make the most of network effects is the bundling of a set of goods or services as a package (Nalebuff 2004; Eisenmann et al. 2011). Both matching and bundling rely on network effects between two groups, but they differ in the way in which the two groups overlap. Matching works well if the two groups do not have any overlaps. For example, if the providers and consumers of a given product are different groups and they do not simultaneously produce or consume the said good, firms can exploit the network effects by matching. But if the two groups mostly overlap—for example, in the case of cell phones and portable audio devices—matching is rather useless, while bundling is an effective strategy to make the most of network effects. Smartphones, which bundle together cell phone and portable audio features, are a good choice for users in situations of considerable overlaps.

If two products are functionally complementary, there will be network effects between the two groups of users. And if there are major overlaps between the said two groups, most users will benefit from the two functions combined. With functional complementarity, meaning the network effect between two products, the utility of a package bundling two functions together exceeds the sum of the utility of the two individual functions. The utility gap between the package and the sum of individual goods widens as functional complementarity becomes stronger.

Bundling works as a kind of lock-in or closure strategy. The firms that bundle two functions in a package enclose the network effects into their products, so that they can benefit from them while other firms do not. For instance, smartphone manufacturers lock in consumers who use cell phones and portable audio. Functional complementarity makes it much harder for firms able to provide one function only to enter the smartphone market. Since network effects derive from functional complementarity between two goods and business ecosystems are characterized by considerable network effects among products, firms have many opportunities to implement bundling strategies.

The three strategic actions described above are quite different from those of product firms operating in ordinary markets, whose main aim is to provide good-quality products at inexpensive prices. Of course, firms in business ecosystems have the same aim, but they cannot disregard the new factor, i.e., network effects for creating value and competing with rivals, which represent a major challenge as well as an opportunity.

In sum, designing of network effects, matching of two markets, and bundling of goods are all strategic actions that exploit the macrostructure of business ecosystems. Since business ecosystems comprise a variety of products that complement each other, network effects will arise as an intrinsic feature that firms can leverage when implementing their strategic actions, which exceed the scope of the strategies that product firms typically have.

These strategic actions are very characteristic of business ecosystems, and the organizations that wish to implement them need strong macroscopic capabilities. With respect to this point, platform firms are able to make the most of these strategies as they do possess macroscopic capabilities. They concern themselves not only with their own products but also with neighboring products, capturing complementary relationships that work as network effects. They also watch the macroelements, such as consortia and open standards, that might be a source of network effects and carefully assess business ecosystem dynamics because these invisible forces affect

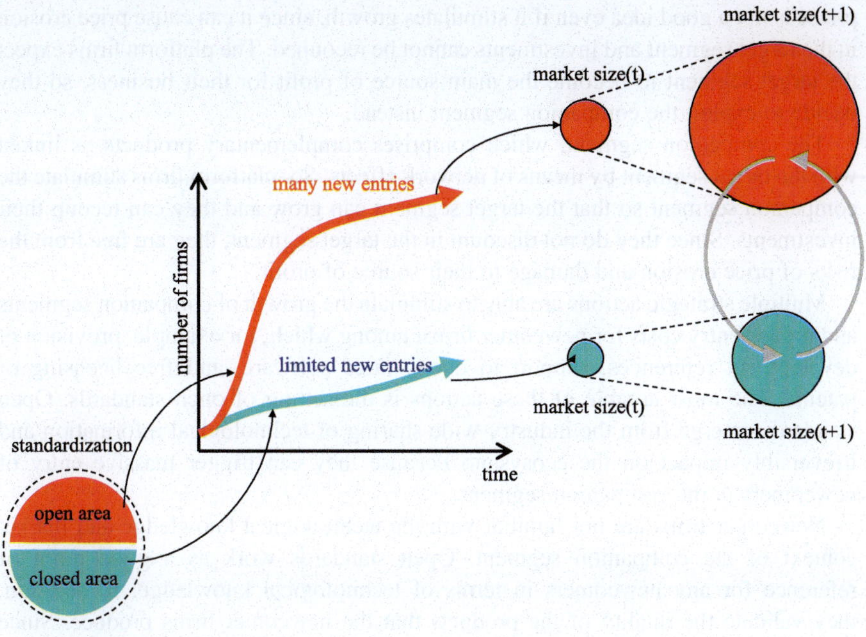

Fig. 1 Growth model of business ecosystem

their business. So, platform firms carefully monitor the dynamics of business ecosystems and exercise strategic actions to exploit them according to their business models.

1.6 Growth Model of Business Ecosystems

As discuss so far, a business ecosystem is an industrial cluster characterized by diversity of firms, complex relationships, and platform firms. Such characteristics come from network effects. Many market segments have indirect relationships that generate network effects, invisible forces affecting the evolution of a business ecosystem. Figure 1 illustrates the growth model of business ecosystems. For the sake of simplicity, we pick two segments that are related by network effects. If one segment grows, the other segment also grows due to network effects. Strong network effects cause a steeper mutual growth of the two segments.

If platform firms, which exploit network effects for their strategic purposes, find that two segments are linked by network effects, they try to leverage this in the most effective way. Typically, they start by setting their strategic framework, picking one of the two segments as the target segment and the other as a companion segment. They sell their own products in the target segment, which they want to grow in order to expand their business. One possible strategy is discounting, but applying cheaper prices is not a good idea even if it stimulates growth, since it can cause price erosion in the target segment and investments cannot be recouped. The platform firms expect the target segment to become the main source of profit for their business, so they decide to exploit the companion segment instead.

The companion segment, which comprises complementary products, is linked with the target segment by means of network effects. So, platform firms stimulate the companion segment so that the target segment can grow and they can recoup their investments. Since they do not discount in the target segment, they are free from the risks of price erosion and damage to their source of profit.

Multiple strategic actions are able to stimulate the growth of companion segments and reduce entry costs for newcomer firms, among which, for example, provision of development references, support to open-source software, and free licensing of patents. The most notable of these actions is the setting of open standards. Open standards emerge from the industry-wide sharing of technological information and irreversibly impact on the ecosystem because they can trigger massive entry of newcomers in the companion segment.

Newcomer firms are not familiar with the technological knowledge and market context of the companion segment. Open standards work as a good point of reference for any newcomers in terms of technological knowledge. In addition, they validate the quality of the products that the newcomer firms produce. Since open standards include explicit compatibility criteria, new entrants can easily adjust their products. Thus, the adoption of open standards compensates for the lack of newcomers' knowledge of the market context.

By learning from open standards, the newcomers rapidly catch up with the incumbent firms in the companion segment. The said segment is thus transformed into an open area for newcomers that would otherwise hesitate to enter. In brief, open standards remove the entry barriers existing in the companion segment. With new entries, market growth in the companion segment is stimulated, supply volumes increase, and prices drop to affordable levels. The size of the target segment increases depending on the number of newcomers entering the companion segment.

The growth of the companion segment also promotes that of the target segment. The network effect, which links the two segments and stems from their complementary relationship, conveys the growth momentum of the companion segment to the target segment. Hence, open standards make both the target and companion segment grow. A noticeable consequence of this process is that the characteristics of the two segments change. The companion segment, for which open standards have been set by the platform firms, becomes an open area for newcomer firms, as entry barriers are removed and opportunities to catch up multiply. Conversely, the target segment remains almost completely out of bounds to newcomers, since entry barrier are still in place. The main reason for this is the fact that platform firms, for which the target segment is the main source of profit, see any newcomers as rivals and strive to keep them out.

The setting of open standards causes business ecosystems to have an open area and a closed area. Newcomer firms enter the open area and stimulate its growth. Due to such increasing entries, the closed area, which is still mostly off limits to newcomers, also grows thanks to the network effect linking the two segments. Therefore, the moment when open standards are set often represents the inflection point at which a business ecosystem opens up one of its segments to newcomer firms and consequently starts to expand.

This is a typical strategy for platform firms because it allows them to achieve two key targets, i.e., ensuring the growth of the business ecosystem as a whole and keeping their own business profitable (Tatsumoto 2017). This so-called platform strategy can be seen in various ecosystems, such as personal computers, digital consumer goods, cell phones, and semiconductor industries, and even traditional industry firms have recently started to express an interest in it. The trend is becoming more and more evident because the industrial structures of various areas are transforming into business ecosystems. For example, in the automobile industry, this trend is often referred to as CASE, which stands for connected, autonomous, shared, and electric. In the case of machine tools, it is called Industry 4.0.

Before exploring this topic further, the next section will illustrate the theoretical framework adopted to analyze these trends, which is mainly based on the architectural theory.

2 Architecture

2.1 Architecture Affects Business Ecosystems

The first section of this chapter provided an overview of the dynamics of business ecosystems. Let us now shed light on the characteristics of business ecosystems, which are deeply connected with the architecture of the products or services provided by firms within them. The concept of architecture refers to the basic configuration design of an artifact, which depends on problem-solving decisions regarding its design factors.

This definition means that the architecture of a product is linked with the way in which a firm solves design problems to manufacture it. In particular, when dealing with large-scale design problems, which require collaboration among multiple firms, product architecture affects the activity of joint problem-solving. Some types of architecture are more suited to a small group of firms jointly dealing with design problems, whereas other types call for a more autonomous, decentralized approach.

With respect to design problems, each architecture has a different approach to developing design solutions. Increasing complexity reinforces this tendency and causes architectural changes, which impact on the industrial structure itself. Business ecosystems, which often arise as a consequence of architectural changes, are a special form of industrial structure. In extreme cases, a slight change in architecture might lead the industrial structure to quickly transform into a business ecosystem, which naturally causes changes in the profitability and competitiveness of firms. Hence, most of the firms who operate within an ecosystem cannot ignore the architecture of its products and how it changes.

In the remaining part of this section, we will examine the architectural characteristics that influence the behavior of firms by laying out the theoretical dimensions of product architecture. We will categorize these characteristics based on the features of the interfaces for which firms have to collaborate with one another and then discuss how and why product architecture affects the shape of the industrial structure.

2.2 Architectural Anatomies of Complex Systems

The study of architecture is rooted in Simon's examination of complex systems (Simon 1969). He found that complex systems share the common feature of having numerous design factors that are mutually dependent. The said design factors and dependencies are the source of complexity that makes product development difficult.

A set of design factors that are mutually dependent can be seen as a design problem of simultaneous equations. During the development process, product designers look for a solution to the design problem. Obviously, as the number of factors and dependencies rises, it becomes ever more difficult to solve the design problem, until a complexity limit is reached and product designers are no longer able to find a solution.

To handle this increasing complexity, product designers encapsulate the design factors into modules (Aoki 2001). A module is an assembled chunk of design factors that are mutually dependent. This design technique is a kind of "divide and conquer" approach to complexity, and it is usually referred to as modularization. Through modularization, a complex system comes to have relatively more dependencies within modules and, at the same time, relatively fewer dependencies across modules. Thus, modularization reduces the complexity of design problems because it divides the large simultaneous equations into small independent ones, so that product designers can handle them individually.

The concept of modularization is quite important, especially in the practical development process, which is often affected by interferences ranging from a simple telephone call to economic turbulence. To reduce this *noise*, product designers use the modularization technique and divide the design factors into modules. A set of design factors and dependencies within a module is unaffected by external noise, and, if the said noise damages a module, the other modules will remain safe. In this way, only one module is lost, and the product designers just need to replace the damaged module with a new one.

Therefore, modularization causes a complex system to become a composite of modules, which Simon (1969) referred to as a nearly decomposable system (NDS), and its behavior can be approximately represented by the combination of its modules' behaviors. Modern products, which are likely to have numerous design factors for many functions, are often complex systems, and we can regard them as a composite of individual modules.

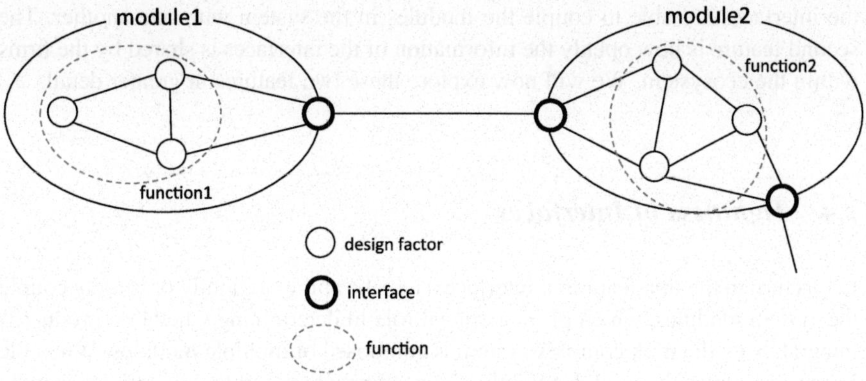

Fig. 2 Design factors, modules, and interfaces

2.3 Features of Module Interfaces

A module contains multiple design factors that realize its functions. Designing a function means solving simultaneous equations involving a set of design factors. Product designers need to identify appropriate parameters for all the design factors, and they usually adopt a trial and error approach to come up with solutions. Once a solution is found, the set of design factors will work well as a function of the system. That is, a set of design factors denotes a function in a module.

Figure 2 illustrates the concept of complex system as a composite of modules. Module 1 contains three design factors, which perform function 1. Similarly, module 2 consists of three design factors and fulfils function 2. Moreover, there are special design factors creating connections across the modules. These design factors, which are called interfaces, are gateways to the design factors within each module. The interfaces connect the modules and the composite of the modules makes up the system.

The combination of the individual modules' behaviors approximately represents the behavior of the system. From the users' point of view, behavior means function. In this interpretation, designing a complex product is an effort to combine modules. Yet, a further complication lies in the fact that the combination of modules produces interaction effects, which might appear negligible at first or immediately come across as difficult to eliminate. Even when they are initially weak, they might stack up and become collectively stronger. This is why those involved in designing complex systems tend to adopt a trial and error approach.

When combining modules, product designers use interfaces. Some interfaces might work well in combining modules, but they might not make the most of the modules' characteristics. Other interfaces might be difficult to use because they require information that is not shared. Hence, interfaces work as a constraint for design problems, and complex systems heavily rely on their characteristics.

Prior research on product architecture has shown that two features of interfaces particularly affect the designing activities. The first feature is how tightly or loosely the interfaces are able to couple the modules in the system with one another. The second feature is how openly the information of the interfaces is shared by the firms within the ecosystem. We will now explore these two features in greater detail.

2.4 *Tightness of Interfaces*

Let us turn to the first feature of interfaces, i.e., the ability to tightly or loosely couple the system modules, which plays a crucial role in determining whether a product is integral or modular. A complex system is composed of multiple modules. When the interface couples the modules tightly, connections are possible only among specific modules. By contrast, when the interface couples the modules loosely, they can easily connect with one another.

The reason for coupling modules is that a function in a module depends on another function in another module. Hence, product designers use interfaces for module coupling to connect a function to another function across modules. If a function depends solely on another function in the same module, the use of an interface is not necessary.

Based on this functional dependency across modules, the use of interfaces reflects the mapping pattern of functions and modules. Figure 3 shows the correspondence between interfaces and architecture.

In the left part of Fig. 3, (i) and (ii) show how the interface couples the modules: (i) refers to the case of tight coupling, whereas (ii) refers to the case of loose coupling of the modules.

In the right part of Fig. 3, (a) and (b) show the types of architecture through bipartite graphs illustrating how the modules and functions are connected. Prior studies have often used bipartite graphs to describe the mapping pattern of modules and functions in product architecture (Ulrich 1995). (A) describes a situation in which a function in a module heavily depends on other functions in other modules; this is the typical mapping pattern of integral architecture. Conversely, (B) describes a situation in which a function in a module does not heavily depend on other functions in other modules; this is the typical mapping pattern of modular architecture.

In the case of integral architecture, where functions heavily depend on one another across modules, a slight modification to function 1 affects both module 2 and module 3 exactly because of mutual dependencies. Since it is difficult to make changes, integral architecture cannot have many design options. Instead, when the architecture is modular, as in situation (B), slight modifications do not affect other modules, and design changes are easily implemented because mutual dependencies across modules are weak.

As for the relationship between interface and architecture, the arrows in Fig. 3 show that tight coupling corresponds to integral architecture and loose coupling to modular architecture.

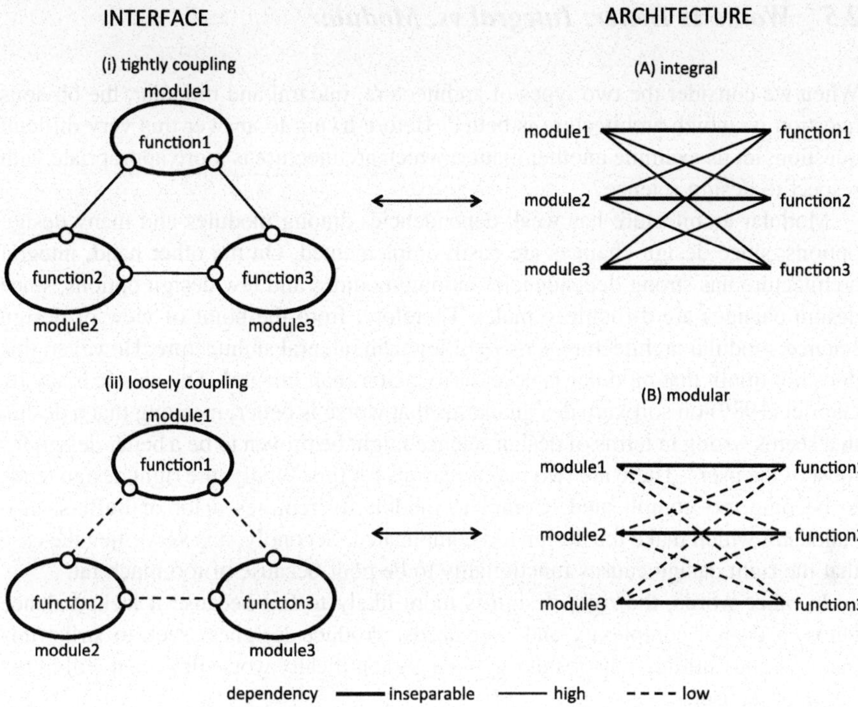

Fig. 3 Interfaces and architectures

In part (i) of Fig. 3, the interfaces tightly couple modules 1, 2, and 3. This means that function 1 in module 1 heavily depends on functions in modules 2 and 3. (A), the graph referring to integral architecture, expresses such dependency pattern in the form of mapping between functions and modules. Function 1 has inseparable mutual dependencies with module 1, which means that a set of design parameters of function 1 forms module 1. Function 1 also has strong dependencies with modules 2 and 3 because it depends on functions 2 and 3. These strong dependencies among functions cause the interfaces to tightly couple the modules. This is the reason why tight coupling corresponds to the case of integral architecture.

By contrast, loose coupling corresponds to modular architecture. In part (ii) of Fig. 3, the interfaces loosely couple modules 1, 2, and 3, which means that function 1 in module 1 does not heavily depend on the functions in modules 2 and 3. This pattern of functional dependencies corresponds to modular architecture (B).

The mapping pattern leads to differences in architectural characteristics. In some cases integral architecture yields better deigns, while, in other cases, modular architecture works more effectively. We will now discuss this point further.

2.5 *Worse Is Better: Integral vs. Modular*

When we consider the two types of architecture, integral and modular, the obvious question is: which architecture is better? Before trying to answer this very difficult question, let us examine another matter: which architecture is more appropriate with respect to design science?

Modular architecture has weak dependencies among modules and many design options, since design changes are easily implemented. On the other hand, integral architecture has strong dependencies among modules and few design options, since design changes are difficult to make. Therefore, from the point of view of design science, modular architecture is more fitting than integral architecture. However, this does not imply that modular is necessarily better than integral. The classic essay by Gabriel (1989) on software design claims that worse is better, meaning that a design that seems wrong in terms of design science might be proven to be a better design for most of the users. There are two major reasons for this. Firstly, the right design tends to become too complicated. Complete modularity requires a lot of buffers, also called *fat*, which make the design too complicated. Secondly, it is very often the case that the right design causes functionality to be poor because of too much fat.

In other words, the right design is more likely to fail because it lacks balance between design complexity and user needs. Product designers seek to strike this balance, and, although the result might be a scientifically worse design, it will fit the needs of its users.

Figure 4 explains the balance between functions and modules based on the type of architecture. Suppose that the architecture is redundant, like in case (b). Modular system 1 has fat (or buffers) in its modules, introduced by product designers to obtain weak dependencies across modules. In other words, it is the fat that ensures the modularity of the system. If product designers remove the fat, the system loses its modularity and becomes lean, but it still performs the required functions. When a system increasingly loses its buffers, its interfaces couple the modules more and more tightly, and its architecture eventually becomes integral.

Now let us consider another redundant system, shown as (c). The functions in modular system 2 have reduced functionalities. Product designers may opt for reduced functionalities because they do not want a function to be heavily dependent on other functions in other modules. In other words, they want to keep the system modular. Nonetheless, in some cases, they need to increase the functionalities to meet the users' needs and have no choice but to break the modularity of the system. They redesign it so that the function in question heavily depends on other functions in other modules. Through this process, the system architecture becomes integral.

Although modular architecture should be preferable from the point of view of design science, it may not be better in terms of balancing functions and modules. Modular systems are naturally characterized by redundancy, which is necessary for design changeability that makes the system, as a product, more appealing to the

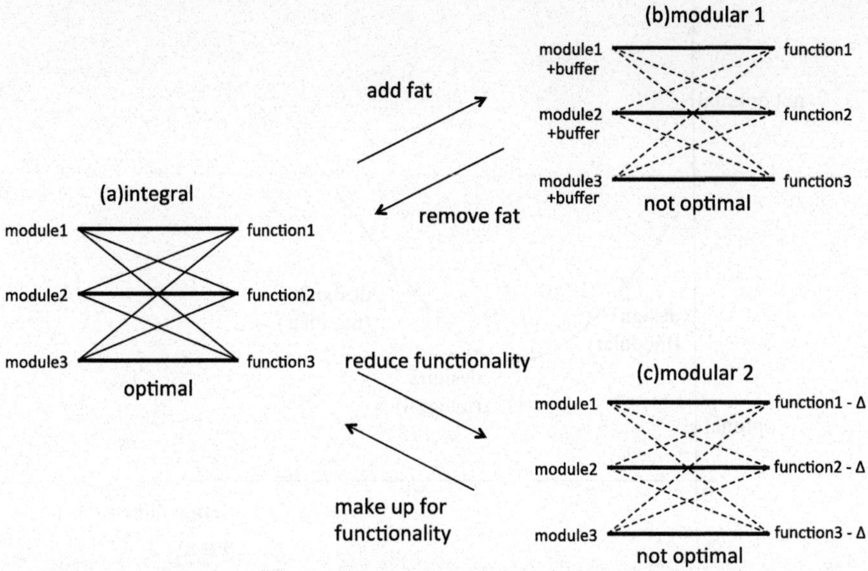

Fig. 4 Worse is better

users, but this upsets the balance between functions and modules. If the system redundancy is removed, its architecture becomes integral. This is the main reason why many successful products are of the integral architecture type. As Gabriel (1989) insisted, product architecture that is appealing to most users is often worse in terms of design science.

2.6 *Design Changeability of Modular Architecture*

Integral architectures strike an optimal balance under the design dimensions given, which consist of vectors of design parameters. However, integral architecture does not always mean better design, as modular architecture might yield superior results in some situations. Design changeability is the relevant concept here. Figures 5 and 6 show when modular architecture is better than integral architecture.

Figure 5 illustrates the relationship between architecture and optimality. The horizontal axis represents the design dimension, i.e., a set of design parameters for the given modules. The vertical axis indicates the system error, which denotes optimality of functions and modules. System error ε is the difference between maximum system utility, u^*, and expected system utility for a given design parameter d, $u(d)$.

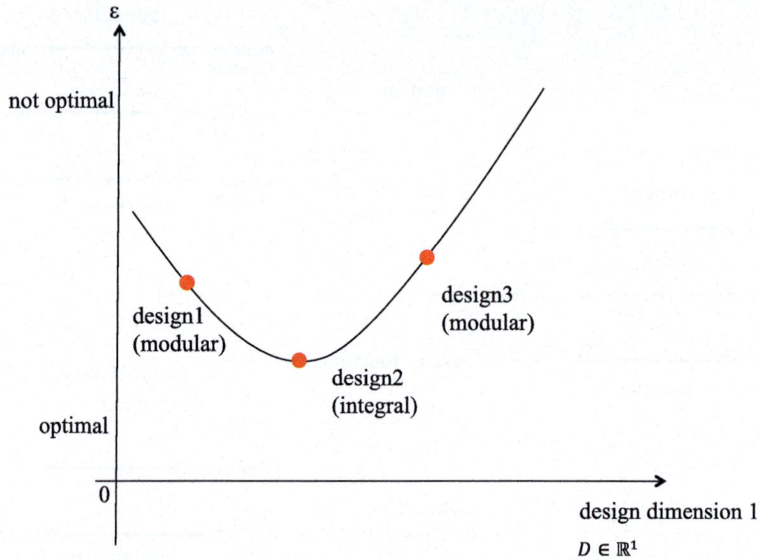

Fig. 5 Architecture and optimality

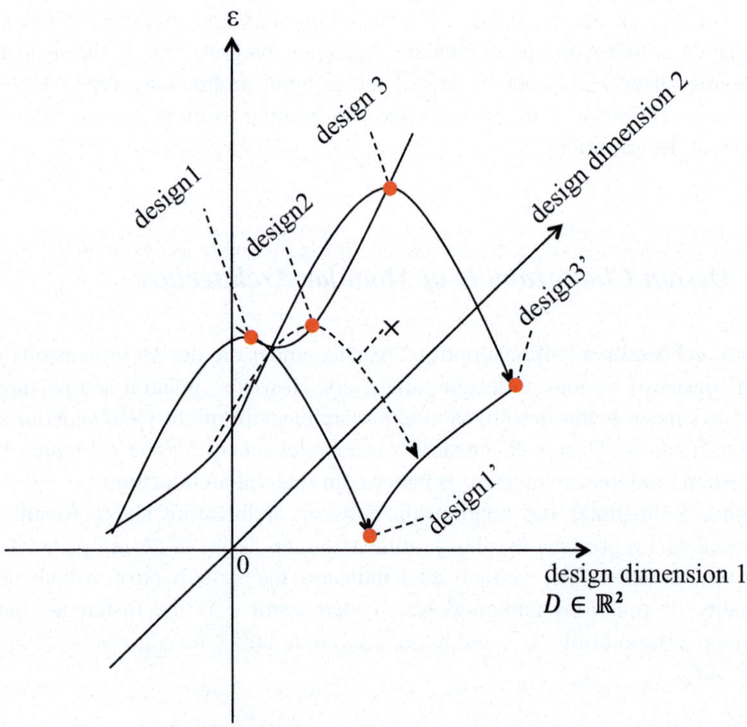

Fig. 6 Expansion of design dimensions

System error ε is defined as:

$$\varepsilon = u^* - u(d) \geq 0$$

$$\begin{aligned} &\text{s.t.} \\ &D \in \mathbf{R}^m \\ &d \in D \\ &u : D \to \mathbf{R} \\ u^* &= \max\{u(d) | d \in D\} \end{aligned}$$

where D is a set of design parameters, m is the number of design dimensions, d is an element of D, u is a utility function, and u^* is maximum utility.

Roughly speaking, expected system utility $u(d)$ means how much utility we expect the system to achieve for a given set of design parameters, d. When the system is optimal, system error ε is zero. Optimality increases moving up along the vertical axis from optimal to not optimal.

In Fig. 5, designs 1 to 3 refer to designs with given sets of design parameters. Each design realizes the system utility and generates the system error. As explained above, the system error is low when the system utility achieved is high. Design 1 is characterized by fat in its modules, which means that some design parameters, packed into modules, do not contribute to realizing the system functions but are simply buffers. In terms of the balance between functions and modules, design 1 is not an efficient design.

On the other hand, design 3 has low functionality. It fully uses the design parameters but still realizes only a small fraction of the needed functions. In other words, design 3 is redundant in its functions. It appears to have many but, in fact, the number of unique functions is small. Just like design 1, design 3 is not an efficient design.

Compared to designs 1 and 3, design 2 achieves greater utility because it has a good balance between functions and modules. It is a lean system under design dimension 1. With respect to architecture, design 2 is an integral system, while designs 1 and 3 are modular systems.

Both designs 1 and 3 have a high system error, which means that a given set of design parameters achieves only low utility, far from optimal. The reason for this is the lack of balance between functions and modules in the system. Too many functions yield a redundant system, while too many modules yield a fat system. Thus, designs 1 and 3 are not optimal, whereas design 2 strikes the optimal balance between functions and modules. In conclusion, integral architecture is the optimal design under design dimension 1.

However, integral architecture has a by-product. When the architecture of a product becomes integral, its interfaces tighten to couple modules, and the addition of other modules becomes problematic. That is, in integral architecture, it is difficult to expand the design dimensions by adding new modules. Conversely, the advantage of modular architecture is that new modules can be added without complications. Modular architecture easily expands the design dimensions by adding modules. If expanding the design dimensions brings major advantages, then modular architecture might be better suited to realizing the design than integral architecture.

Figure 6 shows the relationship between design dimensions and system utility. Under design dimension 1, design 2, which is integral, realizes optimal system utility, while designs 1 and 3, which are modular, are not optimal.

If product designers add another module, the design space expands to design dimensions 1 and 2. Designs 1 and 3 can use the design parameters in design dimension 2. Since their architecture is modular, another module can be easily added. By using the design parameters in dimension 2, designs 1 and 3 can achieve greater utility than design 2. For example, product designers can modify design 3, turning it into design 3′, to improve its optimality. The utility of design 3′ exceeds that of design 2, so that modular architecture makes for better design than integral architecture in this case.

As for design 2, it is not so easy to use the design parameters in dimension 2, since the architecture of this design is integral and its interfaces tightly couple the modules. Thus, adding another module is difficult. If product designers want to use the parameters in dimension 2, they need to loosen the interfaces of design 2, making it more similar to design 1 or 3, and they then can use the design parameters in dimension 2. However, this is a rather tough design process.

In sum, modular architecture is not optimal under the design dimensions given. But, by adding new modules, it easily expands its design space. Under the new dimensions, modular architecture can achieve better designs than integral architecture. With regard to design changeability, modular architecture is superior to integral architecture.

2.7 *Openness of Interfaces*

Another key architectural characteristic is that of the openness of interfaces. The openness or closedness of interfaces refers to the extent to which technological information is shared with other firms. In the case of openness, product designers from one firm can share interface information with other designers from other firms. Conversely, closedness indicates a situation in which product designers can share the interface information only with a few people who have been given permission. These usually range from a limited number of colleagues to selected representatives of the customers or suppliers.

From an industry-wide perspective, information sharing across firms is deeply affected by the degree of standardization of interfaces. Standardization is a coordination process for firms participating in an ecosystem. Open standards, which are the outcome of standardization, represent interface information to be shared for compatibility and can be used by all the firms in the ecosystem. Hence, standardization is an industry-wide activity to enhance information sharing.

Nowadays, most products are complex systems with many interfaces, requiring collaboration among numerous firms. Since the sharing of information is essential

Table 1 Three approaches for open standards

	Setting of standards	Diffusion of standards	Open areas
De facto standards	Market process (dominant firm)	Market process (consumer preference)	Narrow
Consensus standards	Nonmarket process (consortium)	Market process (consumer preference)	Wide
De jure standards	Nonmarket process (committee)	Nonmarket process (often mandatory)	Narrow

for their design tasks, standardization is a key activity for any firm developing complex systems, as it leads to open standards, which are the basis for such collaborative development. The process through which standardization is achieved affects the features of the interfaces of complex systems. Every firm carefully watches the standardization process and its outcomes, since the open standards will influence the growth of the business ecosystem and the competitive advantages gained by firms. Indeed, the main feature deriving from standardization is openness or closedness, which greatly depends on the standardization approaches. The firms that set the standards choose the appropriate approach to standardization according to their business strategies.

The standardization process includes two main phases: setting of standards and diffusion of standards. The setting of standards refers to the designing of interface protocols, while the diffusion of standards is an activity aimed at promoting information sharing about the interfaces. The combination of these two aspects leads to the classification of open standards into three types: de facto standards, de jure standards, and consensus standards. Table 1 provides a summary of three approaches for open standards.

With respect to openness, the three approaches to standardization produce different outcomes. Openness is measured by the size of the open area, i.e., the market segment that does not have entry barriers and is easily accessible to newcomer firms. De facto and de jure standards are likely to produce narrower open areas than consensus standards.

These three approaches are clearly different, especially for what concerns the abovementioned phases of setting and diffusion of standards. However, researchers may confuse them, and, in particular, consensus standards are often mistaken for other standards. This is mainly because consensus standards have appeared more recently and are relatively new to researchers. In addition, consensus standards are partially similar to de facto and de jure standards.

De facto and consensus standards differ in the setting phase. De facto standards are set by a single firm, as there is the risk of infringing antitrust laws if multiple firms gather together to set de facto standards. If they wish to avoid this risk, they should opt for the setting of consensus standards, for which the firms legitimately form a consortium open to third parties. Due to this openness, consensus standards are likely to produce wider open areas than de facto standards.

Turning now to the comparison between de jure and consensus standards, the difference here is in the diffusion phase. De jure standards are ordained by law and are thus often mandatory. Consequently, promoting the diffusion of open standards is not the chief concern in this case. By contrast, consensus standards are voluntary rather than mandatory and are likely to produce much wider open areas for easy diffusion.

There is another reason why consensus standards are characterized by wider open areas. Since they are voluntary, multiple consortia will probably set similar standards and compete in the market. All the consortia care about standards diffusion, which determines the winner of the competition. The wider open area favors the diffusion of standards because it is attractive for newcomers. With the risk of standards wars, consensus standards provide a wider open area through standardization.

In sum, consensus standards usually produce a wider open area than both de facto and de jure standards and are a relatively new approach, whereas de facto and de jure standards represent the old approach to standardization. Consensus standards have been spreading since the mid-1980s through a general relaxation of antitrust laws. This has encouraged the joint development of consensus standards and the creation of wide open areas in business ecosystems. There are many examples of consensus standardization, such as the USB forum for open PC and peripherals interfaces, the AUTOSAR consortium for the car electronics area, W3C for open Internet protocols, as well as Industrial Internet Consortium and OpenFog Consortium for the IoT area.

2.8 *Architectural Convergence: Open-Modular and Closed-Integral*

So far, we have discussed the key features of interfaces. Firstly, we spoke about tight or loose coupling. A system with tightly coupled interfaces often corresponds to integral architecture, whereas loosely coupled interfaces are likely to go with modular architecture. The second aspect is that of openness or closedness. A system that allows the open sharing of interface information will probably have an open architecture; otherwise, its architecture will be closed.

It can be deduced from the above discussion that system architecture can be described along the two dimensions of integral/modular and open/closed. This yields four possible types of architecture: integral or modular in coupling modules and open or closed in sharing information. Clearly, the two architectural dimensions have interactions because both depend on the features of the interfaces in the system, which reflect the mechanisms of collaborative problem-solving among firms. By virtue of these interactions, the four types of architectures are reduced to two combinations: closed-integral and open-modular architectures.

With respect to joint problem-solving, closed-integral architecture is characterized by a consistent approach. As the interfaces are closed, their information is shared among a limited number of firms. Consequently, each of the modules in the system can connect only with a few other modules. However, the large amount of detailed information shared through intense collaboration helps improve the system and optimize its functions. These design changes sacrifice module compatibility to achieve optimal performance and product integrity by means of tight coupling.

On the other hand, open-modular architecture has a different design approach. Its open interfaces enable product designers to collaborate with several other firms, and, since the latter also include newcomers, new module combinations are easily developed and work as a source of new value for the product. Product designers expect high interface compatibility and can make the most of the open interfaces by experimenting with new combinations.

As far as the design logic is concerned, both closed-integral and open-modular architectures are characterized by consistency. But what about open-integral or closed-modular products? If designers opt for an open-integral architecture, which is an integral architecture with open interfaces, then inconsistency occurs. The interface protocols are often modified to optimize system performance, and this drastically reduces open compatibility, which forms the basis of open interfaces for third parties. Integral architecture does not go well with open interfaces in the design process.

Similarly, a closed-modular architecture, i.e., a modular architecture with closed interfaces, usually proves to be a poor choice. In the modular architecture, product designers can experiment with different combinations to try and create new value in use, but, if closed interfaces are chosen, this prevents them from exploring all the possible module combinations. Hence, they do not make use of modular architecture with closed interfaces because it lacks open compatibility with products from other firms.

As mentioned above, in the closed-integral and open-modular types, architectural convergence comes from consistent design logic. In the closed-integral architecture, top system performance is realized through frequent interface adjustments. Detailed information about the interfaces is shared within a small, tight-knit group of product designers that work on customization for optimal performance. Through this process, the system eventually becomes a closed-integral architecture. By contrast, the open interfaces of open-modular architecture allow product designers to share interface information across firms and try various module combinations to assess new value in use. Modularity is crucial for this trial and error process because it ensures compatibly among modules. Through this process, the system eventually becomes an open-modular architecture. Familiar examples of this architectural convergence are automobiles, a notable case of closed-integral architecture, and personal computers, with typically open-modular architectures. The next section discusses these two architectures in relation to industrial dynamics.

3 Architectural Change and Business Ecosystems

3.1 Digitalization: Trigger of Architectural Change

So far we have discussed business ecosystems and product architecture. A business ecosystem is a form of industrial cluster, and product architecture is the basic design of module configuration. Industrial clusters and product architectures are linked by the design logic according to which firms collaborate to solve design problems. Each type of architecture has its different design logic. Therefore, changes in architecture lead to changes in the industrial clusters.

Two main factors that cannot be ignored in the innovation environment and affect product architectures in most industries are digitalization and globalization. Let us look at digitalization first.

Compared with analog technologies, digital technologies are critically different in the design of system interfaces. Analog technologies are based on the laws of nature and cover mechanics, chemicals, and any other areas relying on physical phenomena. By contrast, digital technologies are based on logical rules, and software and electronics are their main applications. Since they are free from the constraints of physical laws, there is great flexibility in their architectures. This means that product designers can easily decouple one system into multiple modules according to their purposes, such as reducing complexity and redesigning by trial and error. Such decoupling of complex systems is not possible in analog technologies. Moreover, digital technologies are much more versatile than analog ones in the creation of system interfaces and in the organization of their interactions.

Interface design flexibility provides different options for firm collaboration. Closed interfaces are often the result of collaboration within a single firm, while open interfaces are ideal for collaboration across firms. Complex systems require many firms to collaborate, so the interfaces linking their modules must be open, to allow for changes and improvements through trial and error. These changes lead the system to become increasingly open in its architecture. In other words, digitalization spurs a shift toward open architecture.

In addition, digitalization affects the firms' business model. With digital technologies, product designers easily make open interfaces, and third-party firms propose new business models based on the flexible combination of modules. Thanks to the existence of open standards for the interfaces, several firms can supply compatible components. More options for module combination enhance the opportunities for new business models, which often take advantage of network effects. Open interfaces generate network effects, and open standards make for even stronger network effects. Indeed, the platform business is a major example of how to make the most of the network effects of open interfaces and standards.

3.2 *Globalization: Amplifier of Architectural Change*

Open architecture is a system with many open interfaces whose information is shared by firms to collaborate in design tasks. Since complex systems require the collaboration of product designers from many different firms, they are likely to have open interfaces. Firms also set open standards, such as protocols or references for open architecture, because open interfaces might not be enough for easy interfirm collaboration. Indeed, firms need efficient measures to share information for joint tasks, and open standards are a powerful tool for this purpose. By providing the formal protocols of interfaces, they enable effective cooperation, so that firms can complete their joint tasks. This is why the development of complex systems is often accompanied by open standardization.

As mentioned above, there are three main approaches to open standardization: de facto, de jure, and consensus standards. De facto and de jure standards are traditional approaches, whereas consensus standards have developed more recently. These three approaches ensure flexibility in standardization, and firms choose the most appropriate one according to their strategies.

As well as supporting firm collaboration, open standards also play a key role in industrial clusters, since they remove barriers and stimulate the entry of newcomer firms. This is because open standards consist of explicit information regarding technological knowledge and industrial contexts, so newcomer firms can easily apply it to their products. As explicit information removes the entry barriers that come from implicit knowledge in the industry, open standards stimulate new entries and increase collaboration among firms.

Globalization amplifies the stimulating effect of open standards. Back in the 1980s, firms based in the developed countries were in a dominant position. However, from the 1990s onward, globalization brought about significant changes, as newcomer firms from developing countries entered the global market. Soon, they faced difficulties in joining pre-existing networks of firms because they were not familiar with the technological knowledge and reference industrial context. Yet, open architecture gave them the opportunity to access explicit information, thanks to open interfaces regulated by protocols released to the public by standardization consortia. Thus, open standards removed entry barriers and provided a common base for firm collaboration.

Newcomer firms from developing countries joined the networks of firms with the aid of open standards. Personal computers are a well-known example. In the 1990s Korean and Taiwanese firms entered the personal computer industry, until then dominated by firms from developed nations, such as the US, Japan, and the EU countries. Due to the availability of open standards for PC components, such Korean and Taiwanese firms soon achieved huge success in the global market. Korean firms aggressively invested in the production of memory semiconductors, which had open standards for mutual compatibility, and quickly reached extremely high production capacity. In a relatively short time, they overtook the US and Japanese manufacturers of memory semiconductors and came to satisfy most of the worldwide memory demand. On the other hand, Taiwanese firms successfully developed a new business model known as ODM, or original design manufacturing (Kawakami 2007), which is a special form of contract manufacturing. ODM firms design and manufacture new products based on the specifications of their client firms, such as computer brand firms in the USA. The clients then rebrand the products and sell them as their own. The ODM business model relies on open standards. By using them, Taiwanese firms learnt to flexibly combine various modules and develop new configurations for product design. By the end of the 1990s, Taiwanese firms accounted for around 90 percent of the global production of notebook computers.

Open architecture products fully enjoy the benefits of globalization. The open standards set for their interfaces stimulate the entry of new firms from developing countries and allow international collaboration among firms from developing and developed economies. From a macroscopic perspective, open standards promote the international division of labor.

Product architecture and industrial clusters are linked by the design logic according to which firms collaborate to solve design problems. In today's globalized world, newcomer firms from developing countries need open interfaces and open standards to develop their products in collaboration with other firms. They choose the open architecture model, which often leads to the formation of industrial clusters as business ecosystems, because it generates network effects that allow business ecosystems to grow rapidly and increase their share in the global market. Personal computers are a good example of the linkage between product architecture and industrial cluster affected by globalization. The interplay of architectural and industrial dynamics is analyzed in the next section by comparing automobiles and personal computers.

3.3 Comparison Between Automobiles and Personal Computers

A comparison between automobiles and personal computers is a good way to show how product architecture influences the shape of the resulting industrial cluster. From the point of view of product architecture, automobiles are typically closed-integral, whereas personal computers are open-modular (Fujimoto 2007; Baldwin and Clark 2000). As for the tightness of interfaces, automobiles adopt an integral architecture in which the interfaces tightly couple the modules. By contrast, the interfaces of personal computers loosely couple the modules. Regarding the openness of interfaces, in automobiles the interfaces are mostly closed and accessible only to a limited number of supplier firms, while personal computers have open interfaces, and supplier firms can freely use them.

Table 2 Comparison between automobiles and personal computers

	Product architecture	Closed-integral	Open-modular
	Example	Automobile	Personal computer
Architecture	Characteristics of products	High integrity and optimal performance	Variety of uses with several module combinations
	Features of the interfaces	Interfaces tightly couple modules	Interfaces loosely couple modules
		Small groups of firms share the information	Firms openly share the interface information
	Design logic among firms	Mutual coordination based on relation-specific assets	Autonomous coordination based on open standards
Industrial cluster	Form of industrial cluster	Cohesive clique	Business ecosystem
	Entry barrier	Relation-specific assets work as entry barriers	Open standards remove barriers and allow entry to many newcomers
Recent impact factors	Digitalization	Interface design flexibility gives new opportunities for open collaborations among firms	More open interfaces allow new module combinations for experimenting new value in use
	Globalization	Entry barriers prevent new entries into the global market	Globalization boosts newcomer entries from developing countries
		Through FDI, the network of firms expands globally, but its speed is slow	Business ecosystems quickly grow to global scale

As discussed above, system architecture may be categorized depending on the features of the interfaces and the approaches to coupling modules and sharing information. This corresponds to the two main dimensions of modular vs. integral and open vs. closed, which have interactions that eventually reduce the architectural types to closed-integral and open-modular. These two architectures affect the development of the industrial structures, as illustrated in Table 2, which also reports the effects of digitalization and globalization.

The first column of Table 2 illustrates the closed-integral case. A closed-integral architecture achieves integrity within the system, the interfaces of which tightly couple specific pairs of modules. Intense collaboration among selected firms allows product designers to share all the relevant information, so that they can calibrate the interfaces and attain optimal performance. This design process causes the interfaces to couple the modules more tightly. Due to information sharing at a deep level, firms gain relation-specific capabilities or assets for mutual problem-solving that give them the ability to efficiently develop the products. On the other hand, these relation-specific assets act as deterrent for those wishing to enter the suppliers network, because they are a collection of implicit knowledge and are not imitable by newcomer firms. So, the industrial structure behind closed-integral systems is a form of cohesive clique, characterized by limited membership and deep information sharing.

Automobile is a well-known product with closed-integral architecture. Automakers and suppliers work together on the designing of cars and often change the interfaces between modules to improve performance. Information sharing helps in joint problem-solving, and relation-specific assets allow the existing suppliers to efficiently collaborate with automakers. Yet, third parties do not have access to this information, which represents a deterrent for newcomer firms (Dyer and Nobeoka 2000). Through joint development, the relation-specific assets make the industrial cluster of automobiles a cohesive clique, which is an efficient network of firms with deep knowledge sharing as well as a closed network with barriers to entry.

Turning now to the second column of Table 2, let us examine the open-modular case. In open-modular systems, interface compatibility makes it possible to combine modules in different ways to find the best setup and achieve maximum value in use. The interfaces are open to third parties and firms often set open standards that further reinforce compatibility by providing protocols and references for compatibility confirmation.

Open standards enable supplier firms to access interface information, so that modules are developed in a decentralized manner. By following open protocols, design problems are jointly solved through autonomous coordination, and supplier firms do not depend on relation-specific assets based on deep knowledge sharing and mutual coordination. The coordination mechanism of open-modular architecture, relying on open standards rather than relation-specific assets, significantly differs from that of closed-integral architecture in that it makes it easier for newcomers to enter the market and join the suppliers network. The industrial cluster of open-modular systems consequently becomes a business ecosystem, in which modules with open interfaces are developed through collaboration based on open standards and there are no barriers to the entry of newcomer firms.

If we look at the case of personal computers, we can easily understand why researchers often refer to them as typically open-modular systems. Personal computers have open interfaces based on open standards, and this enables a variety of module combinations that might produce new value in use. In addition, open standards help newcomers enter the PC market. As the number of newcomers increases, module combinations multiply, generating further value in use. Indeed, the entry of newcomers acts as an engine for the growth of business ecosystems, and, as pointed out in previous studies, new firm entries and new module combinations have worked as the main engine for the growth of personal computers (Baldwin and Clark 2000).

The two architectures described here, open-modular and closed-integral, have different principles and mechanisms for collaboration among firms. Open-modular architectures use autonomous coordination based on open standards, while closed-integral architectures use mutual coordination based on deep knowledge sharing. This translates into differences in the shape of the industrial clusters for open-modular and closed-integral products. Through autonomous coordination, open-modular architectures cause industrial clusters to become business ecosystems, while closed-integral architectures, based on mutual coordination, cause industrial cluster to become cohesive cliques. The comparison between automobiles and personal computers clearly shows how different architectures lead to different forms of industrial clusters according to the coordination mechanisms used.

3.4 Interaction Between Digitalization and Globalization

The previous sections illustrated the features of closed-integral and open-modular architectures, which use different approaches to joint problem-solving and yield different forms of industrial structures: cohesive cliques for closed-integral and business ecosystems for open-modular. This helped us shed light on the fact that, when we investigate an industry, we have to take the architecture of its products into account, because the design logic adopted connects its products and the forms of its industrial clusters.

This connection also means that changes in product architecture modify the form of the relevant industrial cluster. Before comparing automobiles and personal computers, we discussed two critical factors altering product architecture: digitalization and globalization. Since the 1990s, their impact has been growing stronger, causing changes in product architectures which, in turn, have impacted on the shape of industrial structures in different ways.

First of all, let us examine the impact of these two factors on the automobile industry. Within the design process, digitalization makes it easier to set the interfaces, because the rules to be followed are only logical and not physical ones, which is particularly useful when the complexity of the system increases. In order to reduce complexity, product designers often decouple the system into multiple modules and adjust the interfaces. They then reconnect the modules in more effective ways to realize the system functions. This design trend is particularly evident in automotive electronics and software, in which the increasing number of interfaces triggers architectural changes toward the open-modular type. As product architecture evolves, so does the form of the industrial cluster, and, with the development of electric vehicles, autonomous driving, and MaaS (mobility as a service), the pressure spurring this change has grown significantly in the developed countries.

On the other hand, globalization affects the automobile industry by increasing the presence of emerging economies, where consumers tend to choose vehicles having the traditional architecture, i.e., closed-integral, so that the market for closed-integral cars is growing in the developing countries. Through foreign direct investment, or FDI, automobile manufacturers from developed countries try to boost their production capacity by collaborating with local automotive manufacturers and suppliers. The resulting industrial cluster is still a cohesive clique. As FDI increases, the cohesive clique grows globally, thanks to new entries, but this industrial change is relatively slow due to the barriers characterizing this business model that act as a deterrent for newcomer firms.

In brief, the cohesive clique of the automobile industry is experiencing a complex transformation, which is a mixture of disruptive change and robust growth. Digitalization mainly drives the disruptive change, while globalization contributes to the robust growth. In developed countries, the pressure linked to this architectural change is on the rise due to digitalization. Nevertheless, vehicle architecture is stable in the emerging countries, whose role is ever more important thanks to globalization. Both digitization and globalization exert pressure on the architecture of automobiles, but in clearly opposite directions, so that automobile manufacturers have to make tough decision to manage this challenging situation.

Now, let us turn to the case of personal computers and how they are affected by digitalization and globalization. As discussed above, PCs can be described as open-modular systems, and the relevant industrial structure is the so-called business ecosystem. Open-modular systems have open interfaces, and, due to digitalization, the number of such interfaces increases, so that open innovation intensifies. Open standards, created by firms to promote collaboration, make the interface knowledge explicit and formal, stimulating the entry of newcomer firms. These newcomer firms rely on these open standards, since they do not have sufficient implicit knowledge of the segment and of its industrial context. The entry of newcomers expands the size of the business ecosystem. Globalization, which enables the creation of a global pool of newcomer firms from developing countries, accelerates the expansion of the business ecosystem at the global level and, in turn, this helps the international division of labor among firms from developed and developing countries.

The case of personal computers shows the connection between architecture and shape of the industrial cluster in open-modular systems. Since the 1990s, digitalization has impacted on personal computers through the setting of open interfaces. Open standards, which guarantee interface compatibility, have generated numerous combinations of modules and realized new value in use while also removing barriers to entry and, consequently, accelerating the growth of the business ecosystem. Among newcomers, memory chip manufacturers in Korea and ODMs in Taiwan have made the most of the open interfaces and standards to earn a prominent place within the global ecosystem of personal computers. Their rapid growth stems from the very features of the business ecosystem, whose open interfaces allow newcomers to create compatible modules and new module combinations for greater value in use. Thus, the personal computer market has become a global ecosystem due to digitalization and globalization.

The consequence of the above has been not only the expansion of the business ecosystem but also the emergence of a new business model, i.e., the so-called platform business, to take advantage of the network effects generated within the ecosystem. Major platform firms, like Intel and Microsoft, have exploited the network effects to the fullest, and their presence has also boosted the business of newcomer firms from emerging economies. Further entries by newcomers have accelerated the expansion of the business ecosystem internationally, thus forming a global ecosystem (Tatsumoto et al. 2009).

3.5 Revisiting Architectural Dynamics

As explained above, digitalization and globalization strongly affect product architecture and the shape of the reference industrial cluster. Digitalization is likely to cause architectural change, and globalization spreads this impact on a global scale. In other words, digitalization is a trigger of architectural change, while globalization is an amplifier of its impact.

For some time, the automobile industry has been showing major signs of architectural change. These trends, referred to as CASE (connected, autonomous, shared, and electric vehicles), are likely to shift the industrial structure of the automobile sector from cohesive clique to business ecosystem. When estimating their impact, it is important to consider where they happen in the architectural hierarchy. The above four trends may be architecturally divided into two categories: changes occurring inside the core layer and changes occurring outside such core layer.

Changes inside the core layer imply architectural changes of the automobile itself and concern innovations such as autonomous driving and electric vehicles. To deal with them, automobile manufacturers and suppliers need to adjust their structures by investing considerable time and effort in organizational reorientation. This is particularly true for large incumbent firms, whose legacy assets are based on traditional architectures. If they fail to keep up with the changes, they run the risk of losing market share as newcomers quickly catch up and gain prominence.

Changes outside the core layer in which automobile manufacturers usually do their business are likely to have an equally large impact and revolve around connectedness and sharing of vehicles. The former feature means that cars are connected to the Internet and have new functions or services based on big data, while the latter means that car users change their behavior from owing to sharing vehicles. These changes affect the upper layers of the whole system, in which the car itself is just one element, and may lead to the creation of new business models that will replace the existing one. For instance, connectedness relies on peripheral services using big data to offer navigation support and customized recommendations. This model profits from displaying ads and matching services. The second aspect, car sharing, might change the landscape of the automotive industry even more drastically. Indeed, if this becomes a mainstream trend, car users will stop purchasing their own vehicles, causing the total disruption of the traditional automobile business model.

Moreover, the above changes may have interactions and become considerably stronger. Indeed, for the time being, they are mainly occurring in developed countries, but, if they spread to the developing countries as well, they will affect the global system. This may come to pass because automobile manufacturers in developing countries have no legacy assets and because car users in those areas might prefer products and services provided by the new business models as they develop their own consumer culture. Hence, electric vehicles and vehicle sharing might prevail in the developing countries.

Although the architecture of automobiles has been stable for years, the CASE trend seems to suggest that the industry is about to undergo a major shift, likely to be accelerated by the emergence of new data technologies, including IoT, big data, and AI. In recent years, these data technologies have had an impact on all the industries and have become a common technological framework (The Economist 2017). They use data resources coming from the products and realize new value in use, thus intensifying the architectural changes in the automobile industry.

Product architecture is dynamic, and, although it appears stable, dramatic changes can occur at any time, as confirmed by the evolution of the personal computer industry. The architectural change from mainframes to personal computers utterly transformed the landscape of the computer business. Although automobile manufacturers have enjoyed the benefits of a stable architecture for years, recent signs of architectural change are forcing them, especially incumbent firms, to find new approaches.

Nonetheless, we are still in the early stages of this shift, and for now it is impossible to determine with any degree of certainty whether the automobile industry has reached an inflection point. Digitalization has accelerated the above shift and might bring about new business models, but so far globalization has supported the stable growth of the existing architecture. Yet, car users in developing countries might decide to choose products and services that disrupt the traditional business model of the automobile industry, and automobile firms need to learn to cope with this difficult situation.

4 Conclusion and Further Insights

This chapter illustrates the concept of business ecosystem as a special form of industrial cluster and explains how the design logic connects product architecture and the shape of the industrial cluster. It then discusses architectural changes and their impact on the shape of industrial clusters by drawing a comparison between automobiles and personal computers.

Industrial clusters are likely to become cohesive cliques when the product architecture is closed-integral. Conversely, business ecosystems are likely to emerge when the product architecture is open-modular. Business ecosystems have a special growth model because they contain many network effects, and firms seek new business models to exploit them. The platform business is a notable model among them that has exploited network effects as a source of new value in use to grow and reach a global scale.

Industrial clusters are shaped by the design logic adopted to develop the architectures of their products, which relies on different types of collaboration among firms. So, architectural changes affect the form of an industrial cluster, and, if its product architecture changes from closed-integral to open-modular, the cluster will shift from cohesive clique to business ecosystem.

Product architecture can be modified by various factors, and most industries have recently been affected by two phenomena that they cannot ignore: digitalization and globalization. Thanks to digitalization, product designers can now flexibly set interfaces with open standards in order to reduce product complexity. These open interfaces can shift the product architecture toward open-modular. Globalization, characterized by the strong presence of emerging economies, causes the architectural changes to reach global scales. Digitalization works as a trigger for architectural change and globalization acts as its amplifier.

The dynamics of product architecture are constantly evolving, and this is why architecture can drastically change even when it seems stable. Due to digitalization, the automobile industry is now showing signs of major architectural shifts. The pressure is stronger than ever before. Firms need to focus on macro aspects because changes are taking place both inside and outside the core layer where automobile manufacturers mainly do their businesses. Changes inside the core layer, such as autonomous driving or electric vehicles, bring about architectural changes that might transform the shape of this industrial cluster. On the other hand, changes outside the core layer, such as connected cars or vehicle sharing, create new business models that disrupt the traditional practices of the automobile industry. Globalization might amplify the impact of these architectural changes on the global market, since firms from developing countries, which are newcomers in the global economy and have no legacy assets, may choose to adopt new architectures. In addition, car users in developing countries might prefer product or services based on these new architectures closer to their own consumer cultures. To deal with these architectural dynamics, it is crucial to understand the connections between product architecture and industrial cluster.

Acknowledgment This work was supported by KAKEN Grant Number JP16K03850.

References

Aoki M (2001) Towards a comparative institutional analysis. MIT Press, Cambridge, MA

Baldwin CY, Clark KB (2000) Design rules: the power of modularity. MIT Press, Cambridge, MA

Dyer JH, Nobeoka K (2000) Creating and managing a high-performance knowledge-sharing network: the Toyota case. Strateg Manag J 21:345–367

Eisenmann T et al (2011) Platform envelopment. Strateg Manag J 32:1270–1285

Evans DS et al (2006) Invisible engines. MIT Press, Cambridge, MA

Fujimoto T (2007) Architecture-based comparative advantage: a design information view of manufacturing. Evol Inst Econ Rev 4:55–112

Gabriel R (1989) The rise of "Worse is Better". https://www.jwz.org/doc/worse-is-better.html. Accessed 9 Apr 2018

Gawer A, Cusumano MA (2002) Platform leadership: how Intel, Microsoft, and Cisco drive industry innovation. Harvard Business School Press, Boston

Grove AS (1996) Only the paranoid survive. Doubleday Business, New York

Iansiti M, Levien R (2004a) The keystone advantage: what the new business ecosystems mean for strategy, innovation, and sustainability. Harvard Business School Press, Boston

Iansiti M, Levien R (2004b) Strategy as ecology. Harv Bus Rev 82. https://doi.org/10.1108/eb025570

Kawakami M (2007) Competing for complementarity: growth of Taiwanese notebook PC manufacturers as ODM suppliers. In: Sato Y, Kawakami M (eds) Competition and cooperation among Asian enterprises in China. Institute of Developing Economies, Chiba. http://www.ide.go.jp/library/English/Publish/Download/Report/2006/pdf/2006_04_30_04.pdf

Nalebuff B (2004) Bundling as an entry barrier. Q J Econ 119(1):159–187

Paine RT (1966) Food web complexity and species diversity. Am Nat 100:65–75

Rochet J, Tirole J (2003) Platform competition in two-sided markets. J Eur Econ Assoc 1 (4):990–1029

Simon HA (1969) The sciences of the artificial, 1st edn. MIT Press, Cambridge, MA

Tatsumoto H (2017) Platform kigyou no global senryaku, Yuhikaku, Tokyo (in Japanese)

Tatsumoto H et al (2009) Platforms and the international division of labor: a case study on Intel's platform business in the PC industry. In: Gawer A (ed) Platforms, markets and innovation. Edward Elgar, Northampton, pp 345–369

The Economist (2017) Data is giving rise to a new economy. The Economist, Fuel of the Future May 6, 2017. Available at https://www.economist.com/news/briefing/21721634-how-it-shaping-updata-giving-rise-new-economy

Ulrich KT (1995) Product architecture in the manufacturing firm. Res Policy 24:419–440

Part II
Multi-layer Capability Building in Response to Global Competition

After having illustrated the theoretical frameworks and concepts used here to analyze the evolution of industries and firms in terms of manufacturing capabilities and product/platform architectures, let us try to apply them to some of the industrial phenomena of the late twentieth century and early twenty-first century. We have chosen two main aspects for this purpose: (1) Capability building for global competition, and (2) Demand creation in the digital industries.

In Part II of the present book, we shed light on the first aspect, i.e., capability building of manufacturing firms and sites in response to the increasingly intense global competition of recent years. We argue here that efforts for manufacturing capability building, to ensure better flows of value-carrying design information to customers in the market, occur at multiple levels of industries, firms, and sites when global competition is intense. That is, capability-building competition takes place at the level of national industries, global firms, local manufacturing sites, work organizations within them, and so on.

Chapter "Evolution of Organizational Capabilities in Manufacturing: The Case of the Toyota Motor Corporation" explores the evolution of manufacturing capabilities at one of the manufacturing firms with relatively and consistently higher competitiveness and evolutionary capability—the Toyota Motor Corporation. Chapter "The Nature of Surviving Japanese Factories in the Global Competition: An Empirical Analysis of Electrical and Electronics Factories" investigates the level of local factories (i.e., manufacturing sites) in the consumer electronics industry. Chapter "The Effectiveness of Group Leaders in the Lean Production System: Time Study and Agent-Based Model of Leaders' Behavior" deals with work organizations within a factory and their group leaders. Chapter "The Diversity and Reality of Kaizen in Toyota" focuses on Toyota to analyze continuous improvement (i.e., *kaizen*) activities at multiple levels, including work organizations, shop floors, and the firm itself. Chapter "Balancing Standardization and Integration in ICT Systems: An Architectural Perspective" looks at information technologies (IT) that may accelerate manufacturing capability building at the level of both firms and sites.

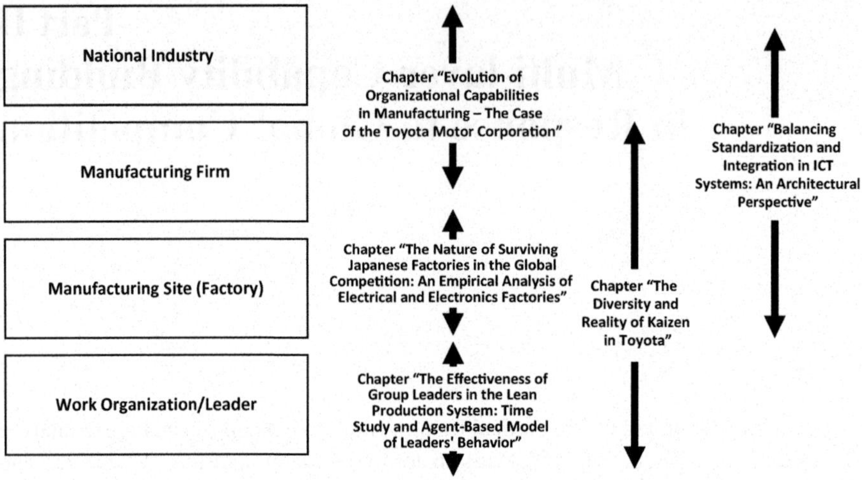

Fig. 1 Multi-layer capability building

Thus, in Part II of this book we argue that firms and industries strengthen their capability building across multiple layers when global competition intensifies. As a result, manufacturing capability evolves over time at several levels, including national industries, individual firms, factories, work organizations, and so on (Fig. 1).

Evolution of Organizational Capabilities in Manufacturing: The Case of the Toyota Motor Corporation

Takahiro Fujimoto

Abstract This chapter focuses on the formation and evolution of organizational capabilities in manufacturing at the individual firm level. Empirically, we look at the emergence of the manufacturing system at the Toyota Motor Corporation, the Toyota Production System, between the 1930s and the 1990s.

We propose to link the resource-capability view of the firm with the evolutionary framework in social sciences or a dynamic perspective that can separately explain an observed system's survival (i.e., functional logic) and its formation (i.e., genetic logic). Within this evolutionary framework, two main concepts are proposed: *multipath system emergence*, for analyzing the complex variations in manufacturing system changes, and *evolutionary learning capability*, for explaining why certain firms can develop competitive manufacturing capabilities faster than their competitors.

We apply these concepts to a historical analysis of the manufacturing system at Toyota. More specifically, we investigate the origins of several major organizational routines of the Toyota Production System and show that they emerged through the unpredictable patterns of various evolutionary paths, including rational calculation, random trials, environmental constraints, entrepreneurial visions, and knowledge transfer, i.e., through multipath system emergence.

It follows from this that Toyota, as a consistently competitive manufacturing firm, possesses not only (1) routinized (static) manufacturing capability and (2) routinized learning (continuous improvement) capability but also (3) evolutionary learning capability, which is a firm's dynamic capability-building capability to improve productive performance in the long run in a situation of multipath system emergence. In other words, a firm's evolutionary capability, or the capability of building capabilities despite a situation of unpredictable multipath system emergence, is critical to its long-term survival and growth, particularly in industries where competition is intense, market/technology environments are uncertain, and products/processes are complex.

T. Fujimoto (✉)
Graduate School of Economics, The University of Tokyo, Bunkyo-Ku, Tokyo, Japan
e-mail: fujimoto@e.u-tokyo.ac.jp

T. Fujimoto, F. Ikuine (eds.), *Industrial Competitiveness and Design Evolution*, Evolutionary Economics and Social Complexity Science 12,
https://doi.org/10.1007/978-4-431-55145-4_6

1 Introduction

1.1 Evolution of Manufacturing Capabilities for Integral Products

Based on the capability-architecture framework proposed in this book, the present chapter explores a framework to analyze the evolution of the organizational capability of certain industrial sites or of a firm as a set of such sites. After discussing generic and specific evolutionary analysis frameworks in natural and social sciences, we apply our framework to one of the most consistent and durable manufacturing capabilities in manufacturing industries worldwide, the evolution of Toyota's automobile manufacturing system (Fujimoto 1999).

This is a case of evolution of a manufacturing firm and its industrial sites in which the product is complex, with highly integral architecture (coordination-intensive), so coordination-rich capabilities are a key requirement. The pattern of industrial competition is rather straightforward, i.e., product versus product—in which the products' market performance and the sites' productive performance matter directly—unlike many digital products and information technology services with open-modular architectures, in which firms may be involved not only in product-versus-product competition but also in platform-versus-platform competition.

In industries with closed or integral product architectures, a firm's productive resources and organizational capability in manufacturing in a broad sense (flow management of value-carrying design information to the market, including production, product development, and procurement) have been regarded as major sources of competitive advantage and above-normal profit. A number of studies have investigated the structures and functions of this operation-based competency, including lean production system and total quality management (Womack et al. 1990; Clark and Fujimoto 1991). And yet, academic research on how a firm develops this competency better than its rivals has so far been rather scarce.

By complex and coordinative manufacturing capability, we mean a firm's specific system of many interrelated organizational routines (Nelson and Winter 1982) in value-creating operations (e.g., production and product development) that results in consistently better productive performance (e.g., productivity and lead time) than that achieved by its competitors. Examples include the respective production systems of Ford and Toyota in the twentieth century. Existing research has revealed that this competency tends to be developed through long-term cumulative processes, as opposed to one-time projects or investments (Hounshell 1984; Cusumano 1985; Fujimoto 1999). The formation of an operation-based competency has also been described as a complex interplay of planning and chance, visions and imperatives, creations and imitations, and trial and error. If this is the nature of the complex operation-based competency in question, what kind of conceptual framework should we adopt to better understand it?

Existing management studies tend to avoid answering this particular research question explicitly. For example, the resource-based view of strategic management explains that a firm's resources or capabilities create above-normal profit, but why this phenomenon occurs in certain firms and not in others is seldom considered. Research on dynamic capability tries to answer this question, but it has not covered the long-term development of complex operation-based competence. The theory of organizational learning analyzes the processes through which a firm changes its organizational routines, but it does not deal with the evolution of routine changes within a firm over time. The literature on operations management explores routines for manufacturing and continuous improvement, but the historical emergence of such routines themselves is beyond its research scope. Process innovation has been a popular topic in technology management literature, but its research agenda has mostly been limited to either each individual innovation or a series of innovations at the industry level. Thus, despite its importance in industrial competition, the long-term formation of an operation-based competency that takes shape within a single firm has so far been a rather uncultivated field in management studies.

1.2 Outline of the Evolutionary Framework

Against this background, the purpose of this chapter is to explore a conceptual framework that may explain how a firm gains competitive advantage by developing and maintaining a complex operation-based competency. The framework that I propose here is an application of the evolutionary theory of social systems to the issue of intra-firm development of the said complex operation-based competency. I chose this evolutionary framework based on my view that it can most consistently explain the dynamic process of complex system formation. More specifically, this chapter puts forward two main arguments. First, it proposes the concept of *multipath system emergence* as a way to explain the process of change of a complex operation-based competency. Second, it argues that a firm gaining operation-based competitive advantage needs three layers of organizational capabilities that include *evolutionary learning capability*. In the empirical part of this chapter, I will apply the abovementioned framework to the case of the Toyota-style manufacturing system in the second half of the twentieth century.

The present chapter tries to reinterpret the evolutionary theories of the social sciences in the context of industrial competition through organizational capability building. Generally speaking, an artificial system, which looks as if it were deliberately designed as a rational system in terms of competitiveness or survival, may have been formed through a complex dynamic process that cannot be reduced to ex ante rational planning alone. When we observe an ex post rational object that may not have been formed in an ex ante rational way, a certain *evolutionary framework* can often be effectively applied to such a case. By evolutionary framework I mean a dynamic perspective that separately explains the survival (i.e., functional logic) and formation (i.e., genetic logic) of the system under observation. For example, a prevalent neo-Darwinian (or synthetic) theory of biological evolution assumes natural selection to explain a living system's survival and random variation to account for its origin. Indeed, a few studies have applied some sort of evolutionary approach to dynamic analyses of biological, social, or economic systems of this kind.

Thus, this chapter aims to propose an evolutionary framework that may be applicable to an artificial system that, I believe, is ex post rational: the abovementioned operation-based competency, such as that of the Toyota Motor Corporation. Although my analysis is by no means a direct application of neo-Darwinism or biological models, it is still evolutionary, given that I separate the functional logic and the genetic logic of the manufacturing system at Toyota.

The next section will present an evolutionary framework for social systems that may be applicable to certain types of process innovations and compare it with existing evolutionary theories in biology and the social sciences. In the third section, starting from the said general evolutionary framework, I will propose a new framework to explore the evolution of organizational capabilities of manufacturing firms and sites. More specifically, the concepts of *multipath system emergence* and *evolutionary learning capability* will be illustrated in greater detail. In the fourth section, these concepts will be redefined operationally and applied to the case of Toyota-style manufacturing capability, which is a system of many interconnected organizational routines that control and improve flows of design information to the customers. Based on this historical-empirical analysis, I will argue that the ultimate core capability of this high-performing manufacturer in the last half of the twentieth century is likely to be its evolutionary learning capability.

2 An Evolutionary Framework for Social Systems

2.1 General Scheme of the Evolutionary Framework

Since C. Darwin's work on living systems, various types of evolutionary frameworks have been applied to social, economic, and managerial systems. The notion of evolution, however, has been quite equivocal, creating misunderstandings among researchers from different fields. In order to avoid conceptual confusion, I distinguish two levels of an evolutionary framework: a general scheme and a specific scheme. The present evolutionary approach shares its basic logical structure with many other evolutionary theories of biological and social systems at the general level. Yet, at the second level, it is also rather specific to the present analytical purpose, i.e., empirical research on manufacturing systems and process innovations.

Let us start from the general level. By evolutionary framework at this level, I do not mean specific theories of biological or social evolution, but a general logical scheme that such theories may share, a common denominator for any model or theory that can be called *evolutionary*. At this level, the present framework shares basic logical patterns with the contemporary synthetic (neo-Darwinian) theory of biological evolution, as well as with the evolutionary theory of firms, technologies, organizations, and strategies. At the said general level, we may call a framework evolutionary if the conditions listed below apply:

1. *Variety and stability*: The framework's main purpose is to explain why we observe a certain variety, or difference, of stable patterns (e.g., species) in the objects concerned (e.g., living systems).
2. *Ex post rationality*: The objects observed behave so functionally that they *look as if* someone had purposefully designed them for survival, regardless of whether such purposive motivation actually existed beforehand.
3. *History*: The present pattern of the objects in question is conjectured to have been formed historically along a certain path over a long period.
4. *Genetic and functional logic*: The framework provides three complementary explanations for a given dynamic phenomenon: the logic for system *variation* (generation of a variety of patterns), *selection* (elimination of low-performing patterns), and *retention* (preservation of the remaining patterns). In other words, the evolutionary perspective offers *genetic* and *functional* explanations of the same object separately: the former shows how it has evolved into what we see now, whereas the latter demonstrates how it behaves effectively for higher performance or survival rate.
5. *Anti-teleology*: Because of the above logical separation, this framework does not have to depend on ex ante rational foresight of omnipotent decision-makers to explain the formation of the ex post rational system. In other words, the evolutionary logic denies completely depending upon teleology.

The evolutionary framework of this chapter (see also Fujimoto 1999) shares the above logical scheme with many other evolutionary theories for biological and social systems.

Note, again, that the generic evolutionary framework separates genetic analysis (i.e., how the system was created and has changed to yield its present form) from functional analysis (i.e., how a system's structure has contributed to its survival and growth), which is the core of evolutionary thinking.

2.2 *Specific Scheme of the Evolutionary Framework*

At the second level, the evolutionary framework is applied specifically to the case of a complex operation-based competency within a single manufacturing firm, like Toyota. In other words, the present framework deals with intra-firm evolution of organizational routines.

Considering the nature of this chapter's topic (i.e., long-term changes in operational routines within a single surviving company), I assume that individual firms (e.g., Toyota) can adapt their internal systems to the environment and thereby survive at the firm level. As Barnett and Burgelman (1996) point out, one may also conceive of another type of evolutionary model, in which individual firms are unable to adapt their internal structures to the selection environment and firms with ineffective routines are simply weeded out: thus, it is the population of firms as a whole that adapts (e.g., organizational ecology by Hannan and Freeman 1989). The present chapter does not adopt this neo-Darwinian (or population ecology) version of evolutionary models, though. It instead assumes that individual firms may internally select their manufacturing routines through what Robert Burgelman would call an *intraorganizational ecological process* (Burgelman 1994), before being selected by means of an external ecological process.

The specific scheme for cumulative operational capability building can be summarized as follows (see Fujimoto 1999 for further details):

1. *Retention:* Based on the author's view that production is transmission of product design information from the process to the product (Fujimoto 1999), the present framework assumes that what is retained in a manufacturing system is a stable pattern of *information assets and flows* that collectively influence manufacturing performance. This informational pattern, or *gene* of the manufacturing system, may also be called *routines*, *productive resources*, or routinized capabilities. In other words, the present study reinterprets manufacturing routines as a detailed pattern of stocks and flows of value-carrying design information.
2. *Variation*: The present framework treats changes in manufacturing routines as a multipath system emergence, a complex and irregular combination of rational plans, entrepreneurial vision, historical imperatives, pure chance, and so on. In this sense, our evolutionary model is not neo-Darwinian, given that neo-Darwinism only assumes pure randomness as the ultimate source of genetic variation, excluding the possibility of feedback from the environment to the genes (routines). The present model is rather Lamarckian, in that it recognizes an individual firm's efforts to adapt its routines to the environment, although imperfectly. Multipath system emergence also differs from intentional *search* activities, assumed by some evolutionary economic theories, in that the former may include unintended trials. System emergence is obviously akin to the concept of *emergent strategy* (Mintzberg and Waters 1985), but the former assumes a situation in which managers do not know if the next system change will be driven by a deliberate strategy or by an emergent strategy in the first place.
3. *Selection*: The present framework assumes a lenient selection mechanism. As mentioned earlier, selection of routines may occur when firms with low-performing routines are immediately eliminated through market competition—a situation that the so-called organizational ecology may assume (Hannan and Freeman 1989). This kind of harsh selection seldom happens in today's automobile industry.[1] For our empirical analysis, a more realistic assumption is that firms with lower performance can survive for some time, but competitive pressure from best-practice rivals tends to force them to select and change their

[1]Population ecology models may be applied more effectively in the case of earlier phase of automobile industrial evolution, in which many births and deaths of individual automobile manufacturers were observed. See, for example, Abernathy (1978) and Carroll et al. (1996) for the case of the US auto industry.

routines in the long run. In other words, the selection environment is generous enough to allow automobile firms with different performance levels to survive. This study does not deal with direct selection of individual firms by the market environment but analyzes selection of routines within a surviving firm (i.e., Toyota). The market does function as a routine selection mechanism, but its impact on most of today's automobile firms is at best indirect (i.e., existing firms may switch routines in response to signals from the market). For long-term survival, what matters as a market signal is relative performance. The survivor's routines do not need to be optimal but simply better than the competitors' in the long run. To sum up, the selection process of the present framework is less severe than that of neo-Darwinian models of biology or population ecology of organizations or of equilibrium models in microeconomics. Accordingly, our framework regards relative manufacturing performance as a surrogate indicator of each individual firm's *probability of future survival.*

2.3 Comparison with Neo-Darwinian Biological Models

The evolutionary framework of the present chapter is by no means new, but it is clearly different from some other interpretations of evolutionary models (Fujimoto 1999). For instance, whereas the present framework shares its general logical scheme with today's synthetic theory of biological evolution, it is not a direct application of the biological model at the specific level (Table 1).

The prevalent paradigm in biology theorizes that variation (mutation as random changes in genetic information or DNA), selection (natural selection caused mostly by different propagation rates), and retention (reproduction of genetic DNA information in and across individuals) jointly create changes and diversification of genetic information, which then materialize in living systems through adaptation to changing environments. For the purposes of microscopic empirical analyses of operational-technical systems, however, I do not follow the biological model at the specific level.

In the context of a social system, *variation* is explained not through the purely random process that neo-Darwinism[2] may find in a biological system but through the complex interaction of forces ranging from purely random to purely purposeful changes, i.e., multipath system emergence.

For what concerns *selection*, as mentioned earlier, the present framework assumes a rather lenient selection environment, in that relatively weak organizations, with lower competitive performance, can survive at least for a while. Hence, it is possible to observe significant cross-organizational differences in performance for survival at a certain point in time (Clark and Fujimoto 1991; Womack et al. 1990).

[2]Note that I use the term *neo-Darwinism* rather broadly here, as synonymous with *Modern Synthesis*, the prevalent theory in biological evolution that includes revised Darwinism and Mendelian genetics.

Table 1 Comparison of evolutionary frameworks

	Framework of this book	Neo-Darwinian (synthetic) theory in biology
Object	Manufacturing systems	Living systems
Criteria of ex post rationality	Relative manufacturing performance	Survival/reproduction
Logic of retention	Object to be retained = manufacturing routines as informational patterns	Object to be retained = genes as information
	Routines are stored in the firm; they may be diffused across firms	Genetic information is stored in the organism; it may be reproduced across generations
Logic of variation	Emergent processes change routines	Random chance changes genetic information
	Feedback from the environment may trigger routine changes (Lamarckian)	No feedback from the environment to the gene
Logic of selection	Long-term elimination of low-performing routines by either market or organization	Long-term elimination of low-performing genotypes by the environment
	A rather generous selection environment is assumed	A rather harsh selection environment is assumed
	Individual firms may select high-performing routines	Individual organisms cannot select high-performing genes
Separation of functional and genetic explanations	Genetic explanation = emergent processes result in changes in manufacturing routines	Genetic explanation = random DNA variations result in variations of phenotypes
	Functional explanation = certain routines result in relatively high performance (static and improvement capability)	Functional explanation = certain phenotypes result in higher performance for survival in the environment (natural selection)
Anti-teleology	Rejects the idea that omnipotent decision-makers create the entire system through perfect foresight	Rejects the idea that the omnipotent creator made all the living systems through predetermined plans

Also, the framework may include certain internal selection mechanisms within the organization that preselect the routines which have higher probabilities of survival in the external selection environment.

Finally, I only assume an incomplete mechanism for *retention* and duplication of organizational knowledge or information in a social system, unlike the relatively strict mechanism of gene or DNA duplication in the case of living systems. As anyone who has worked in a company knows, organizational routines and memories erode quickly. Also, they are often difficult to imitate by other organizations.

To sum up, the specific framework of this chapter (see also Fujimoto 1999) is indeed evolutionary, but it is not neo-Darwinian, since the latter approach assumes random variation and harsh selection, while the former features emergent variation and lenient selection. And, quite obviously, I reject crude *social Darwinism*, which may be used to justify the elimination of the weak.

My framework may also disagree with some other specific interpretations of existing evolutionary models applied to social phenomena. For instance, I do not assume *progressivism*, or the doctrine that system evolution causes constant progression toward something inherently valuable or supreme. What the evolutionary process tends to bring about is not progress but simply adaptation to environmental requirements.

By proposing an evolutionary framework, I also do not mean a *linear stage model*. Distinctive stages may be identified in ex post historical analyses, but I do not assume such a predetermined sequence of regular stages.

Neither do I assume that system changes are always incremental (i.e., evolutionary rather than revolutionary), although I do emphasize cumulative aspects of system changes in the present empirical analysis. As said, the above are specific interpretations of the evolutionary perspective, but the current framework does not follow such specific versions of evolutionary theories.

2.4 Comparison with Existing Evolutionary Theories of Firms

Finally, let us discuss the relationship between the present evolutionary framework and existing evolutionary theories and models of business firms. Although our research is empirically motivated, some existing theories are more suited than others to explaining its data and material: *evolutionary theories of the firm*, *resource-capability-based approaches*, and the concept of *emergent strategy formation*. This chapter adopts some (but not all) aspects of these theories. In this sense, much of my analysis is based on previous work by economists, historians, and business researchers, including Penrose, Nelson, Winter, Dosi, Rumelt, Teece, Chandler, and Mintzberg.[3]

Both resource-capability-based approaches to strategic management and evolutionary theories of the firm have attracted much attention among business academics and practitioners in recent years. Although they come from different academic traditions, these theories tend to share a common view of the company as a collection of firm-specific and difficult-to-imitate resources, organizational routines, capabilities, or competencies. Their proponents expect such resources, especially when they represent a stable pattern of performance and associated behavior, to account for competitive differences among firms, as well as for the evolution of business

[3]Penrose (1959), Nelson and Winter (1982), Dosi (1982), Chandler (1990), Teece et al. (1994)

enterprise systems.[4] The current framework also follows this routine-capability-based view of manufacturing firms in a broad sense.

Nevertheless, at the specific level, our evolutionary framework is not a direct application of these theories. A certain amount of reinterpretation and modification is needed for a detailed empirical analysis of automobile manufacturing competency, such as that of Toyota.

While most of the existing resource-capability literature—found in the fields of strategic management, applied economics, and business history—analyzes the dynamics of the overall systems of multi-product firms, these studies have not been designed for a detailed competitive analysis of production and product development systems at the level of a single plant or project.[5] Previous research in technology and operations management has done a much better job of exploring the specifics of manufacturing systems, but it tends to lack either a total system perspective or a long-term historical perspective.[6]

It is also important to distinguish the evolutionary framework adopted in the present empirical analysis from the population ecology version of evolutionary models (e.g., Hannan and Freeman 1989). The question here is whether firms can revise their routines for better chances of survival. As already mentioned, the population ecology versions of evolutionary theories hypothesize that firms are unable to adjust their routines in response to signals from the environment—a neo-Darwinian interpretation of evolution. According to this assumption, existing routines may persist within a firm, or they may change in a purely random way, but it is the external environment, as a selection mechanism, that creates a nonrandom distribution of high-performing routines in a given population of firms. Firms cannot select effective routines; the environment selects the firms. Note that, if numerous random variations and strict selection of optimal routines are assumed, the neo-Darwinian model can be compatible with equilibrium models in microeconomics (see Alchian 1950; Hirshleifer 1977).

Our version of the evolutionary framework does not adopt this aspect of the neo-Darwinian view. It assumes that firms are able to change their routines in a nonrandom way for better survival, although imperfectly and slowly, in response to signals from the environment. In other words, the evolutionary model of this chapter (Fujimoto 1999) emphasizes internal selection processes, whereas the ecology

---

[4]For the concepts of resource, organizational routine, capability, and competence, see, for example, Penrose (1959), Nelson and Winter (1982), Dosi (1982), Barney (1986), Rumelt (1984, 1991), Wernerfelt (1984), Itami (1984), Chandler (1990, 1992), Prahalad and Hamel (1990), Grant (1991), Leonard-Barton (1992), Teece et al. (1992), Kogut and Kulatilaka (1992), Iansiti and Clark (1993), and Teece et al. (1994). For evolutionary aspects of the organization and its strategies and technologies, see also Weick (1979), Nonaka (1985), Mintzberg (1987), and Burgelman (1994).

[5]Much of the recent literature, such as Chandler (1990), Prahalad and Hamel (1990), and Teece et al. (1994), mainly analyzes multi-product or multi-industry situations.

[6]Abernathy (1978), Hayes and Wheelwright (1984), and Hayes et al. (1988) are among the rare cases that include both a total system perspective and a dynamic approach, but they do not establish any explicit connections with evolutionary theories of firms or organizations.

version emphasizes external selection. As stated above, however, my framework does not say that such changes in routines are always preceded by rational plans or foresight; it features system emergence instead.

To sum up, the present evolutionary framework shares its basic logical structure with neo-Darwinian models of biological evolution and their application to theories of the firm at the general level, but it is by no means neo-Darwinian at the specific-scheme level. Neo-Darwinism emphasizes random variations and external selection, recognizing the latter as a dominant force in system formation, whereas the approach adopted here focuses on emergent variations and internal selection, highlighting the interactions between the firms' distinctive capabilities and historical imperatives.

3 System Emergence and Evolutionary Capability

3.1 Multipath System Emergence

Based on the evolutionary framework explained in the previous section, I will explore two main concepts for the study of operation-based capability building: multipath system emergence and evolutionary learning capability. The former argues that a firm experiencing evolution of its manufacturing capability (i.e., a system of manufacturing routines) tends to face ultimately unpredictable situations, so much so that it does not even know whether its evolutionary path is deliberate and based on ex ante rational decision-making or an emergent process with unintended results. Evolutionary learning capability emphasizes a firm's capability-building capability, thanks to which it outperforms its rivals in creating certain ex post rational manufacturing systems even in the abovementioned situation of ultimate uncertainty.

First, let us consider a conceptual framework that may explain the long-term formation of a firm's operation-based competency. For example, the overall Toyota-style manufacturing competency gradually evolved through the second half of the twentieth century as the cumulative result of changes in individual routines. Yet, unlike the case of continuous process improvements (i.e., *kaizen*), it is difficult to find common traits among these *histories of routines* in terms of patterns and timing. Instead, the overall process of competence development is seen as a complex network of events that include random chance, ex ante rational decisions, environmental constraints, unintended successful trials, unsuccessful trials, and so on. Such social system changes are mainly driven by intentional human actions, but they may have unintended consequences.[7] The process of change may be explained after it happens, but it is difficult to predict its pattern beforehand. I call this complex and dynamic phenomenon *multipath system emergence.*

[7]See Merton (1968) for this notion, which is related to his discussion of *latent functions* and *dysfunctions*.

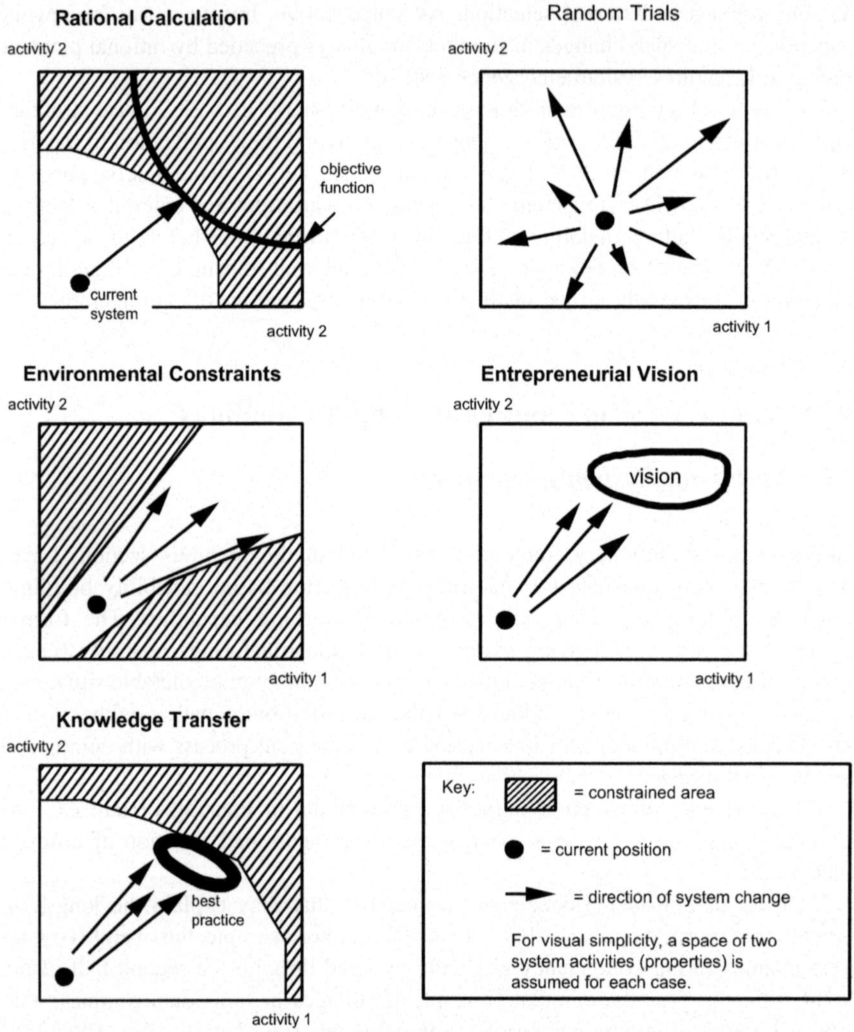

Source: Modified and adopted from Fujimoto (1995) "A Note on the Origin of the 'Black Box Parts' Practice in the Japanese Motor Vehicle Industry." In Shiomi, H., and Wada, K. Fordism Transformed. Oxford University Press.

Fig. 1 Multipath system emergence

Generally speaking, new manufacturing routines gradually emerge as a result of complex interactions between firms and environments, in which firm-specific capabilities may play only a partial role. In other words, system emergence can occur along a number of different paths, and combinations of them may be needed to explain a particular system change. My evolutionary framework to analyze manufacturing firms includes the following paths (see Fig. 1).

Random Trials Those who follow this path believe that an organization's trials are a matter of pure chance. Firms that are lucky come up with a better system, while unlucky organizations create a poor one. This is why you might as well try everything.

Rational Calculation Decision-makers deliberately choose a new course of action that satisfies or maximizes an organization's objective function; they examine feasible alternatives based on their understanding of environmental constraints and capability limits.[8] This is the ex ante rational problem-solving that many managers see as the only way to create successful change.

Environmental Constraints Decision-makers detect certain constraints imposed by the environment and voluntarily prohibit certain sets of actions. The constraints may be objective (e.g., laws and regulations), or they may be *self-restraints*, based on the managers' perception of the environment (e.g., perceiving that the market is diversifying rapidly, planning ambitious product proliferation to match this perception, and facing the constraint of shortage of engineers).

Entrepreneurial Vision A desirable set of activities is willingly chosen by the entrepreneurs, based on their vision, philosophy, or intuition, without any extensive analysis of organizational capabilities and constraints.

Knowledge Transfer A certain pattern is transferred from another organization to the one in question. The transfer may happen within an industry (competitor, supplier, customer) or across industries. Moreover, it may be of the *pull* type, in which the adopter-imitator of the system takes the initiative, or of the *push* type, in which the source organization is the driving force behind the said transfer.

In brief, multipath system emergence indicates a situation in which the operators of the system in question cannot predict which of the above paths will emerge in the next system change. It is a complex and irregular combination of both intended and unintended system changes, as opposed to the dichotomy between pure randomness (i.e., neo-Darwinism) versus pure foresight (i.e., Christian teleology) in the case of biological evolution.

The words *emergence* and *emergent* have been used by a number of scholars in biological evolution (e.g., Lloyd Morgan), sociologists and general system theorists (e.g., Parsons 1937; Weinberg 1975), and business academics (e.g., Mintzberg and Waters 1985).[9] Across these various disciplines, emergence implies a certain system

---

[8]The neoclassical decision theory further assumes that economic actors are equally capable and face an identical environment.

[9]See, for example, Weinberg (1975). For an application of the concept of emergent process to organizations and management, see Mintzberg and Waters (1985). In the natural sciences, the so-called chaos theory is a similar attempt to explain apparently disorderly or irregular phenomena through subtle interactions between deterministic processes and random processes (see, e.g., Hall 1991). However, the current chapter does not try to apply this stream of research to social systems directly.

Table 2 Three layers of manufacturing capability

	Basic nature	Influence on	Interpretation
Routinized manufacturing capability	Static and routine	Level of competitive performance (in stable environments)	Firm-specific pattern of steady-state information system in terms of efficiency and accuracy of repetitive information transmission
Routinized learning capability	Dynamic and routine	Changes in or recovery of competitive performance	Firm-specific ability to handle repetitive problem-solving cycles or a routinized pattern of system changes
Evolutionary learning capability	Dynamic and nonroutine	Changes in patterns of routine capability	Firm-specific ability to handle system emergence or the nonroutine pattern of system changes in building routine capabilities

trait that, due to its complexity, cannot be explained through the behavior of its constituent parts alone or predicted from previous states of the system.[10] Its nuances may be somewhat different case by case, but my notion of *system emergence* shares a common feature with the above disciplines: it rejects the rational optimism that system change can be entirely controlled by means of purposeful plans existing prior to any change; it also denies the cynical notion that social system change is merely a stochastic process, a long-term accumulation of random accidents with little connection to human efforts.

3.2 Three Layers of Organizational Capabilities

In applying the concept of organizational capability to the case of operation-based competencies in manufacturing systems (e.g., at Toyota), I propose the following three-layer conceptual framework: *routinized manufacturing capability*, i.e., a set of organizational routines that affects the level of competitive performance in a steady state; *routinized learning capability*, i.e., a set of organizational routines that affects the pace of continuous performance improvements (as well as recovery from frequent system deterioration or disruptions); and *evolutionary learning capability*, i.e., a nonroutine ability that affects the creation of the above capabilities themselves[11] (Table 2). The

[10]According to von Bertalanffy (1968) and Weinberg (1975), the complexity of system behavior stems from interactions among a medium number of elements in the system, rather than from a random process involving a large number of objects or a mechanistic process affecting a small number of objects.

[11]Similar concepts include static versus dynamic routines (Nelson and Winter 1982), absorptive capacity (Cohen and Levinthal 1990, 1994), as well as dynamic capability (Teece and Pisano 1994). The concept of evolutionary capability adopted here is different from these concepts, in that the former emphasizes the nonroutine and emergent nature of the process for creating routines.

first two are organizational routines that have been analyzed by much of the past literature, but the third category is a new and nonroutine concept.

3.2.1 Routinized Manufacturing Capability

The first layer of this framework, i.e., routinized manufacturing capability, refers to a stable pattern involving a set of productive resources, as well as their repetitive interactions, which creates firm-specific advantages in competitive performance at a given point in time. For instance, if a factory operates with a consistently lower number of defective parts per million compared with its competitors, and if certain defect-preventing routines, such as *poka-yoke*, *jidoka*, *andon*, and 5-S, are implemented more thoroughly by this factory than by others, we can infer that this set of routines is the factory's routinized manufacturing capability.[12] In an industry where repetitive production tends to be the dominant mode, a stable level of manufacturing performance in stable environments implies that such a set of organizational routines exists.

Further, by applying the informational point of view mentioned earlier (Fujimoto 1999), we can reinterpret routinized manufacturing capability as a pattern in a steady-state information processing system that allows given product design information to be repetitively transmitted in a more effective, accurate, efficient, and/or flexible manner than what happens in competitor firms. For instance, poka-yoke, jidoka, andon, and 5-S can be reinterpreted as a set of routines that jointly enhance accuracy of repetitive information transmission on the shop floor from the production process to the products. In other words, routinized manufacturing capability refers to a firm-specific set of organizational routines that control information assets in the manufacturing system, as well as repetitive patterns of information transmission among them. In a static environment, where it is possible to ignore changes in market needs and internal system disruptions over time, strong routinized manufacturing capability would be a sufficient condition for a stable level of high manufacturing performance. However, a static environment may not be a realistic assumption.

Routinized manufacturing capability can be defined for each dimension of competitive performance: factor productivity, throughput time, design quality, and so on (Fujimoto 1999). Yet, a firm's ability to consistently achieve high performance in multiple dimensions *at the same time* may be even more important for its survival and growth. For example, as William J. Abernathy notes, some firms may improve productivity by sacrificing flexibility; others may improve conformance quality

[12]These routines may not only yield lower levels of in-process defects and field defects but also facilitate problem recognition and thereby trigger continuous improvement (kaizen) activities, which is an aspect of the improvement capability discussed later. Thus, the two types of routine capability tend to overlap in real shop floor settings.

while lowering productivity.[13] When dilemmas concerning trade-offs between two dimensions are commonly observed in an industry, a firm that successfully reduces or eliminates them may be able to outperform its competitors in both competitive dimensions at the same time.

3.2.2 Routinized Learning Capability

The second layer of the framework, i.e., routinized learning capability, refers to a firm's specific ability to change its manufacturing system in a frequent and regular manner to improve functionality. Frequent incremental changes to a firm's products or processes (which allow it to compete more effectively) imply that the firm has a certain routinized learning capability, which may be defined as an organization's ability to conduct kaizen, or continuous improvement activities. And when a manufacturing firm faces frequent deterioration or disruptions in its products or processes but is able to recover from such problems more effectively and swiftly than its competitors, we can infer that the said firm does indeed possess this capability.

From an information point of view, routinized learning capability is essentially a set of organizational routines to renew a firm's information assets for better adaptation to a dynamic environment, one in which product obsolescence and production process disruptions are common. For instance, suppose that defect rates at one factory, measured in parts per million, decrease continually and at a higher rate than among its rivals, over an extended period, and there is evidence that the factory has implemented TQC (total quality control) and TPM (total productive maintenance) programs more effectively than others over the same period. One may infer from these observations that the manufacturing firm in question has a certain routinized learning capability.

Again, by applying the information processing perspective, we can refer to this capability as organizational problem-solving, or the firm-specific ability to perform routinized problem-solving cycles more effectively than competitors. In the context of manufacturing systems, a standard problem-solving cycle refers to a heuristic routine that converts problem information (input) into solution information (output). Generally speaking, a firm's routinized learning capability consists of the following sub-capabilities[14]:

Problem Identification A system's ability to reveal and visualize problems, to diffuse problem information among problem-solvers, and to make individual members aware of problems, as well as its organizational members' willingness to accept higher performance goals

[13]See Abernathy (1978) for the concept of *productivity dilemma*.

[14]A standard linear model of problem-solving is applied here (e.g., Simon 1969, 1976; March and Simon 1958).

Problem-Solving The ability to effectively find, simulate, and evaluate alternatives; to coordinate knowledge, skills, responsibility, and authority for solving problems; to diffuse tools for problem-solving throughout the organization; to circulate knowledge of alternative action plans and their effects; and to share evaluation criteria

Retention of Solutions The ability to formalize and routinize new solutions in standard operating procedures both quickly and accurately, maintaining and retrieving them effectively, thereby providing stability for organizational members who internalize said solutions[15]

In an industry such as automobile manufacturing, in which repetitive production of multiple products is common, a firm's routinized learning capabilities can be observed in at least two areas. First, continuous process improvements occur on the factory shop floor. For example, a problem-solving routine can be applied to a standard case of total quality control (TQC), kaizen program, and Toyota Production System (TPS). Say, a Toyota Group parts supplier has a factory in which just-in-time and visual management are thoroughly implemented, so that workers and supervisors can easily find defect problems in the production process. In the same factory, workers, supervisors, and engineers consistently follow standard steps for incremental improvements (sometimes called "the quality control story"): problem definition (theme setting), root cause analysis, shop floor experiments for evaluating alternatives, decentralized selection and implementation of the revised procedures, and standardization and countermeasures to make sure that the old ways are abandoned. The company may have a standard format for kaizen report sheets, which workers fill out for each of the steps following the standard sequence mentioned above. Standardized analytical tools may also be used regularly by the shop floor people (e.g., histograms to identify the critical problem, cause-effect diagrams to find root causes, scatter diagrams to evaluate the effects of alternative solutions, and so on).

Second, we can observe product development improve through design quality and product mix. Other things being equal, model renewal may occur at a higher pace. In this case, we can interpret each of the product development stages (i.e., concept generation, product planning, product engineering, and process engineering) as a distinctive problem-solving cycle.[16] At the product engineering stage, for example, informational outputs from the product planning stage, including product specifications, styling, and layout, become goals (goal setting). Product designs are then developed (alternative idea generation); prototypes are constructed according to the designs (model building); they are tested at laboratories and proving grounds (experiment); and the cycles are iterated until a satisfactory result is achieved, when the final engineering drawings are chosen as a solution (selection).

---

[15]Retention of solutions may also be regarded as an essentially static capability, since it enables repetitive activation of the same information.

[16]The various problem-solving cycles are linked to one another, so that solutions from the upstream cycles become goals for the downstream cycles (Fujimoto 1989; Clark and Fujimoto 1989, 1991).

One could argue that the reality of shop floor management and product development is much more complicated, ill-structured, less streamlined, and more ambiguous than what the above standard model of linear problem-solving cycles assumes.[17] In order to explain a company's routinized learning capability, however, the standard problem-solving routine remains relevant. Although this linear problem-solving model may not reflect the actual messiness of the shop floor, an organization that imposes a problem-solving framework on this confusing reality is still likely to improve its performance faster than other organizations that have not established such routines. In other words, a firm possessing a consistent problem-solving routine throughout its organization is likely to achieve better results in terms of continuous improvements. Hence, the core element of a firm's routinized learning capability is a routine for standard problem-solving cycles.

3.2.3 Evolutionary Learning Capability

The third layer of this framework, i.e., evolutionary learning capability, is not easily observed in the everyday workings of a company. It refers to a firm's distinctive ability to create a set of effective (i.e., ex post rational) organizational routines in the long run faster and earlier than its competitors. Thus, to the extent that changes in organizational routines can be regarded as multipath system emergence, evolutionary learning capability indicates *the firm-specific ability to cope with multipath system emergence*, or a complex process of capability building, which is neither totally controllable nor wholly predictable.

Note here that evolutionary learning capability is not a routine itself.[18] That is, while both dynamic, routinized learning capability and evolutionary learning capability should be distinguished from each other. The former deals with routinized or regular patterns of system changes, whereas the latter is related to higher-order

[17] Alternative frameworks, for example, include the *garbage can* model (March and Olsen 1976; March 1988). See also von Hippel and Tyre (1993).

[18] There is always the risk that such logic may lead to an infinite chain of backward explanations (capability of capability building and so on). The present framework with three layers of capabilities tries to avoid this by giving each construct a concrete definition, rather than by simply calling them meta-routines. Thus, improvement capabilities manage repetitive routine changes, while evolutionary capabilities cope with nonroutine emergent changes. Also, in practical terms, it is rather meaningless to discuss the capability to build evolutionary capabilities, because the creation of an evolutionary capability itself is likely to be a unique series of historical events, whose stable pattern cannot be analyzed through a meaningful hypothesis-testing process. Being aware of this problem, in the present chapter, we will not try to go further backward and explain explicitly why a company like Toyota was historically able to build a certain evolutionary capability. It would be impossible, in the first place, to explain such rare events through the concept of organizational capability.

system changes, which are in themselves rather irregular and infrequent and often connected with rare, episodic, and unique historical events.[19]

To the extent that firm-specific patterns of productive performance and operation-based capability are observed, *differences* in each firm's evolutionary learning capability do matter. In particular, we can identify the following two aspects of the evolutionary learning capability:

1. Intentional (pretrial) learning capability: A firm's ability to find, experiment with, and acquire new organizational routines more effectively than its competitors. This may include the ability to rationally work out potentially effective trials or an entrepreneur's ability to intuitively envision effective trials.
2. Opportunistic (ex post) learning capability: But what if trials for a new capability are made inadvertently and turn out to be effective in terms of competition? In this case, a firm with a strong ex post learning capability can still generate specific advantages for itself, through its ability to grasp the potential consequences of unintentional trials on competition and to routinize and retain successful trials.

Even when competing companies do not differ in their intentional (pretrial) learning capability or in their ability to problem-solve, one of them may still be able to outperform the others by possessing better opportunistic (ex post) learning capabilities.

To sum up, a firm's evolutionary learning capability is its ability to manage the multipath system emergence processes of routine-capability building better than its competitors. As such, it is a nonroutine dynamic capability embedded in the organization.

[19]The idea of multilayer structures in organizational capabilities, routines, programs, knowledge, learning, etc. is not particularly new in the literature on organizational studies. For instance, the concept of *initiation* as the creation of new programs (March and Simon 1958), *structuration* as conditions governing continuity or transmutation of the structures of rules and resources (Giddens 1984), *double-loop learning* (Argyris and Schön 1996), and *higher level learning* (Fiol and Lyles 1985) all assumes a multilayer structure. Our definitions of improvement versus evolutionary capabilities are somewhat different from the above concepts, as they emphasize the distinction between repetitive regular changes and emergent irregular changes within the system in question. Also, the distinction between improvement capability and evolutionary capability is different from the traditional distinction between the ability to handle incremental innovations and that needed for radical innovations (Abernathy and Utterback 1978, Hayes and Wheelwright 1984). The evolutionary capability discussed here is not the ability to perform a one-off, major system change, but the ability to cope with an emergent process over an extended period.

4 Application: Operation-Based Competency at Toyota

4.1 *Basic Facts About the History of Automobile Manufacturing*

So far, I have proposed the concepts of multipath system emergence and evolutionary learning capability to explore dynamic aspects of a firm's operation-based competency. Now I will try to apply them to the history of manufacturing routines in the automobile industry.

Let us start by summarizing some key historical phenomena of the world automobile industry as stylized facts. First, by the 1980s a group of Japanese firms had demonstrated significantly higher levels of manufacturing performance in the world auto industry. Second, high-performance Japanese manufacturers, like Toyota, had improved certain aspects of their performance faster and more consistently than other firms during an extended period prior to the 1980s. Third, the said high levels of performance and pace of continuous improvements appeared to stem from an overall manufacturing system, rather than individual techniques or practices. Fourth, the Toyota-style system was not created all at once, but it had gradually and cumulatively evolved mainly between the 1940s and 1980s (Cusumano 1985; Fujimoto 1999).

The literature on operations and productive performance in this industry has also identified universally prevalent, region-specific, and firm-specific patterns of manufacturing capability that all exist at the same time (Clark and Fujimoto 1991). That is, all the following conditions affect the evolution of operation-based competence:

1. *Universally prevalent patterns of practice* may emerge when rational problem-solvers share identical objectives and constraints worldwide, when best practices have been transferred to everyone, and when severe selection environments allow only a particular pattern to survive.
2. *Region-specific patterns of capabilitie*s may emerge when firms face regional environmental constraints or objectives or when knowledge transfer occurs only within each region.
3. *Firm-specific patterns* may emerge when each company is allowed to take a *random walk* in changing its systems, when each individual firm faces different environmental constraints, when each firm is led by a different entrepreneurial vision, when firms have varying levels of problem-solving capabilities, or when knowledge transfer among firms is limited.

Thus, although pure chance and historical imperatives often play an important role in system emergence and capability-building processes, a company may still be able to build certain manufacturing capabilities faster and more effectively than any competitors through its strong evolutionary learning capability. For example, historical imperatives may explain why the Japanese automakers in general acquired

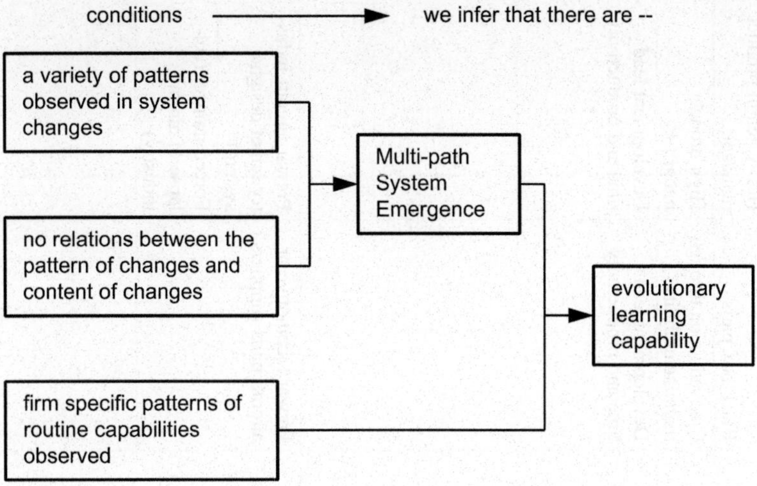

Fig. 2 Operational definition of system emergence and evolutionary capability

certain region-specific capabilities, but they do not explain why certain Japanese companies, like Toyota, developed better capabilities than others.

4.2 *Operational Definitions*

Having laid out the above basic empirical observations and key concepts, we can now apply the present framework to the actual case of evolution of the Toyota-style manufacturing capability (Fujimoto 1999). For this historical analysis, first of all, the concepts of *multipath system emergence* and *evolutionary learning capability* need to be defined operationally (Fig. 2).

That is, (i) when a wide variety of evolutionary paths for system changes (e.g., rational calculation, environmental constraints, entrepreneurial vision, knowledge transfer, and random trials in Fig. 1) are historically observed in each case of the main organizational routines making up the manufacturing capability in question (ii) and when the pattern of the aforementioned evolutionary paths cannot be explained logically by the nature of the organizational routine in question (i.e., there is no predictable relation between the types of paths and the content of each individual routine), we regard the system in question to be an outcome of *multipath system emergence* (see Table 3, which is shown later, as an example of the Toyota's manufacturing system).

In addition, when we also observe that the resulting system (i.e., a set of organizational routines) demonstrates stable and firm-specific competitive performance, we infer from these facts that the firm possessed a certain *evolutionary learning capability*.

Table 3 Summary of the evolution of selected production-development capabilities

	Just-in-time	Multitasking with product-focus layout	Jidoka and flexible equipment	Kaizen and TQC	Black box parts	Heavyweight product manager
Competitive effect (rationality)	Creating pressure for productivity improvements Throughput time Inventory cost	Productivity improvement	Pressures for quality improvement Flexibility	Quality improvement Productivity improvement	Cost reduction through manufacturability Development lead time and productivity	High product integrity Development lead time and productivity
Entrepreneurial vision	Kiichiro Toyoda, 1930s ("just-in-time" slogan) Taiichi Ohno, 1940s–1950s (system building)	Kiichiro Toyoda, 1945 (vision of rapid productivity catch-up without economy of scale)	Kiichiro Toyoda, 1931 (vision of high productivity with small volume production)			
Transfer from other industries	Textile (benchmarking of Nichibo) Prewar aircraft production	Textile: multi-machine operations in spinning (through Ohno)	Textile: Sakichi Toyoda's automatic loom	TQC was established in other industries (e.g., process industry)	Prewar locomotive or aircraft parts suppliers	Prewar aircraft industry (chief designer system) Forced transfer (collapse of aircraft industry)
Transfer from Ford system	Synchronization idea from Ford (invisible conveyor lines) Kanban as incomplete synchronization	Productivity benchmarking with Ford Modified Taylorism	Adoption of Detroit-type automation when feasible U-shaped layout as "incomplete transfer machine"	Suggested system from Ford Training within industry Statistical quality control		

Imperative of forced growth with resource shortage		Limitation of permanent workforce after the 1950s strikes Forced productivity increase in the 1960s	Shortage of investment resources: low-cost automation had to be pursued	Shortage of supervisors replacing craftsmen-foremen = need for TWI	High production growth and model proliferation created pressures to subcontract subassembly and design	
Imperative of forced flexibility with a small and fragmented market			Forced flexibility of equipment due to small volumes		Product proliferation in the 1960s created pressures to subcontract design jobs out	Product proliferation with limited engineering resources created pressures toward compact projects
Imperative of shortage of technology	Lack of computer technology for production control in the 1950s–1960s		Lack of adaptive control automation: jidoka needs human intervention		Lack of electric part technology at Toyota in 1949 (separation of Nippondenso)	
Ex post capability of the firm		Flexible task assignment and flexible revision of work standards to better exploit productivity increase opportunities		Toyota maintained momentum for TQC by creating organizations to diffuse it to suppliers	Toyota institutionalized a version of the black box part system that could better exploit competitive advantages	Only Toyota adopted the heavyweight product manager system from the aircraft industry as early as the 1950s

4.3 Histories of the Routines: A Summary

Based on the above operational definition of system emergence and evolutionary learning capability (see Fig. 2 again), the author explored the evolution of various core elements of the Toyota-style system (Fujimoto 1999), or operational routines. Just-in-time, mechanisms for productivity improvement, multitasking, flexible production, total quality control, suppliers' design capability, and heavyweight product manager system are examples of such routines, whose histories may be investigated one by one.

Although a full historical analysis is omitted in this chapter (see Fujimoto 1999 for further details), some results may be summarized with regard to the manufacturing routine patterns mentioned earlier as stylized facts, i.e., the coexistence of universally adopted routines, region-specific routines, and firm-specific routines.

1. *Factors Affecting Universally Adopted Routines*

   1.1. *Pressure from International Competition*: Toyota's routine-capability building was consistently motivated, since the 1930s, by perceived competitive pressure from US mass producers, particularly Ford. Even with a strongly protected domestic market between the 1930s and 1950s, Toyota's perception of imaginary competitive pressures persisted.

   1.2. *Direct and Indirect Adoption of the Ford System*: Partly motivated by the above perception of international competition, Toyota adopted, mostly indirectly, many elements of the Ford system and of the American mass production system, including moving conveyors, transfer machines, product and component designs, the Taylor system, supervisor training programs, and statistical quality control. A pure dichotomy between the Ford system and the Toyota system is therefore misleading.

2. *Factors Affecting Region-Specific Routines*

   2.1. *Benefits of Forced Growth*: Some of the region-specific historical imperatives that all the Japanese firms faced during the postwar period almost *forced* them to make certain responses, a number of which turned out to contribute to their competitive advantage. Many such responses were not recognized as competitive weapons when the firms first adopted them. For example, the imperative of forced growth in both production and product development, with limited supply of production inputs and fear of labor conflicts, turned out to facilitate capability building for productivity improvements through avoidance of intra-firm overspecialization, division of labor between assemblers and suppliers, as well as by limiting the excessive use of high-tech equipment on the shop floor.

   2.2. *Benefits of Forced Flexibility*: The imperative of forced flexibility in a fragmented market also benefited the Japanese firms. This is partly because of region-specific patterns of industrial growth: *rapid production growth accompanied by rapid product proliferation*. When their capabilities were first built, the flexibility that the firms acquired tended to be recognized as a necessary evil to cope with a fragmented market, rather than as a measure for international competition. Furthermore, as the comparison between the Japanese and UK production systems clearly shows, fragmented markets do not automatically create effective flexibility.

2.3. *Benefits of Lack of Technology*: While the excessive use of high-tech automation equipment often became an obstacle to productivity improvements for many Western automakers in the 1980s, the effective Japanese makers apparently avoided the problem. This may be partly because the latter consciously rejected the temptation of overspecialization, but it is also due to the fact that high technology was not there in the first place. To the extent that this can be caused by certain region-specific technology gaps, the lack of technology may bring about unintended competitive benefits for firms in a given region.

2.4. *Benefits of Intended Knowledge Transfer*: Region-specific patterns of capabilities may also emerge when intra-regional knowledge transfer is more intense and frequent than inter-regional knowledge transfer. The supplier network shared by the Japanese firms was one such transfer instrument. Intense competition among domestic makers in the 1960s and 1970s may have also facilitated their efforts in learning from domestic competitors.

2.5. *Benefits of Unintended Knowledge Transfer*: As seen in the case of engineers from the prewar aircraft industry, the *push* type of knowledge transfer, which the receivers did not intend to initiate, brought about rapid advances in automobile technologies and product development systems in the Japanese postwar automobile industry.

2.6. *Benefits of Incomplete Knowledge Transfer*: Although the Japanese auto firms tried to adopt many practices and techniques used by US mass producers, some of these processes were incomplete, due to the historical imperatives mentioned above and imperfect firm absorption capacity. In this sense, the *kanban* system may be regarded as an incomplete version of the conveyor system, the U-shaped machine layout as an incomplete transfer machine, and the *jidoka* as an incomplete adaptive automation. The very incompleteness of the transfer may have facilitated its subsequent diffusion through the entire system.

3. *Factors Affecting Firm-Specific Routines*

3.1. *Benefits of Self-Fulfilling Vision*: Firm-specific entrepreneurial vision sometimes played an important role in building distinctive manufacturing capabilities. This was particularly true when an apparently unrealistic vision triggered self-fulfilling efforts to achieve bold objectives. In the 1930s and 1940s, Kiichiro Toyoda played a pivotal role in advocating cost reductions without economies of scale, catching up with Ford, and the just-in-time philosophy. In those days, Nissan did not have an equivalent figure.

3.2. *Benefits of Linkages with Other Industries:* Toyota's inherent connections with the textile industry may have facilitated knowledge transfer from it (particularly through Taiichi Ohno) and helped create the automaker's competitive advantages in production control techniques.

3.3. *Advantages of Opportunistic (Ex Post) Learning Capability*: Although hardly any firms recognized the potential competitive advantage of a new system when they first tried it, some were still able to develop firm-specific advantages through their *opportunistic (ex post) learning capability*. They did so by recognizing the potential competitive advantage of a new system, modifying it to exploit it to the full, institutionalizing it, and retaining it until the advantages were realized. For example, while all the Japanese automakers faced similar environmental pressures to adopt the black box parts system in the 1960s, seemingly only Toyota created a system that could fully exploit the potential advantages of this practice. Moreover, although all the Japanese auto manufacturers accepted to take on aircraft engineers after the war, Toyota was the only company that institutionalized the heavyweight product manager system, which was prevalent in the aircraft industry. Thus, even when all Japanese firms faced certain historical imperatives that facilitated new practices, only some of them made the most of this potential luck through their firm-specific evolutionary learning capability.

4.4 Interpretation: Toyota's Evolutionary Learning Capability

Following the operational definitions of multipath system emergence and evolutionary learning capability mentioned in Fig. 2, the above results can be summarized in Table 3, in which Toyota's capability-building cases are classified according to types of routines and types of paths.

This analysis clearly shows that:

1. There were several system change paths for each main component of the Toyota-style manufacturing system (see the variety of explanations in each column of Table 3).
2. There was no clear correlation between the nature of the routines and the types of paths (compare the patterns of explanations across the columns of Table 3).

Therefore, by applying the operational definition specified in Fig. 2, I argue that Toyota's routine capability building can be described as *multipath system emergence* and that, as it created distinctively competitive routines though the emergent process, Toyota possessed an *evolutionary learning capability*.

Hence, the evolution of the Toyota-style manufacturing system can be characterized by multipath system emergence. Although Toyota's official corporate history portrays the firm's success as a combination of entrepreneurial vision and rationally controlled follow-through, its ability to seize the day when the unforeseen occurred—its superior evolutionary learning capability—may prove more important in the long run (Fujimoto 1999).

5 Implications and Conclusions

5.1 Evolutionary Learning Capability and Related Concepts

After having looked at the concept of evolutionary learning capability and its application to an actual case of operation-based competency, let us link the present framework to related concepts for explaining the emergence of operation-based competency.

Emergent Strategy Mintzberg's emergent strategy (as well as Quinn's logical incrementalism) is a powerful concept that can explain a significant portion of my historical material (Mintzberg and Waters 1985; Quinn 1978). Indeed, an essential aspect of the ex post evolutionary capability is the ability to realize an emergent strategy better than other firms. Note, however, that the concept of multipath system emergence is broader than that of emergent strategy, as it also includes a deliberate strategy as its possible path.

Organizational Learning Evolutionary learning capability is also closely related to the concept of organizational learning.[20] Yet, there are some important differences between the two. Organizational learning implicitly assumes regular patterns of learning (problem-solving routines), repetitions (learning by doing), or the prior existence of an overt intention to learn (learning from others through benchmarking). The concept of evolutionary learning capability does not assume these conditions, since it includes irregular, non-repetitive, and unintended learning factors. Evolutionary learning capability, in this context, implies the ability to acquire effective routines in any number of ways, although it is hard to predict what types of learning opportunities will emerge and when. These opportunities may include organizational learning from deductive theories, learning from others, and learning by intentionally doing, but they may also include *unintentional* learning through inadvertent actions.

Problem-Solving The process of multipath system emergence does include the standard problem-solving cycles discussed earlier, but problem-solving heuristics cannot always explain system emergence and a firm's evolutionary learning capability. The regular sequence of problem identification-solution-retention may not exist, and, in many cases, trials of solutions precede problem recognition, as James G. March and Johan Olsen point out in relation to their *garbage can model.*[21] Solutions to certain noncompetitive problems may subsequently and inadvertently become solutions to competitive problems. For instance, Toyota reliance on its

[20]To the extent that organizational learning is "encoding inferences from history into routines that guide behavior" (Levitt and March 1988) or "improving actions through better knowledge and understanding" (Fiol and Lyles 1985), evolutionary learning capability may overlap the concept of a certain higher-order learning ability to change routines for learning or values (Argyris and Schön 1996; Fiol and Lyles 1985). For concepts and definitions of organizational learning, see, for example, Fiol and Lyles (1985), Levitt and March (1988), and Argyris and Schön (1996).

[21]See March and Olsen (1976) and March (1988).

suppliers' engineering activities increased in the 1960s, which managers and engineers at the time believed to be a solution to the problem of mounting workload for Toyota's in-house engineers. Yet this specific solution to workload problems had an unintended impact on competitive performance. Indeed, relying on its suppliers' engineering activities had become a major competitive weapon for Toyota and other Japanese automakers by the 1980s, because it facilitated cost saving through component design for manufacturing (DFM) and product development cost/time saving.

Rationalism An evolutionary perspective recognizes that ex ante rational human actions (as described by either the *perfect rationalism* of neoclassical economics or the *bounded rationalism* of Herbert Simon[22]) may trigger organizational changes, but in a complex system, such an approach only represents one of the many possible paths. Entrepreneurial vision, which business historians tend to emphasize, plays an important role—yet so do environmental constraints and the many external forces that can affect system changes for better or worse.

5.2 The Remaining Questions: Where Did Evolutionary Capabilities Come from?

The present chapter interpreted a firm's evolutionary learning capability as its ability to perform both ex ante (intentional) learning and ex post (opportunistic) learning. I paid particular attention to the latter: a company with strong ex post (opportunistic) learning capability, like Toyota, converts miscellaneous existing solutions, many of which are unpolished, into a set of distinctive routine capabilities in product development, production, and purchasing.

What are, then, the organizational features that may facilitate the solution-refinement cycle? Again, systematic empirical research will be needed to shed light on this. However, after many contacts with Toyota employees, it is my impression that they view new situations in daily life—whether new problems, solutions formulated for other problems, partial solutions to present problems, or chance events—as potential opportunities to improve competitiveness more often than employees in other firms. Original trials may be linked to pure luck or unintended consequences, rather than to intended and realized success, but a firm with many *prepared* people, who associate every element with its competitive effects, may be able to recognize the competitive value of such trials and exploit them for capability building more effectively than its rivals (Cohen and Levinthal 1990, 1994).

Other questions follow naturally from the above discussion: How is the evolutionary learning capability formed in the first place within a particular company? Where did Toyota's company-wide *competition-consciousness* or *prepared mind*

[22] Simon (1969, 1976)

come from? These are intriguing points, but there are no clear explanations. Some say that the spirit of Kiichiro Toyoda or even of Sakichi Toyoda was retained in the employees' attitudes for a long time; others claim that the distinctive culture of Eastern Aichi Prefecture (traditionally called Mikawa) shaped their tenacity and concentration. Both interpretations are plausible, but neither is convincing on its own. After all, the formation of the evolutionary learning capability itself is so rare that it is extremely difficult to analyze it as a social science subject. For now, all we can say is that historical circumstantial evidence leads us to infer that Toyota had distinctive evolutionary learning capabilities, at least up to the 1980s.

5.3 Conclusions: Fortune Favors the Prepared Organizational Mind

This chapter presented an evolutionary framework to analyze a certain type of long-term process innovation. By introducing such concepts as multipath system emergence and evolutionary learning capability, it argued that the cumulative process innovation of the manufacturing system at the Toyota Motor Corporation may be explained more persuasively by applying the present framework, which does indeed rely on multipath system emergence and evolutionary learning capability.

Empirical and historical research on the world automobile industry shows that a wide variety of system performances and practices exist in different firms and that these systems have changed in nonroutine ways over time. Thus, the evolutionary framework proposed in this chapter may provide additional insights into why certain manufacturing practices have emerged at Toyota.

A company's decision-makers should certainly attempt to solve problems rationally, but they should not assume that rational plans always solve problems. The actual process of system change is essentially emergent. And, no matter how successful a company has been, it needs to develop an organizational culture of *preparedness*, by converting both the intended and unintended consequences of its actions, the lucky breaks and the well-laid plans, and the temporary successes and failures into a long-term competitive advantage. This is the key ingredient of the effective evolutionary learning capability. After all, *fortune favors the prepared organizational mind.*[23]

---

[23]The original sentence is a famous statement by Louis Pasteur: "Fortune favors the prepared mind". Its relevance was pointed out to me by both David A. Hounshell (Carnegie Mellon University) and the paper by W. M. Cohen and D. A. Levinthal "Fortune Favors the Prepared Firm" (1994).

References

Abernathy WJ (1978) The productivity dilemma. Johns Hopkins University Press, Baltimore
Abernathy WJ, Utterback JM (1978) Patterns of industrial innovation. Technol Rev 80(7):2–9
Alchian AA (1950) Uncertainty, evolution, and economic theory. J Polit Econ 53(3):211–221
Argyris C, Schön DA (1996) Organizational learning II: theory, method, and practice. Addison-Wesley Publishing Company, Reading
Barnett WP, Burgelman RA (1996) Evolutionary perspectives on strategy. Strateg Manag J 17:5–19
Barney JB (1986) Strategic factor markets: expectation, luck and business strategy. Manag Sci 32:1231–1241
Burgelman RA (1994) Fading memories: a process theory of strategic business exit in dynamic environments. Adm Sci Q 39:24–56
Carroll GR et al (1996) The fates of De Novo and De Alio producers in the American automobile industry 1885–1981. Strateg Manag J 17:117–137
Chandler AD (1990) Scale and scope. Harvard University Press, Cambridge
Chandler AD (1992) What is a firm? Eur Econ Rev 36:483–492
Clark KB, Fujimoto T (1989) Lead time in automobile product development: explaining the Japanese advantage. J Technol Eng Manag 6:25–58
Clark KB, Fujimoto T (1991) Product development performance. Harvard Business School Press, Boston
Cohen WM, Levinthal DA (1990) Adaptive capacity: a new perspective on learning and innovation. Adm Sci Q 35:128–152
Cohen WM, Levinthal DA (1994) Fortune favors the prepared firm. Manag Sci 40(2):227–251
Cusumano MA (1985) The Japanese automobile industry. Harvard University Press, Cambridge
Dosi G (1982) Technological paradigms and technological trajectories. Res Policy 11:147–162
Fiol CM, Lyles MA (1985) Organizational learning. Acad Manag Rev 10(4):803–813
Fujimoto T (1989) Organizations for effective product development – the case of the global automobile industry. Unpublished D.B.A. Dissertation, Harvard Business School
Fujimoto T (1999) The evolution of a manufacturing system at Toyota. Oxford University Press, New York
Giddens A (1984) The constitution of society. University of California Press, Berkeley
Grant R (1991) The resource-based theory of competitive advantage: implications for strategy formulation. Calif Manag Rev 33(3):114–135
Hall N (ed) (1991) Exploring Chaos: a guide to the new science of disorder. W. W. Norton & Company, New York
Hannan MT, Freeman J (1989) Organizational ecology. Harvard University Press, Cambridge
Hayes RH, Wheelwright SC (1984) Restoring our competitive edge. Wiley, New York
Hayes RH et al (1988) Dynamic manufacturing. Free Press, New York
Hirshleifer J (1977) Economics from a biological viewpoint. J Law Econ 20(1):1–52
Hounshell DA (1984) From the American system to mass production 1800–1932: the development of manufacturing technology in the US. Johns Hopkins University Press, Baltimore
Iansiti M, Clark KB (1993) Integration and dynamic capability: evidence from product development in automobiles and mainframe computers. Harvard Business School working paper 93-047
Itami H (1984) Shin Keieisenryaku no Ronri (the logic of business strategy: revised). Nihon Keizai Shinbunsha, Tokyo (in Japanese)
Kogut B, Kulatilaka N (1992) What is a critical capability. Paper presented at the Joseph A. Schumpeter Society, Kyoto, August 1992
Leonard-Barton D (1992) Core capabilities and core rigidities: a paradox in managing new product development. Strateg Manag J 13:111–125
Levitt B, March JG (1988) Organizational learning. Annu Rev Sociol 14:319–340
March JG (1988) Decisions and organizations. Basil Blackwell, Oxford
March JG, Olsen JP (1976) Ambiguity and Choice in Organizations Universitetsforlaget, Bergen
March JG, Simon HA (1958) Organizations. Wiley, New York

Merton R (1968) Social theory and social structure. Free Press, New York
Mintzberg H (1987) Crafting strategy. Harvard Business Review July–August 1987, pp 66–75
Mintzberg H, Waters JA (1985) Of strategies, deliberate and emergent. Strateg Manag J 6 (3):257–272
Nelson RR, Winter SG (1982) An evolutionary theory of economic change. Harvard University Press, Cambridge
Nonaka I (1985) Kigyo Shinkaron (the theory of corporate evolution). Nihon Keizai Shinbunsha, Tokyo (in Japanese)
Parsons T (1937) The structure of social action. Free Press, Glencoe
Penrose ET (1959) The theory of the growth of the firm. Basil Blackwell, Oxford
Prahalad CK, Hamel G (1990) The core competence of the corporation. Harv Bus Rev 68(3):79–91
Quinn JB (1978) Strategic change: 'Logical Incrementalism'. Sloan Manag Rev 1(20):7–21
Rumelt RP (1984) Towards a strategic theory of the firm. In: Lamb RB (ed) Competitive strategic management. Prentice-Hall, Engelwood Cliff
Rumelt RP (1991) How much does industry matter? Strateg Manag J 12:167–118
Simon HA (1969) The science of the artificial. MIT Press, Cambridge
Simon HA (1976) Administrative behavior. Free Press, New York
Teece DJ, Pisano G (1994) The dynamic capabilities of firms: an introduction. Ind Corp Chang 3 (3):537–556
Teece DJ et al (1992) Dynamic capabilities and strategic management. Revised, June 1992. Working Paper, University of California, Berkeley
Teece DJ et al (1994) Understanding corporate coherence: theory and evidence. J Econ Behav Organ 23:1–30
von Bertalanffy L (1968) General system theory: foundations, development, applications. George Braziller, New York
von Hippel E, Tyre M (1993) How 'Learning by doing' is done: problem identification in novel process equipment. M.I.T. Sloan School Working Paper
Weick K (1979) The social psychology of organizing. Addison Wesley, Reading
Weinberg GM (1975) An introduction to general systems thinking. Wiley, New York
Wernerfelt B (1984) A resource-based view of the firm. Strateg Manag J 5:171–180
Womack J et al (1990) The machine that changed the world. Rawson Associates, New York

The Nature of Surviving Japanese Factories in the Global Competition: An Empirical Analysis of Electrical and Electronics Factories

Mitsuhiro Fukuzawa, Nobuyuki Inamizu, Junjiro Shintaku, Kodo Yokozawa, and Nobutaka Suzuki

Abstract During the last few decades, Japanese electrical and electronics companies have been struggling to compete against rivals from emerging countries, such as South Korea and China. As a result, they have lost global market share. However, are Japanese firms really losing their competitiveness? This chapter investigates the following questions: (1) Are Japanese factories maintaining competitiveness today? (2) If so, which are the main characteristics of these competitive organizations? 3) Which aspects are important for Japanese factories to maintain competitiveness in the future? Interviews with eight successful Japanese electrical and electronics companies and a survey of 97 business units were conducted. Our findings show that Japanese electrical and electronics companies are still competitive at the shop floor level, with higher performance than their overseas transplant, except in terms of cost. Some common features of these factories include (1) carrying out sales and promotion activities targeting their own headquarters or other companies in order to find new business opportunities and (2) gathering various functions in one place, ranging from product development and design to production and sales. However,

This chapter is based on Fukuzawa (2015a), Fukuzawa and Inamizu (2017), Inamizu and Fukuzawa (2017), and Shintaku et al. (2014).

M. Fukuzawa (✉)
Faculty of Economics, Seikei University, Tokyo, Japan
e-mail: mfukuzawa@econ.seikei.ac.jp

N. Inamizu · J. Shintaku
Graduate School of Economics, University of Tokyo, Tokyo, Japan

K. Yokozawa
Graduate School of International Social Sciences, Yokohama National University, Yokohama, Japan

N. Suzuki
Graduate School of Information and Management Systems Engineering, Nagaoka University of Technology, Nagaoka, Japan

T. Fujimoto, F. Ikuine (eds.), *Industrial Competitiveness and Design Evolution*, Evolutionary Economics and Social Complexity Science 12,
https://doi.org/10.1007/978-4-431-55145-4_7

key issues to ensure future competitiveness still need to be addressed, such as personnel development and transmission of skills caused by unbalanced age distribution.

1 Introduction

In the last few decades, Japanese manufacturing companies have been facing hard times in global competition. Starting from the shift to the floating exchange rate system in 1971, the yen began to appreciate against the US dollar and, after the Plaza Accord was signed in 1985, the exchange rate of the US dollar versus the yen declined by 51% in 2 years. Consequently, Japanese firms' painstaking efforts to cut costs proved vain.

When the Cold War ended in the early 1990s, China swiftly established itself as a huge labor supplying country, with 1.2 billion workers whose wage was only one twentieth of that in Japan. In terms of costs, Japanese factories had no chance to compete against China, no matter how hard they strived to achieve cost reductions, and some even started to claim that it was pointless to try and maintain manufacturing bases in Japan.

Furthermore, the Japanese economic bubble burst in the 1990s, causing sluggish demand in most industries. This downturn in the domestic market damaged Japanese manufacturers, which had been losing export competitiveness due to the appreciation of the yen and the emergence of Chinese factories with lower-wage labor. To make matters worse, especially for Japan's electrical and electronics sectors, many emerging enterprises came to the fore as powerful rivals in the 2000s, including Samsung and LG from Korea; TSMC, AUO, and Quanta from Taiwan; and Haier and Gree from China.

Under these circumstances, several Japanese electrical and electronics companies found a way to survive by transferring their mass production bases overseas, for instance, to China and the ASEAN countries, and the production scale of their domestic plants was consequently reduced. At the same time, the number of employees and new recruits went down, and the use of non-regular employment increased. The economic situation showed some signs of improvement in the early 2000s, but the financial crisis of 2008 and the unprecedented Great East Japan Earthquake and tsunami resulted in even greater difficulties for the yen and a rise in electricity prices. Even in such adverse conditions, Japanese electrical and electronics companies still tried to hold out.

In this chapter, we will attempt to answer the following questions: (1) Are Japanese factories maintaining competitiveness today? (2) If so, which are the main characteristics of these competitive organizations? (3) Which factors are critical for Japanese factories to maintain competitiveness in the future?

To explore these matters, we will analyze shop floor level competitiveness, organization of manufacturing sites, and shop floor communication in Japanese electrical and electronics factories. Our aim is to provide some insights for the effective management of domestic and foreign business divisions and factories.

2 Methods

2.1 *Exploratory Case Study*

From August to December 2013, an exploratory case study was conducted in eight well-performing factories in Japan's electrical and electronics industries. Its purpose was to investigate the key reasons why some factories were able to maintain their competitiveness, while many others suffered severe losses or even went bankrupt as a consequence of the difficult economic situation.

In each case, we visited the factories and conducted the interviews in person. The interviews began by speaking about the background of the factory, for about 30 min to an hour. Then a factory shop floor visit was carried out. Further questions, a list of which had been sent to the interviewees before the visit, focused on functions in the factory, positioning in relation to all other group factories, shop floor activities, and human resources development. The interviewees included the plant manager, production department staff, plant management staff, and union staff. If the factory also had product development functions, staff members from the development department were asked to participate too.

After our visits, we held a discussion and identified some common features displayed by competitive factories: (1) being a self-driven factory, (2) gathering functions in one spot, (3) combining human skills and machines, (4) being a *fighting mother factory*, (5) making strenuous efforts for kaizen, and (6) overcoming unbalanced age distribution.

2.2 *Questionnaire-Based Survey*

From December 2013 to January 2014, a questionnaire-based survey was carried out in factories with production functionality and employing a minimum of 200 members of the Japanese Electrical Electronic and Information Union. Our exploratory case study, existing research on manufacturing strategies and organizations, and feedback from practitioners were taken into consideration to draft the questionnaires.

Depending on the respondents' position in the organizational hierarchy, three surveys were developed: the factory survey (Study A), the workplace (i.e., shop floor) leader survey (Study B), and the worker survey (Study C):

Study A (Factory Survey) Respondents comprised plant and administrative managers, who understood the overall state of the factory well. Questions were asked about product configuration, performance (market and productive performance), and the human resource policies of the company. There were 97 respondents (response rate 59.5%).

Study B (Workplace Leader Survey) Each firm selected three manufacturing sites, and the leaders of those sites participated in the survey. Questions were asked about workers' skills, coordination with other departments, attitudes toward work, and organizational climate. In total, 354 leaders responded (response rate 79.4%).

Study C (Worker Survey) All the workplace leaders mentioned above distributed this questionnaire to ten regular workers under them. In total, 3116 workers responded (response rate 78.1%).

3 Findings

3.1 Case Analysis

Let us now look in greater detail at the key common features of competitive factories listed in the previous section: (1) being a self-driven factory, (2) gathering functions in one spot, (3) combining human skills and machines, (4) being a *fighting mother factory*, (5) making strenuous efforts for kaizen, and (6) overcoming unbalanced age distribution.

3.1.1 Being a Self-Driven Factory

In Japanese firms, it is common practice for the headquarters to assign specific productions to each factory, and it is the factories' responsibility to manufacture the products assigned and enhance productivity. However, regardless of the factories' efforts, the headquarters may always decide to stop production due to negative changes in the environment, such as unfavorable market conditions or currency exchange.

Most of the factories that we analyzed were not just sitting still, waiting for new business to materialize. Rather, they actively carried out promotional and sales activities, targeting both their own headquarters and other companies, in order to maintain production and employment.

One of the factories investigated faced potential closure more than once after the 1990s, when the headquarters decided to alter the main product which it had been manufacturing for a long time. The factory reacted by presenting the strengths of its specialized product planning area to the development department at headquarters. In addition, its managers worked as sales personnel and tried to secure orders from third parties in sectors where its specialized technology could prove valuable.

In another case, the core business of the factory—production of a scanning device produced on a small scale and already in the maturity stage of product lifecycle—was moved elsewhere due to restructuring. In order to survive, the firm in question developed a mechanical technology to scan paper of any size and a software technology to accurately digitize the scanned characters and images. Based on these technologies, it launched a new scanner and carried out sales activities that led the new product to be supplied to financial institutions and governments around the world. The firm currently holds the largest market share in its field worldwide, as well as being number one in China. Chinese paper is often of poor quality, but the scanner can read data accurately without jamming, and this is the reason why it was adopted by China's Bureau of Statistics to carry out the national census. It is now advertised with the slogan: "700 million sheets of paper have been scanned."

The group to which this factory belongs also produced scanners at a Chinese plant. However, disaster, floods, and political risks in China caused the need for redistributing production. To secure successful assignment, the manager of the factory in question talked to related companies, such as manufacturers of molded products. He showed them the production costs in China and started an intensive negotiation. Finally, the factory succeeded in taking the work of the Chinese manufacturer. Furthermore, by carrying out kaizen activities, it eventually managed to produce at the same cost as the Chinese factory.

3.1.2 Gathering Functions in One Spot

In order to be self-driven, a factory needs not only production functions but also other functions, such as development and sales. With the production function alone, the scope of its activities remains limited. Yet, this can change if the factory is able to gather several functions in one place, including product development, production facility development, and sales and services.

Gathering functions in one factory does not only help develop and manufacture products, but it also shortens the time needed to launch new products, reduces costs, and improves overall productivity through collaboration among different departments. Furthermore, it makes it possible to listen to their customers' voice and promptly provide feedback, and each department can contribute to solving customers' problems. Such a factory can secure new business.

However, in many cases it is not easy to gather development, sales, and services functions in one place. One of the factories that we visited had the necessary development functions, but development could not proceed smoothly because the quality assurance department and laboratory of the development department were far away. To improve the situation, the departments related to development were all gathered on the 2nd floor of the factory, while the production department was placed on the 1st floor to achieve efficient development and production.

In another case, not only all the functions but also the headquarters were relocated to a specific factory. By concentrating all its functions in one spot, the firm managed to solve issues regarding product development, quality, and production cost through a global approach. Indeed, it has now reduced its production costs to a level that allows it to compete with overseas transplants.

Since the 2000s, the electrical as well as other industries have pursued the objective of making the factory a production subsidiary. Among the eight factories that we investigated, quite a few had become production subsidiaries. This might imply a distant relationship with the headquarters and dealings concerning cost issues only. Nevertheless, the firms analyzed had consciously implemented development and production activities in close collaboration with their headquarters, and this was producing positive results.

3.1.3 Combining Human Skills and Machines

Being self-driven and gathering functions in one spot are important features, but a factory's most important function is production. It goes without saying that employees are at the core of production, and it is only by transferring skills and abilities to the next generation that a factory will grow. Japanese firms have increased their productivity by combining machinery/equipment developed or improved in house with the accumulated skills and abilities of their production sites.

Currently, in overseas factories, for instance, in China, automation is progressing because labor costs are rising. The capability of Chinese production sites is lower than that of Japan, so the latest production equipment is introduced to compensate for it. However, the rate of utilization of said state-of-the-art equipment entirely depends on the skills possessed by the workers at the production sites. If Japanese production sites stopped improving their production capability, competition with overseas factories would only revolve around labor costs and full use of production facilities. This is why a successful factory must build up its strengths in terms of both employees and machinery/equipment.

One of the factories in our sample dealt with products that were difficult to produce and had considerable variations. It addressed these issues by developing its own production equipment from the very beginning and managed to reduce variations through in-house production. In addition, by implementing QC activities, boosting skills at the production site, and making full use of its production facilities, it managed to secure additional business.

In another case, because the quality of the products might impact human safety, the factory producing them was not allowed to fail, had to provide a long warranty period (10–15 years), and was also required to produce many kinds of different products. This factory first switched its production method from line production to cell production, then introduced a traceability system, and clarified who produced what and when. Additionally, it adopted large-scale special equipment, such as testing equipment for thorough inspections. Thanks to this combination of human skills, systems, and facilities, the company now makes extremely reliable products and ranks among the top players in terms of global market share.

3.1.4 Being a *Fighting Mother Factory*

Most of the factories which we visited were mother factories that controlled and integrated the productions of their overseas manufacturing bases. They were no ordinary mother factories either but *fighting mother factories*, committed to maintaining mass production capabilities. Few decades ago, many Japanese companies moved their mass production bases to emerging countries (mainly China and ASEAN countries) seeking cheaper labor. This limited the role of domestic plants to the development of trial products and the launch of mass production lines. Yet, recent data show a rise in labor cost in those emerging countries and increasing demand for higher-quality products. Consequently, overseas transplants have started to ask their mother factories to play the role of *teachers*, i.e., to transfer knowledge and techniques to enhance the quality and productivity of their overseas transplants. However, as the mother factories stopped mass production a long time ago, their capabilities are lost, and many of them find it difficult to take on this teaching role.

The majority of Japanese factories in our study still retain the function of leading mass production, and some have learnt to use this to their advantage. For instance, one of them has a mother factory function controlling eight overseas manufacturing bases. It has been accumulating techniques for quality and productivity improvement under the mass production system and sends experienced workers to its overseas factories to transfer knowledge and skills. It also invites workers over to Japan to provide training regarding the structural and technical characteristics of the products and the relationship between machines and molds. The resulting situation is a fruitful relationship in which both parties can learn from each other through friendly rivalry.

In brief, the Japanese factories that have managed to survive in a harsh economic environment are not those that abandoned the mass production system and specialized only in teaching, but they are what we can call *fighting mother factories*, able to retain and improve their skills under the mass production system.

3.1.5 Making Strenuous Efforts for Kaizen

A further feature of competitive factories is the ability to make strenuous efforts for kaizen through activities such as QC circles and 5S. In the organizational context, *kaizen* is defined as “corporate-wide continuous improvement activities by eliminating waste and using the employees’ wisdom, which in the long run results in major performance improvements.” Kaizen consists of small and steady continuous improvement activities and is often compared with wringing a dry cloth. It is tough and requires a lot of discipline, but, in the long run, it results in higher productivity and safety.

At one of the factories which we visited, everyone was involved in kaizen activities aimed at achieving preset goals. The factory manager made the overall corporate policy more explicit and concrete, and the team leaders further broke it down into more measurable targets. Each worker defined in detail what he/she should do to achieve the set objectives by the end of the fiscal year. Both target and progress information for each worker were posted on a wall on the shop floor, so that everyone could see the others' progress. Improvements were monitored through monthly checks, and the last 10 minutes of each working day were reserved for kaizen activities.

One of our major findings was that factories in rural areas carry out kaizen activities not for the sake of improved profits, following a capitalist logic, but to maintain and increase employment opportunities for the surrounding communities. During our interviews, we repeatedly heard respondents say that they must create and protect jobs through kaizen, which they saw as the main way to become a competitive factory. One factory was putting in a great deal of effort to develop a good relationship with the local community. Even during the recession, it continued to hire human resources from nearby schools and provided steady feedback to the schools about the employment situation.

3.1.6 Overcoming Unbalanced Age Distribution

Its human resources obviously play a critical role in the survival of a factory. However, most of our interviewees mentioned difficulties related to human resources development and transfer of skills caused by unbalanced age distribution in the workforce. This problem was significant in seven of the eight factories analyzed, which had a small percentage of employees in their 20s and 30s and a majority of older workers, aged 40 and upward. In order to prepare for the future, measures were being introduced to support development of younger employees and technology succession.

For instance, one factory identified the core skills need for succession, then set goals for each skill, and paired up experienced and younger employees for skills transmission. By using a skill matrix and identifying clear deadlines, succession proceeded slowly but surely. Another factory set up a so-called *craft workshop*, where employees could meet and develop machines and tools. As a result, within a year, approximately 6000 kaizen ideas were realized, and the gap between older and younger workers, caused by lack of vertical communication, was effectively bridged. Another difficulty was detected in the fact that emails and other digital tools make communication within the firm weaker. Many respondents recognized the importance of analog communication and of revitalizing QC circles as well as organizing social gatherings. These findings suggest that analog tools facilitate vertical communication and, as a result, play a critical role in successful skill succession.

3.1.7 Summary of the Exploratory Case Study

Our case study explored the main reasons behind the success of some Japanese factories in the electrical and electronics sectors. The six common aspects able to improve competitiveness are as follows:

1. Rather than waiting for new business to be provided by the headquarters, successful factories carried out sales and promotion activities targeting their own headquarters or other companies in order to secure new opportunities.
2. Competitive firms gathered various functions in one spot, including research and development, production, and sales. As a result, they were able to expand the scope of their activities and achieve improvements in product development, cost reductions, and higher productivity.
3. Successful factories managed to combine human skills with machines, and continuous interaction led to enhanced capabilities.
4. A winning strategy was to maintain and improve their skills under the mass production system.
5. Even during particularly tough periods, kaizen efforts were never neglected. Their main purpose was to maintain strong connections with the local community.
6. In addition, competitive factories effectively dealt with aspects such as development of young employees and difficult succession of skills caused by unbalanced age distribution.

It ought to be noted that these results refer to best-performing factories and do not reflect the overall situation of the electrical and electronic sectors in Japan. In the next section, we will examine some more general characteristics of Japanese factories using data from our large-scale survey.

3.2 *Survey Results*

3.2.1 Basic Data on Respondents

Of the factories that took part in the survey, 17% had sales below 10 billion yen, 69.1% had sales between 10 billion and 100 billion yen, and 13.8% had sales exceeding 100 billion yen. Also, 53.2% had a gross profit margin of 0–9%, and 22.8% were in the 10–19% bracket. These two groups comprised 75% of respondents overall.

Respondents were asked to specify their business domain (multiple choice allowed). Thirty one percent selected “heavy electric/industrial-use electric equipment”, followed by “electronic components” (23%), “others” (19%), “telecommunications/computers” (17%), “home appliances and audio/video equipment” (12%), and “industrial-use electronic equipment for measuring” (9%).

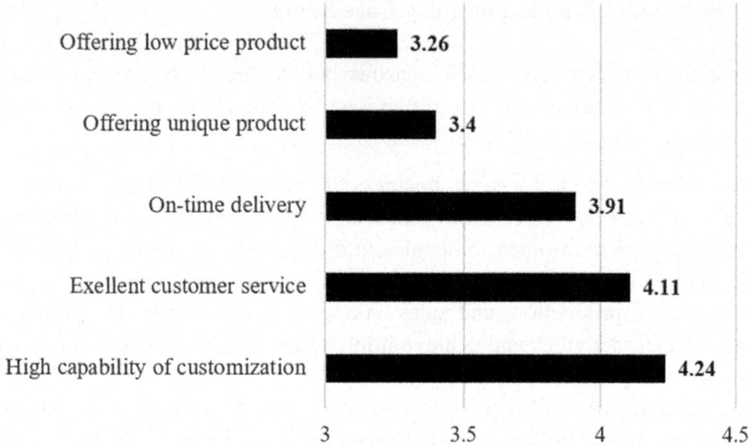

Note : N=87, Based on the mean value of the answers.

Fig. 1 Market competitiveness

3.2.2 Competitiveness of Japanese Factories

Questions on factory competitiveness pertained to (1) market competitiveness and (2) productive competitiveness. These two concepts are based on Fujimoto (2003).

Market Competitiveness As for each factory's market competitiveness in its primary business—in order to explore the reasons for higher ratings by customers vis-à-vis rival firms, both domestic and international—respondents were asked whether they agreed with the five items below. Each statement could be assessed on a five-point scale, with one point corresponding to "completely disagree," three points indicating "neutral," and five points meaning "completely agree." As shown in Fig. 1, market competitiveness derives mainly from excellent customer response capabilities, such as "Responsiveness to customization demands from customers" and "Strong customer service."

MC1: Delivery of low-price products through cost cutting
MC2: Delivery of unique products/services
MC3: Accuracy/shortness of delivery times
MC4: Responsiveness to customization demands from customers
MC5: Strong customer service

Productive Competitiveness Productive competitiveness is particularly difficult to measure for two reasons. First, productivity and cost figures are typically not publicly available. Second, even if they can be collected through a questionnaire, it is meaningless to make a simple comparison based on said data when a company may produce many different products, as is often the case in the electrical industry. Accordingly, relative measures, such as comparisons with other plants making similar products, are more useful than specific and absolute performance figures. Since the responding companies have accurate benchmarking metrics to compare Japanese factories with their overseas transplants (e.g., in China/ASEAN), relative performance evaluations of Japanese and overseas factories producing the same products were used in our survey.

We inquired about commonly used QCDF (quality, cost, delivery, and flexibility) measures, as well as the ability to develop manufacturing and production engineering for long-term competitiveness (Bhamu and Sangwan 2014; Fujimoto 1999, 2003, 2012; Holweg 2007; Holweg and Pil 2004; MacDuffie and Pil 1995; MacDuffie et al. 1996; Schroeder and Flynn 2001; Shah and Ward 2003, 2007; Turkulainen and Ketokivi 2012; Womack et al. 1990). For the most recent year (from October 2012 to the end of September 2013), respondents were asked to compare their own factory with rival plants within the same company producing the same types of products, focusing on the ten items below. Each statement was assessed on a five-point scale: one point indicated that the rival site was superior, three points corresponded to a balanced situation, and five points meant that the respondents' own location was superior. Rival factories included in the comparison were mostly located in China (approximately 62%), followed by ASEAN countries (approximately 13%). Table 1 presents key statistics for these items. As Fig. 2 shows, Japanese factories (i.e., the mother factories) were on average superior to their overseas transplants in all aspects except production cost (e.g., labor cost).

A1: Customer satisfaction
A2: External defect ratio
A3: Production cost (e.g., labor, materials)
A4: Productivity (e.g., man-hours per item manufactured)
A5: Delivery (e.g., number of days from customer order to delivery)
A6: Flexibility in altering production models/volumes
A7: Number of new product introductions (per year)
A8: Development of unique production technology (e.g., production refinement, faster processing)
A9: Rapid mass production start-up
A10: New product proposal and development

3.2.3 Human Resources in Japanese Factories (Study A)

Figure 3 shows the age composition of full-time employees (end of October 2013, $N = 55$). The distribution is clearly quite unbalanced, with few younger workers in their 20s and early 30s and many middle-aged workers in their 40s or older. Here are the major issues arising from this situation: (1) As the workers' age increases, the already relatively high cost of labor will continue to grow in the future; (2) The next-generation shop floor managers (currently in their 30s) do not have sufficient experience to teach younger workers; (3) The number of veteran workers (55 and above), who also happen to

Table 1 Mean, SD, and correlations of the ten items of productive competitiveness

	Mean	SD	A1	A2	A3	A4	A5	A6	A7	A8	A9
A1: Customer satisfaction	4.07	.966									
A2: External defect ratio	4.16	.987	.876**								
A3: Production cost	1.83	1.195	−.161	−.331**							
A4: Productivity	3.90	1.014	.585**	.496**	.016						
A5: Delivery	3.85	.833	.567**	.496**	.033	.515**					
A6: Flexibility	3.78	1.504	.418**	.263*	.150	.430**	.521**				
A7: # of new product introductions	3.59	1.295	.303**	.241*	.048	.259*	.213*	.238*			
A8: Develop. of unique production tech.	4.22	1.066	.530**	.544**	−.243*	.411**	.210*	.304**	.537**		
A9: Mass production start-up	4.17	.979	.484**	.345**	−.037	.333**	.428**	.357**	.528**	.577**	
A10: New product proposal and develop.	4.13	1.086	.418**	.427**	−.171	.397**	.321**	.319**	.671**	.800**	.654**

$N = 82$, ** $p < 0.01$, *$p < 0.05$, + $p < 0.1$

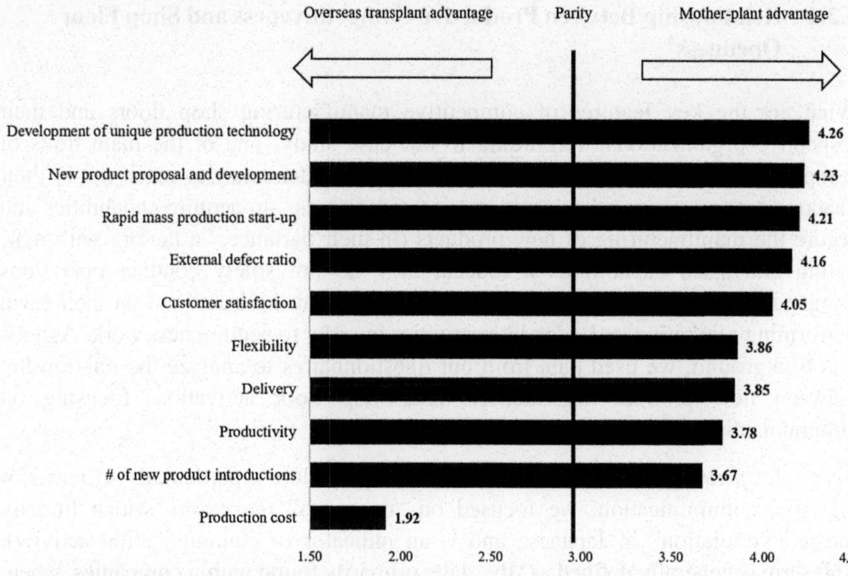

Note: N=73, Based on the mean value of the answers.

Fig. 2 Productive competitiveness

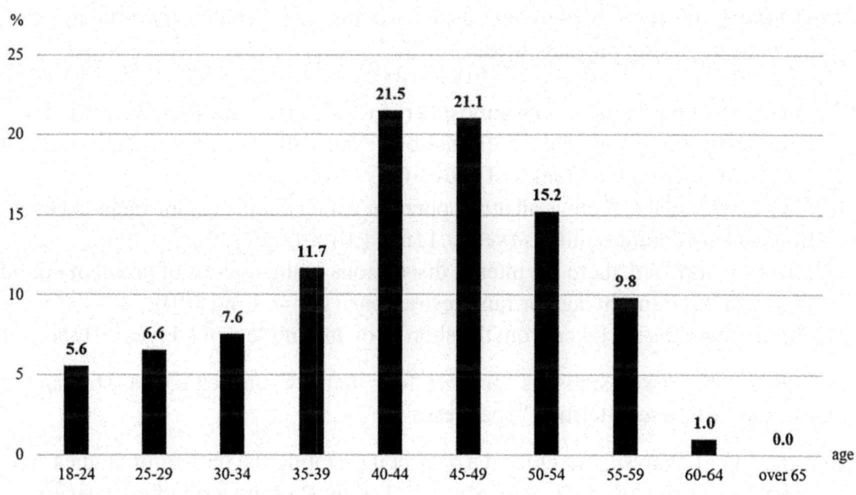

Note: N=55, Based on the mean value of the answers.

Fig. 3 Age composition of regular workers

be the greatest source of advanced skills, will decline; (4) A mechanism must be created to transfer the advanced skills of those with 20 or 30 years of experience to younger workers. These findings are also supported by our case study.

3.2.4 Relationship Between Productive Competitiveness and Shop Floor Openness[1]

What are the key features of competitive manufacturing shop floors and their respective organizations? According to our case study, one of the main traits of competitive factories with active shop floors is their tendency to reach out to their headquarters—or even other companies in some cases—to acquire capabilities and secure the manufacturing of new products (in their parlance, "a factory with marketing and sales capabilities"). Indeed, they do not solely conduct operations assigned to them by their respective headquarters but think and act on their own, performing marketing and sales-like activities in order to acquire new work. Against this background, we used data from our questionnaires to analyze the relationship between new product introductions and shop floor activation, focusing on communication.

Items of Openness As an indicator of active shop floors, particularly in terms of effective communication, we focused on "openness" (*kazetōshi*, which literally means "ventilation" in Japanese and is an indicator of communication activity). This term is generally defined as "the state, primarily found within companies, where there is an environment of free communication of opinions, regardless of hierarchical rank, and where there is a strong flow of communication and information sharing."[2] Indeed, Study C included four items related to the general climate in the workplace,[3] and respondents were asked whether they agreed (two-point, yes-no scale) with the following statements:

C1: In my workplace, there is an atmosphere in which it is easy to make work-related requests or have work-related discussions even with those not directly linked to my chain of command (yes = 1, no = 0).
C2: In my workplace, there is an atmosphere in which people can be included even if they have different opinions (yes = 1, no = 0).
C3: In my workplace, there are intense discussions in the interest of problem-solving that are not bound by age or rank restrictions (yes = 1, no = 0).
C4: Many opinions are taken from the shop floor and implemented (yes = 1, no = 0).

Table 2 shows key statistics for these four items. Cronbach's α is 0.69, and the simple sum of these is termed "openness."

Results In our analysis, we created a dataset integrating the data from Studies A and C under the framework defined by Study B. For most of the individual businesses in our sample, we obtained 1 response to Study A, 3 to Study B (meaning that

---

[1] A part of this section was originally published in Inamizu and Fukuzawa (2017).

[2] Translated from the entry found at http://www.practical-japanese.com (search date 20 February 2017).

[3] In selecting items describing workplace openness, we also considered indices related to "organizational weight," which is said to indicate poor openness (Numagami et al. 2007).

Table 2 Mean, SD, and correlations of the items of openness

	Mean	SD	C2	C3	C4
C1	.59	.493			
C2	.54	.499	.51**		
C3	.43	.495	.30**	.33**	
C4	.43	.496	.29**	.36**	.33**

$N = 3021$, ** $p < 0.01$, *$p < 0.05$, + $p < 0.1$

Table 3 Correlations between competitiveness and openness

	Correlation with
A1: Customer satisfaction	.12
A2: External defect ratio	.14
A3: Production cost	−.14
A4: Productivity	.06
A5: Delivery	.13
A6: Flexibility	.17
A7: # of new product introductions	.28**
A8: Develop. of unique production technology	.28*
A9: Mass production start-up	.16
A10: New product proposal and development	.35**

$N = 82$, ** $p < 0.01$, *$p < 0.05$, + $p < 0.1$

3 workplaces responded per business), and 30 to Study C (meaning that 10 workers responded per workplace, with 3 workplaces for each business). When inserting data from Studies A and C into Study B's dataset, the same data from Study A were used for all the workplaces belonging to the same business. As for Study C's data, mean values were calculated for each workplace.

After creating this integrated dataset, we analyzed only the workplace (shop floor) from top-selling businesses. We did this because it allowed us to conduct an analysis limited to each factory's core workplace. Moreover, when creating this dataset, since we inserted the same data from Study A for workplace at the same factory, we were able to sift through and choose one workplace at each factory.

The process outlined above yielded a total of 81 factories. The sample included 38 cases of 1 workplace selected per factory, 31 cases of 2 workplaces selected per factory, and 12 cases of 3 workplaces selected per factory. Therefore, the overall number of workplaces was 136 ($38 \times 1 + 31 \times 2 + 12 \times 3$). Workplaces with missing values were excluded from the analysis.

Table 3 presents the correlation coefficients between openness and each item of competitiveness. There is no visible correlation between QCDF and openness. However, for items A7 (number of new product introductions) and A10 (new product proposal/development), the correlations are positive and significant, with a 1% significance threshold.

We then concentrated on A7 and A10 for the next steps of our analysis. Figures 4 and 5 show the respective frequency distributions of the two items. About half of the

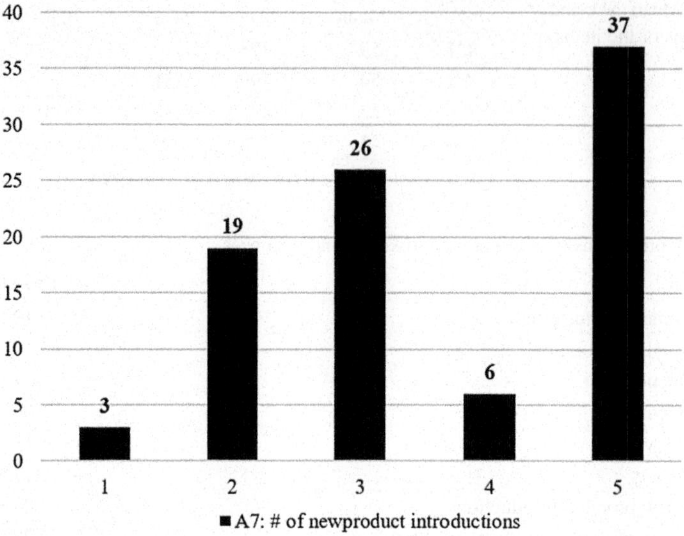

Fig. 4 Frequency distribution of A7

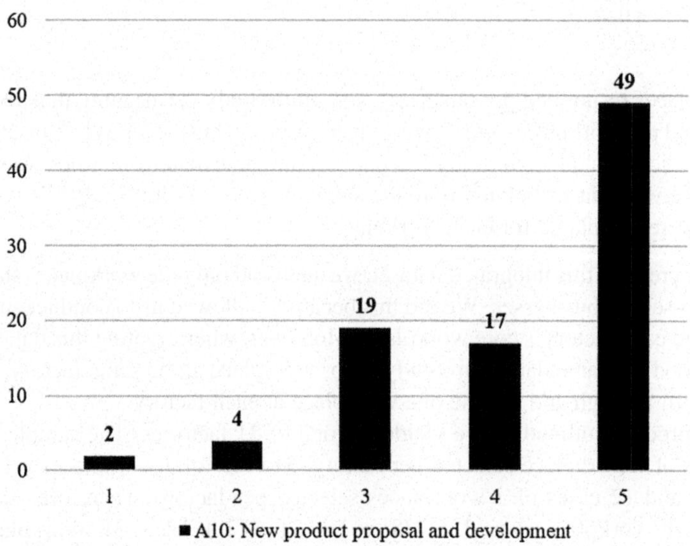

Fig. 5 Frequency distribution of A10

Table 4 Differences in mean openness

	Lower group			Upper group			
	N	Mean	SD	N	Mean	SD	*t* value
A7: # of new product introductions	50	1.65	.724	33	2.01	.676	2.278*
A10: New product proposal and development	39	1.57	.656	44	1.99	.727	2.807**

** $p < 0.01$, *$p < 0.05$, + $p < 0.1$

Fig. 6 Relationships among A7, A10, and openness

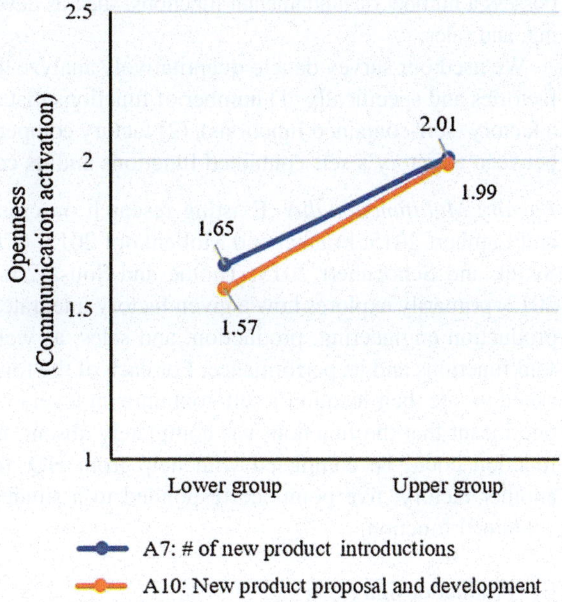

total respondents assigned a score of 5 to both items, with a two-peak distribution. We then split the respondents based on the scores given: those who had given a score of 5 were assigned to the "upper" group, while scores ranging from 1 to 4 were classified in the "lower" group.

Table 4 and Fig. 6 show the results of the analysis aimed at detecting differences in mean openness indices between the upper group and lower group, for items A7 and A10. The results of the *t*-test for A7 (number of new product introductions) revealed a significant difference ($t = 2.278$, $p < 0.05$). A similar *t*-test for A10 (new product proposal/development) also yielded a significant difference ($t = 2.807$, $p < 0.01$). Thus, we can conclude that the openness indices related to A7 and A10 are higher for the upper groups than for their respective lower groups.

3.2.5 Relationship Between Factory Competitiveness and Multifunctionality[4]

As explained earlier, competitive factories in the electrical and electronics industries have multiple functions beyond production, including design and production engineering, and have been able to survive by finding their own new business opportunities and altering product lines and operation structures based on superior productivity and design capabilities. Indeed, in order to secure new work autonomously, a production site must possess a number of fundamental functions, such as development, production engineering, and sales.

We used our survey data to quantitatively analyze the traits of highly competitive factories and specifically (1) number of functions that can be fully completed within a factory (self-contained functions), (2) factory competitiveness, and (3) relationship between a factory's self-contained functions and its competitiveness.

Factory Multifunctionality Existing research on cross-functional integration (Enz and Lambert 2015; Frankel and Mollenkopf 2015; O'Leary-Kelly and Flores 2002; Swink and Schoenherr 2015; Thomé and Sousa 2016; Turkulainen and Ketokivi 2012) primarily explores how a given factory integrates the four functions of design, production engineering, production, and sales, as well as the relationship between said functions and its performance. For each of the four functions, respondents were asked to rate their location's self-containment level on a five-point scale. A score of one meant that the function was completely absent; three points indicated that the function could be completed with help from HQ, the relevant business unit, or another factory; five points corresponded to a situation of fully operational, self-contained function.

F1: Product design
F2: Process design (production engineering)
F3: Production
F4: Sales

Figure 7 displays self-containment data for each of the four aforementioned functions ($N = 93$) across the factories analyzed. As for the design functionality, approximately 54% of the factories are completely or almost independent (scores of 4 or 5), whereas around 20% cannot perform this function at all (scores of 1 or 2), regardless of whether the production functionality was moved overseas. About 90% of the factories have fairly or highly self-contained production functionality (scores of 4 or 5).

Furthermore, we considered each individual factory and assigned a value of 1 to the top scores for each function (score of 5, can be completed independently) and a value of 0 to all other responses (scores between 1 and 4). We then calculated the totals for the four functions, obtaining values between 0 and 4 corresponding to the numbers of self-contained functions. Their distribution ($N = 93$) is shown in Fig. 8.

[4]A part of this section was originally published in Fukuzawa and Inamizu (2017).

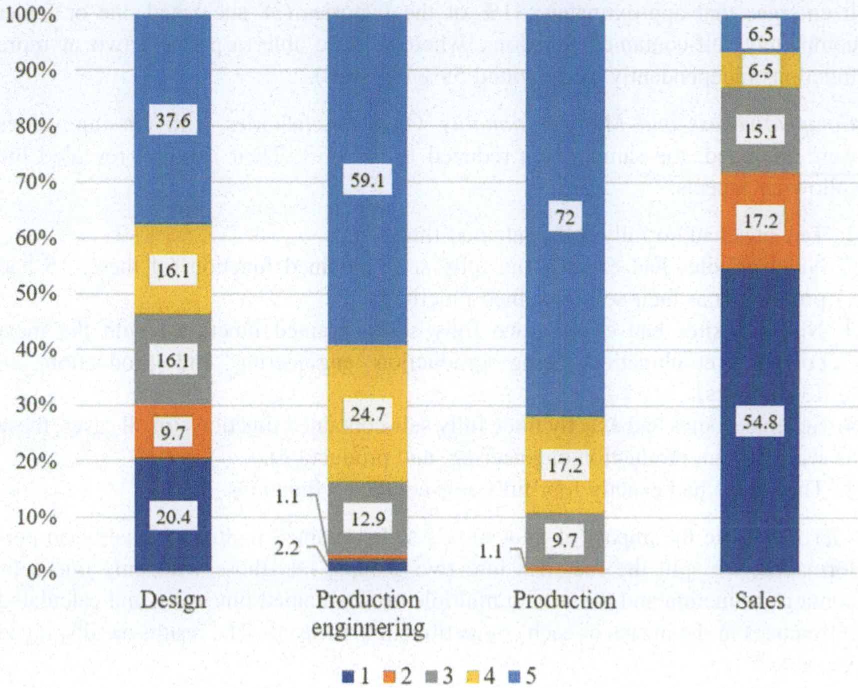

Fig. 7 Self-containment of functions

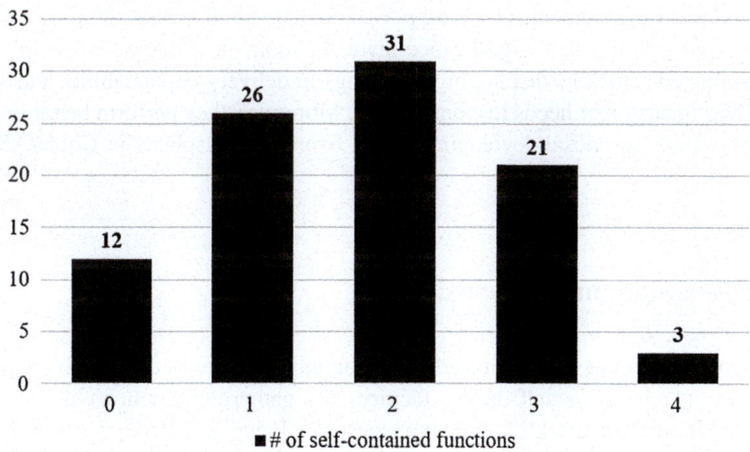

Fig. 8 Frequency distribution of multifunctionality

It emerges that approximately 41% of the factories (38 sites) had one or fewer completely self-contained functions, whereas those able to perform two or more functions independently were around 59% (55 sites).

Competitiveness and Multifunctionality Once the factories with missing values were excluded, the sample was reduced to 68 sites. Their analysis revealed the following aspects:

1. Ten sites had no fully self-contained functions.
2. Nineteen sites had exactly one fully self-contained function (of these, 15 had production as their self-contained function).
3. Nineteen sites had exactly two fully self-contained functions (with the most common combination being production engineering and production, in 13 factories).
4. Seventeen sites had exactly three fully self-contained functions (in all cases, these were design, production engineering, and production).
5. Three sites had exactly four fully self-contained functions.

To evaluate the impact of a location's self-contained multifunctionality on performance, we split the factories into two groups, i.e., those with only one self-contained function and those with multiple self-contained functions, and calculated differences in the means of each competitiveness measure. The results are displayed in Table 5.

The most significant gaps in competitiveness indices are found in MC3 (delivery dependability and speed, $t = 2.36$, $p < 0.5$), MC4 (customer needs responsiveness, $t = 3.6$, $p < 0.1$), MC5 (customer service, $t = 2.1$, $p < 0.5$), and A10 (new product proposal and development, $t = 2.10$, $p < 0.05$). The trend is also significant for A4 (productivity) at the 10% significance level. In short, factories with multiple fully self-contained functions display higher ratings in delivery dependability and speed, as well as in customer needs responsiveness. Moreover, they perform better in terms of new product proposal/development than overseas transplants in China/ASEAN countries.

4 Discussion and Conclusions

Our exploratory case study revealed some common features of competitive Japanese factories: (1) being a self-driven factory, (2) gathering functions in one spot, (3) combining human skills and machines, (4) being a *fighting mother factory*, (5) making strenuous efforts for kaizen, and (6) overcoming unbalanced age distribution.

Our survey shed light on shop floor level competitiveness, organization of manufacturing sites, and shop floor communication across Japanese electrical and electronics factories. Firstly, *Study A* empirically confirmed that, on average, Japanese electrical and electronics factories are superior to their overseas transplants in

Table 5 Differences in means of performance

	Multifunctionality				
	≦1		2≦		
	($N = 29$)		($N = 39$)		
	Mean	SD	Mean	SD	t
MC1: Low price	3.00	0.85	3.36	1.18	1.46
MC2: Uniqueness of product	3.24	0.79	3.54	1.12	1.28
MC3: Delivery dependability and speed	3.62	0.82	4.10	0.85	2.36*
MC4: Customer needs responsiveness	3.76	0.87	4.44	0.60	3.60**
MC5: Customer service	3.79	0.90	4.23	0.78	2.10*
A1: Customer satisfaction	4.03	0.87	4.23	0.90	0.91
A2: External defect ratio	4.21	0.90	4.31	0.89	0.46
A3: Production cost	1.90	1.18	1.77	0.99	−0.47
A4: Productivity	3.59	1.09	4.03	0.99	1.72 +
A5: Delivery	3.79	0.73	4.00	0.89	1.06
A6: Flexibility	3.90	0.90	3.92	1.20	0.10
A7: # of new product introductions	3.59	1.30	3.85	1.39	0.79
A8: Develop. of unique production technology	4.24	0.91	4.46	0.85	1.01
A9: Mass production start-up	4.14	1.06	4.33	0.96	0.78
A10: New product proposal and development	4.02	0.98	4.51	0.85	2.10*

$N = 68$, ** $p < 0.01$, *$p < 0.05$, + $p < 0.1$

all measures of competitiveness except production cost. Research on "reshoring" (Ellram et al. 2013; Gray et al. 2013; Shih 2014; Bailey and de Propris 2014) suggests that American and European companies need to reconfigure their global manufacturing locations especially because of increasing labor costs at their Asian production sites in the 2010s. Recent studies have identified a number of factors that influence the choice of manufacturing locations at the global level, such as cost (labor and material), access to local markets, logistics, currency volatility, country risk, and government trade policies (Ellram et al. 2013). Nonetheless, the issue of productivity stemming from the lean production system has not been sufficiently discussed in these lines of research, and it appears that Japan might be in the same situation as American and European companies for what concerns reshoring.

On this point, Study A suggests that, when deciding to reconfigure their firm's global manufacturing bases, managers need to distinguish between the advantages of *factor cost* (e.g., labor and material cost) in their overseas transplants and the advantages of *productivity* deriving from high-level shop floor capabilities. If wages in emerging countries continue to rise in the future, it will be reasonable to manufacture both in the highly productive Japanese plants (remade in Japan) and at overseas transplants by aggressively transferring lean production know-how from the mother plants in Japan. In order to do so, it is essential to build organizational capabilities for continuous productivity improvements.

Secondly, to analyze the relationship between factory competitiveness and organizational openness at the shop floor level, we used data from our questionnaires

(Study A and Study C). Despite not revealing a correlation between openness and QCDF, the results show a positive and significant relationship between openness and "Number of new product introductions" and "New product proposal/development." Indeed, as our case study suggested, proposing new products to the headquarters and/or to other companies—and the consequent frequent introduction of new products—is likely to lead to the activation of shop floor organizations. The proposal, adoption, and resulting introductions of new products by factory managers lead to additional work. Moreover, the outsourcing of production to overseas factories causes increasingly significant reductions in production volumes at Japanese factories. Given this situation, additional work would help to activate factory shop floors. As it is new work, it would also be unconstrained by the usual processing cycles that the factory is accustomed to, which would have a positive impact on workplace communication not only with other individuals but also with other company divisions. Consequently, this would probably create a stronger feeling of openness in the workplace.

Caution is necessary when considering the relationship between the activation of shop floor organizational communication and QCDF. Workplaces with good levels of openness are also proactively engaged in kaizen-style initiatives, which are generally seen as contributing to better performance in QCDF and mass production start-up (Koike 2012, 2013; Koike et al. 2001). However, the survey conducted as part of this research was unable to find significant relationships among these items. Normally, the start of production of new products would have a negative effect on QCDF,[5] and, even in a high-performing factory with an active shop floor, frequent introductions of new products might represent an obstacle in terms of maintaining superior QCDF measures. Furthermore, active communication alone might not be sufficient to improve QCDF, and some additional factors are likely to be critical for the attainment of those results. All things considered, while the proposal and subsequent manufacture of new products can have a negative impact on QCDF measures, this does improve shop floor organizational openness and can be seen as a rather effective way to activate shop floor communication.

Thirdly, our study suggests that steadily developing manufacturing competences and design/development capabilities at the same time is critical for preserving the superiority of the Japanese factory shop floor in the future. In other words, the key functions of design, production, and marketing/sales should all be present at the same manufacturing site. As Fujimoto et al. (2015) and Nakazawa et al. (2016) explain, to maintain and improve their competitiveness, Japanese companies should aggressively pursue market creation and new business opportunities by taking advantage of their strengths in manufacturing and design activities.

[5]In Table 1, we can see an overall positive and significant relationship with A1–A10. However, there does not appear to be a significant correlation among A7 (number of new product introductions), A2 (external defect ratio), and A3 (production cost). Moreover, no significant correlation emerges among A10 (new product proposal and development), A3 (production cost), and A5 (delivery).

Regarding multifunctionality, domestic production sites that are completely autonomous in multiple functions (design/manufacturing/production) display the following traits: (1) they are superior to their rivals in delivery dependability and speed and in responding to market needs, and (2) they are better than comparable factories in China/ASEAN countries at proposing and developing new products. Our results indicate that, to increase global market competitiveness and retain superiority in new product proposals and development, production sites should be fully self-contained in multiple functions. Multifunctionality at a production site may result in higher costs but also in better dynamic capability (e.g., Eisenhardt and Martin 2000; Helfat et al. 2007; Teece et al. 1997; Zollo and Winter 2002; Fukuzawa 2015b), which is needed to respond to environmental changes in order to create new markets via new product proposals/development.

Finally, our evidence supports the idea that the current level of competitiveness of Japanese factories relies on a large number of skilled workers. However, skill transfer from factory veterans to younger workers is crucial to sustain the competitiveness of Japanese shop floors in the future.

Acknowledgments The authors deeply appreciated the cooperation of the Japanese Electrical Electronic and Information Union.

References

Bailey D, de Propris L (2014) Manufacturing reshoring and its limits: the UK automotive case. Camb J Reg Econ Soc 7(3):379–395

Bhamu J, Sangwan SK (2014) Lean manufacturing: literature review and research issues. Int J Oper Prod Manag 34(7):876–940

Eisenhardt KM, Martin JA (2000) Dynamic capabilities: what are they? Strateg Manag J 21 (10–11):1105–1121

Ellram LM et al (2013) Offshoring and reshoring: an update on the manufacturing location decision. J Supply Chain Manag 49(2):14–22

Enz MG, Lambert DM (2015) Measuring the financial benefits of cross-functional integration influences management's behavior. J Bus Logist 36(1):25–48

Frankel R, Mollenkopf DA (2015) Cross-functional integration revisited: exploring the conceptual elephant. J Bus Logist 36(1):18–24

Fujimoto T (1999) The evolution of a manufacturing system at Toyota. Oxford University Press, New York

Fujimoto T (2003) Nouryoku kouchiku kyousou (Capability building competition). Chuou Kouron Shinsha, Tokyo (in Japanese)

Fujimoto T (2012) Evolution theory of production systems. Ann Bus Adm Sci 11:25–44. https://doi.org/10.7880/abas.11.25

Fujimoto T et al (2015) Nihonno Monodukuri no Sokodikara. Toyo Keizai Shinpousha, Tokyo (in Japanese)

Fukuzawa M (2015a) Competitiveness of Japanese electric and electronics factories. Ann Bus Adm Sci 14:217–230. https://doi.org/10.7880/abas.14.217

Fukuzawa M (2015b) Dynamic capability as fashion. Ann Bus Adm Sci 14:83–96. https://doi.org/10.7880/abas.14.83
Fukuzawa M, Inamizu N (2017) Multi-functional factories: survey study of Japanese electric and electronics companies. Ann Bus Adm Sci 16(5):229. https://doi.org/10.7880/abas.0170421a
Gray JV et al (2013) The reshoring phenomenon: what supply chain academics ought to know and should do. J Supply Chain Manag 49(2):27–33
Helfat CE et al (eds) (2007) Dynamic capabilities: understanding strategic change in organizations. Blackwell Publishing, Malden
Holweg M (2007) The genealogy of lean production. J Oper Manag 25(2):785–805
Holweg M, Pil FK (2004) The second century: reconnecting customer and value chain through build-to-order. MIT Press, Cambridge, MA
Inamizu N, Fukuzawa M (2017) New product introductions and activation of shop floor organizational communication. Ann Bus Adm Sci 16(4):149–163. https://doi.org/10.7880/abas.0170313a
Koike K (2012) Kouhinshitsu Nihon no kigen: Hatsugen suru shokuba wa koushite umareta (The origin of Japanese high quality: how were the workplaces to voice opinions born?). Nihon Keizai Shinbunsha Shuppan, Tokyo (in Japanese)
Koike K (2013) Tsuyoi genba no tanjo: Toyota sougi ga umidashita kyodo no ronri (The birth of a strong Genba (shop floor): cooperative logic generated from Toyota dispute). Nihon Keizai Shinbunsha Shuppan, Tokyo (in Japanese)
Koike K et al (2001) Monozukuri no ginou: Jidosha sangyo no shokuba de (Skills for manufacturing: at workplaces in the automotive industry). Toyo Keizai Shipo Sha, Tokyo (in Japanese)
MacDuffie JP, Pil FK (1995) The international assembly plant study: philosophical and methodological issues. In: Babson S (ed) Lean work: empowerment and exploitation in the global auto industry. Wayne State University Press, Detroit, pp 181–196
MacDuffie JP et al (1996) Product variety and manufacturing performance: evidence from the international automotive assembly plant study. Manag Sci 42(3):350–369
Nakazawa T et al (2016) Monodukuri no Hangeki. Chikuma Shobo, Tokyo (in Japanese)
Numagami T et al (2007) Soshiki no omosa (Organizational weight). Nihon Keizai Shinbun Shuppansha, Tokyo (in Japanese)
O'Leary-Kelly SW, Flores BE (2002) The integration of manufacturing and marketing/sales decisions: impact on organizational performance. J Oper Manag 20(3):221–240
Schroeder RG, Flynn BB (eds) (2001) High performance manufacturing: global perspectives. Wiley, New York
Shah R, Ward PT (2003) Lean manufacturing: context, practice bundles, and performance. J Oper Manag 21(2):129–149
Shah R, Ward PT (2007) Defining and developing measures of lean production. J Oper Manag 25(4):785–805
Shih WC (2014) What it takes to reshore manufacturing successfully. MIT Sloan Manag Rev 56(1):55–62
Shintaku J et al (2014) Denki sangyo no genbaryoku chosa: Nihon no genba no kyosoryoku wo sasaeru shokuba (Genba-capabilities in electric and electronics companies: a survey of factories, group leaders and workers). Akamon Management Review 13(10):371–406 (in Japanese). https://doi.org/10.14955/amr.131001
Swink M, Schoenherr T (2015) The effects of cross-functional integration on profitability, process efficiency, and asset productivity. J Bus Logist 36(1):69–87
Teece DJ et al (1997) Dynamic capabilities and strategic management. Strateg Manag J 18(7):509–533
Thomé AT, Sousa R (2016) Design-manufacturing integration and manufacturing complexity: a contingency investigation of job rotation and co-location. Int J Oper Prod Manag 36(10):1090–1114

Turkulainen V, Ketokivi M (2012) Cross-functional integration and performance: what are the real benefits? Int J Oper Prod Manag 32(4):447–467

Womack JP et al (1990) The machine that changed the world: based on the Massachusetts Institute of Technology 5-million dollar 5-year study on the future of the automobile. Rawson Associates, New York

Zollo M, Winter SG (2002) Deliberate learning and the evolution of dynamic capabilities. Organ Sci 13(3):339–351

The Effectiveness of Group Leaders in the Lean Production System: Time Study and Agent-Based Model of Leaders' Behavior

Nobuyuki Inamizu and Mitsuhiro Fukuzawa

Abstract This study adopts the mixed approach of simulation modeling and field study to investigate the effectiveness of leaders engaged in the role of line chief in lean production. The simulation results show that there is a tipping point at which help from a line chief drastically improves production flow when the workload is relatively high and problem occurrence is rare. However, the efficient production flow thus achieved is vulnerable to slight changes in problem occurrence. The field study conducted in a Japanese automobile assembly plant, especially the time study of group leaders, supports the results of the simulation.

1 Introduction

> A manager said to me, 'What are you doing? Can you supervise the production line even when you are working on the line?' Then, I answered 'What do you mean? We cannot keep up with production if I don't work on the line. You should see this immediately.' (Shimizu 2005, p 574)

In Shimizu (2005), a group leader from Toyota recalls the years of spectacular economic growth in the 1960s. Group leaders were not supposed to work on the production line, but Toyota experienced explosive car sales growth, and its production could not keep up with demand.

Since the 1990s, many studies have examined the *work team* in the production line or on the shop floor. First, the *Toyota Production System* and the *Lean Production System* as sources of strength for Japanese automobile companies attracted widespread attention (Berggren 1992, 1993; DankBaar 1997; Fujimoto 1999; MacDuffie et al. 1996; Ohno 1988; Womack et al. 1990). Then, there was a

N. Inamizu (✉)
Graduate School of Economics, University of Tokyo, Tokyo, Japan
e-mail: inamizu@e.u-tokyo.ac.jp

M. Fukuzawa
Faculty of Economics, Seikei University, Tokyo, Japan

T. Fujimoto, F. Ikuine (eds.), *Industrial Competitiveness and Design Evolution*, Evolutionary Economics and Social Complexity Science 12,
https://doi.org/10.1007/978-4-431-55145-4_8

surge in interest in the team as a new system of work organization (Appelbaum and Batt 1994; Benders and van Hootegem 1999; Cutcher-Gershenfeld et al. 1994; MacDuffie 1995, 2002; Morita 2008).

In general, *lean production* is a production system that responds well to changes in market demand by implementing changes in production volume and variety. It also achieves high productivity and high quality by eliminating intermediate stock and surplus labor (buffer or *muda*) (Ohno 1988; Monden 2012). Lean production, as defined above, is achieved by applying methods that follow four key concepts: (1) just-in-time delivery (producing the necessary units, in the necessary quantities, at the necessary time), (2) autonomation (an autonomous defects control mechanism), (3) flexible workforce (adjusting the number of workers to changes in demand), and (4) creative thinking or inventive ideas (capitalizing on workers' suggestions) (Monden 2012, pp. 7–8). These concepts and methods are similar to the operational constructs of lean production proposed by Shah and Ward (2007).

The lean work team is required to perform a number of tasks, in keeping with the four key concepts. First, it has to (1) perform the standard work, (2) stop defective products from moving to the next process stage (by stopping the production line and immediately dealing with the defective products), and (3) be responsible for the multistage process. Second, the work team has to (1) determine the real causes of the detected anomalies, (2) revise the standard work through *kaizen* activities and *quality circles* (QC), and (3) develop multiskilled workers through job rotation (Monden 2012). Existing studies have pointed out that team leaders play a central role in lean production (Benders and van Hootegem 1999; Berggren 1992; Delbridge et al. 2000).

However, these studies have also noted that not enough research has been performed on the role and behavior of team leaders. For example, Delbridge et al. (2000) states that "...The role of the team leader within teams is typically neglected or treated as unproblematic in debates over team effectiveness or organizational performance. These findings suggest that this is a major oversight" (p 1475). We cannot adequately understand lean production without understanding the details of the leaders' roles and behavior.

In general, there is vertical division of labor in a lean or TPS team. Each worker is required to correctly complete standardized work. Team leaders (TLs) lead teams of approximately five workers, and their tasks include addressing processing line stoppages and other malfunctions (the so-called line chief role), as well as substituting for absent workers. A group leader (GL) usually leads four teams and concentrates on managerial tasks rather than working directly on the line. The main tasks of a GL include managing the production site daily, revising operations standards, and overseeing kaizen activities (Fujimoto 1999).

This chapter focuses specifically on the line chief position, which involves engaging in production line management and maintaining production flow. If abnormalities arise along the production line, the line chief works to solve them. In general, TLs are required to take on the line chief role, which directly affects productivity and quality and enables the work team to adapt to changes in short-term production volume (Koike 2005; Mishina 1995). Yet, despite its importance, the detailed reality of a line chief's role has not been researched systematically.

Here, this issue is addressed through a mixed approach involving simulation modeling and field study. Field studies can describe the actual and detailed behavior of leaders, but their weakness is the poor generalizability of results. In this regard, computer simulation can test the effects of many variables and possibly clarify thc boundary conditions of field study results. Nonetheless, computer simulation has a weakness in the external validity of its results, which can be overcome through the ample empirical data obtained from field studies. Moreover, compared to the mathematical modeling approach, computer simulation can more easily deal with the nonlinear phenomena often seen in organizations and societies (Davis et al. 2007; Harrison et al. 2007).

In particular, we adopt cellular automata (a simple agent-based approach) to develop our model. Traffic flow analysis through cellular automata has advanced over the past decade. Recently, these traffic flow models have been actively applied to production management studies, especially lean production and TPS (Nishinari 2006). Therefore, we developed a simple production flow model using cellular automata to examine the effectiveness of the line chief's behavior. We also conducted a field study in a Japanese automobile assembly plant that exhibited strength in maintaining high productivity, despite the tough economic environment subsequent to the 2008 financial crisis. This high productivity was supported by leaders' (line chiefs') behavior. We then filmed the behavior of its leaders as line chiefs for an hour each day over several days and analyzed the videos through a *time study*.

In the following sections, we will first describe the simulation model, called *lean flow model*, and present its results. Next, we will illustrate the field study results, focusing on the time study. Finally, we will discuss the effectiveness of the leaders' behavior in relation to their roles and responsibilities in lean production or TPS.

2 Lean Flow Model

The primary element of lean production or TPS is "just what is needed, only when needed, only in the quantity needed." A later process takes what is needed from an earlier process, and an earlier process produces only the kind and quantity of parts used by the later process. The second element of the production system is "quality must be built into each process." Workers are careful not to let any defects through to the next process. Whenever a problem occurs, the worker calls the leader through the so-called andon and asks him or her to deal with it. In their relief role, leaders keep the production line flowing smoothly. These elements can easily be described by *Rule 184* of a simple cellular automata model and its revised model (Nishinari 2006; Schiff 2008). The lean flow model is based on these models.

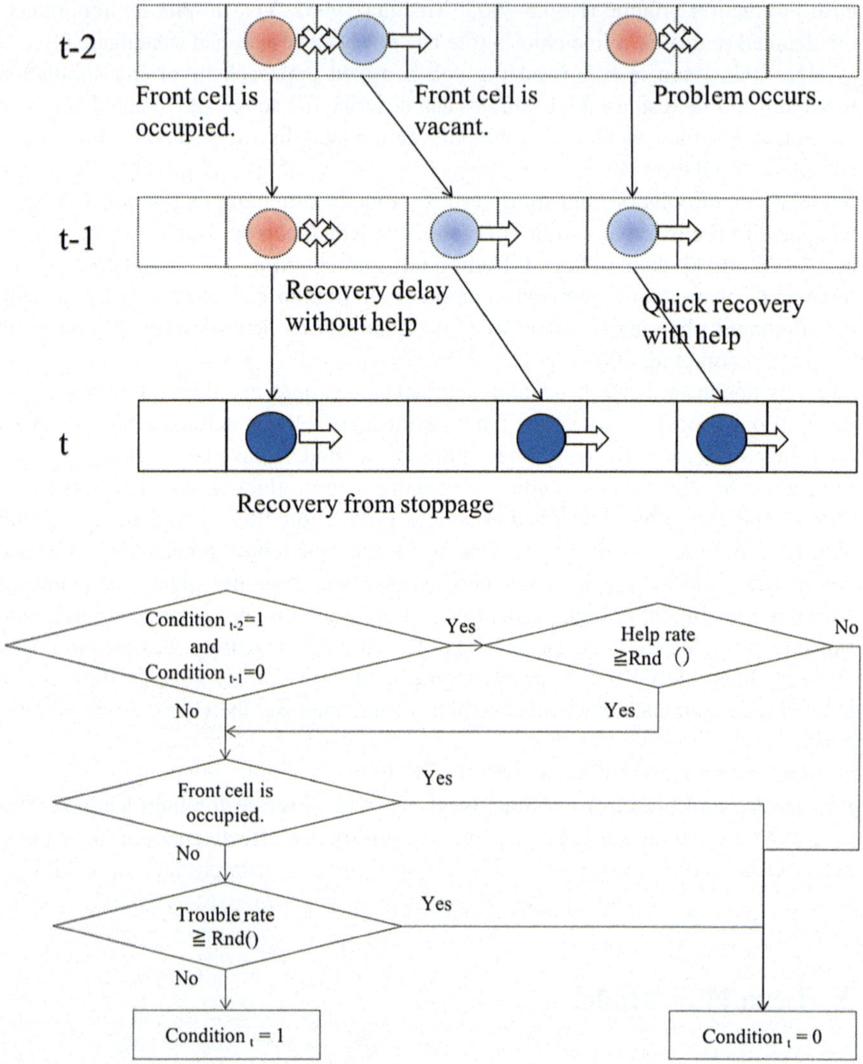

Fig. 1 Lean flow model

2.1 Initial Setting

TASKs are placed randomly in the space where cells are lined up side by side (see Fig. 1).

2.2 *Rules of TASK Flow*

Only Rule 1 is executed at the first period ($t = 1$), only Rule 2 is executed at the second period ($t = 2$), and only Rule 3 is executed after the third period ($t > 2$). Conditiont = 0 holds if the TASK stops at period *t*. $\text{Condition}_t = 1$ holds if the TASK moves at period *t*.

Rule 1: Stop (Conditiont = 0) if the front cell is occupied. Otherwise, stop ($\text{Condition}_t = 0$) if a problem occurs. The probability of a problem occurring depends on the "problem rate." Otherwise, move to ($\text{Condition}_t = 1$).
Rule 2: Remain stopped ($\text{Condition}_t = 0$) if the condition at the previous period was "stop" ($\text{Condition}_{t\text{-}1} = 0$) and if there is no help from a line chief. The probability of help depends on the "help rate." Otherwise, execute "Rule 1."
Rule 3: Execute "Rule 1" if the condition was "stop" two periods before ($\text{Condition}_{t\text{-}2} = 0$). Otherwise, execute "Rule 2."

Rule 3 means that, without help, the stopped but not helped TASKs cannot move until two periods have passed. Rule 2 means that the helped TASKs take only one period to move. As shown in these rules, the line chief's role is built into the Lean Flow Model.

2.3 *Input Parameters and Time Structures*

We ran simulations, fixing the number of cells at 100 and changing each input parameter as follows: (1) number of TASKs = 20–60, (2) problem rate = 0.0001–0.001, and (3) help rate = 0.0–1.0. One period passes when all TASKs execute the above rules. One simulation run consists of 500 periods, and we conducted 50 simulation runs in each setting.

2.4 *Output Data*

"Flow" as output data is calculated using the following equation:

$$\text{Flow} = \text{number of TASKs whose Condition } t \text{ is one}/\text{total number of TASKs}$$

Flow at each period from 401 to 500 is averaged. Then, the mean value of each of the 50 simulation runs under the same setting is also averaged. The obtained value is "average flow." This value indicates the degree of smooth production flow, which is a performance measure of lean production or TPS.

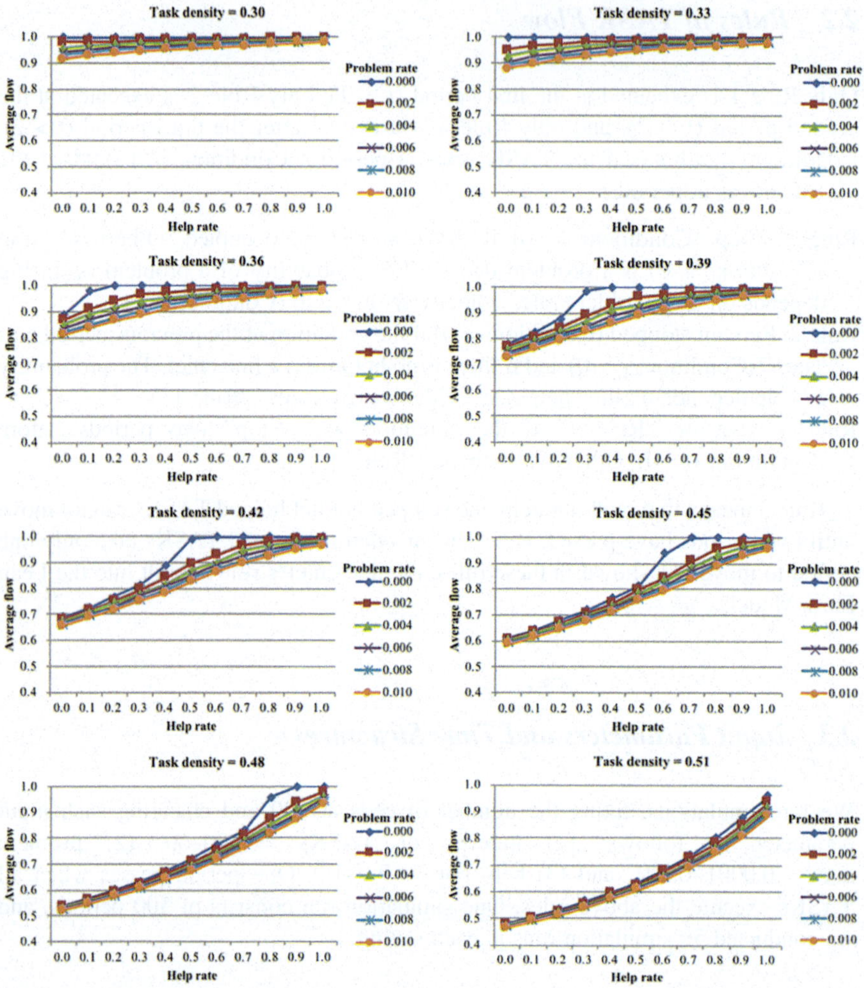

Fig. 2 Simulation results of the lean flow model

2.5 *Simulation Results of the Lean Flow Model*

We drew graphs showing the help rate as the abscissa and average flow as the ordinate (Fig. 2). Each line displays the results under each problem rate. The help rate indicates the degree of help from the line chiefs, while the problem rate indicates the possibility of problem occurrence and work delays. We drew similar graphs for each task density (total number of TASKs/number of cells). The task density, which indicates the assembly line workload, is high when the line speed is high or when there are many elementary tasks occurring at each station.

When the task density is low (task density = 0.30 and 0.33), the average flow is relatively high (more than 0.90) regardless of the help rate and problem rate. Even when the front TASK stops and takes two periods (a long time) to be resolved, the rear TASK does not need to stop because TASKs take place at a distance from each other, which results in sufficient buffer time.

As the task density increases (task density = 0.36, 0.39, and 0.42), the average task flow decreases significantly when there is less help from line chiefs. However, the average flow is greater than 0.9 when the help rate is 1.0.

Interestingly, when the problem rate is 0.0 and task density is 0.39, a slight increase in the help rate, from 0.2 to 0.3, drastically improves the average task flow from 0.88 to 0.99; that is, there is a tipping point at which the rate of help by line chiefs improves the average task flow. At the same time, however, a slight increase in the problem rate sharply worsens the average task flow. For example, when the task density is 0.39 and the help rate is 0.3, an increase in the problem rate from 0.000 to 0.002 decreases the average flow from 0.98 to 0.89; that is, help from line chiefs facilitates a drastic improvement in productivity to an extremely high level, but, at the same time, such help amplifies the instability of production flow.

As the task density increases, the tipping point moves upward. For example, when the task density is 0.36, the tipping point of the help rate is between 0.0 and 0.1, but it moves to around 0.2 under a task density of 0.39 and to around 0.4 under a task density of 0.42. On the whole, when the average task flow is approximately 0.85, additional help drastically improves production flow.

Finally, when the task density is extremely high (task density = 0.51), the tipping point cannot be seen. It is certain that the higher the help rate, the higher the average task flow, but the maximum average task flow is only 0.96.

The simulation results can be summarized as follows:

1. When the task density is moderate (less than 1/3), good production flow can be achieved without help from leaders.
2. When the task density is extremely high (more than 1/2), help from leaders improves production flow, but these improvements are rather minimal. Thus, the task density needs to be lowered (by either enhancing the capability of the production line or facilitating a slowdown in the line speed).
3. When the task density is relatively high and the likelihood of problem occurrence is extremely low, there is a tipping point at which help drastically improves production flow. Nevertheless, in this scenario, the achieved efficient flow is extremely unstable. A slight decrease in help or a slight increase in problem occurrence significantly worsens production flow.

There are two key aspects regarding the third result. First, under certain conditions (a relatively high workload and an extremely low problem occurrence), help from line chiefs drastically improves production flow. Second, this helpful behavior can amplify fluctuations in production flow.

3 Field Study in an Automobile Assembly Plant[1]

3.1 Research Setting

Our research setting was Plant Y of Japanese automobile Company X, which produces small quantities of various models of cars. Since 2007, this plant has been producing three models on one line. We focused on "Group α (chassis process)" in Plant Y, which consisted of 12 workers, two TLs, and one GL. Each team leader managed six workers.

Group α had 12 workstations, divided into two zones. Zone A comprised "a2-2 (tube installation)," "a3 (tank mounting)," "a6-1/16-2 (engine mounting)," and "a9-2 (muffler installation)." Zone B included "b2-1 (bumper installation)," "b3-1 (right-liner installation)," "b5-1 (skid plate installation)," "b5-2 (left-liner installation)," "b7-1 (normal load tightening)," "b8-1 (right-tire installation)," and "b8-2 (left-tire installation)." In managing each zone, the two team leaders were responsible for responding to and solving problems occurring in their respective zones. Thus, the GL was in charge of two team leaders and 12 workers. Two more stations, "a5" and "a7," were added according to 4WD specifications. This group had a U-shaped production line (Fig. 3).

We conducted our research from February to June 2010. We implemented 12 field studies, spending 8 h each day on our investigation. Our research team had approximately 15 members, with about five participating regularly in each field study. We examined the following factors: (1) production line work; (2) skill development of workers, TLs, and GLs; (3) production line performance (e.g., man-hours per vehicle, operational availability, defect rates, time of line stops); and (4) problems and problem-solving, specifically the details of the TLs' (and GL's) relief activities and teamwork.

In particular, we focused on the behavior of the GLs acting as line chiefs (see Fig. 3). We videotaped them for 1 h, twice daily, over several days and analyzed them through a time study. For this analysis, we divided 1 h into 7 major categories (and 27 detailed categories) and compared the proportions of each behavioral category at each time period.

3.2 Performance Analysis

3.2.1 Variations in Production Volume

At the beginning of the 2000s, Plant Y's average monthly production volume was approximately 6000 units. This number grew consistently until 2008, when it exceeded its 10,000-unit production peak of 2007. During this period, production

[1] A part of this section was originally published in Inamizu et al. (2014).

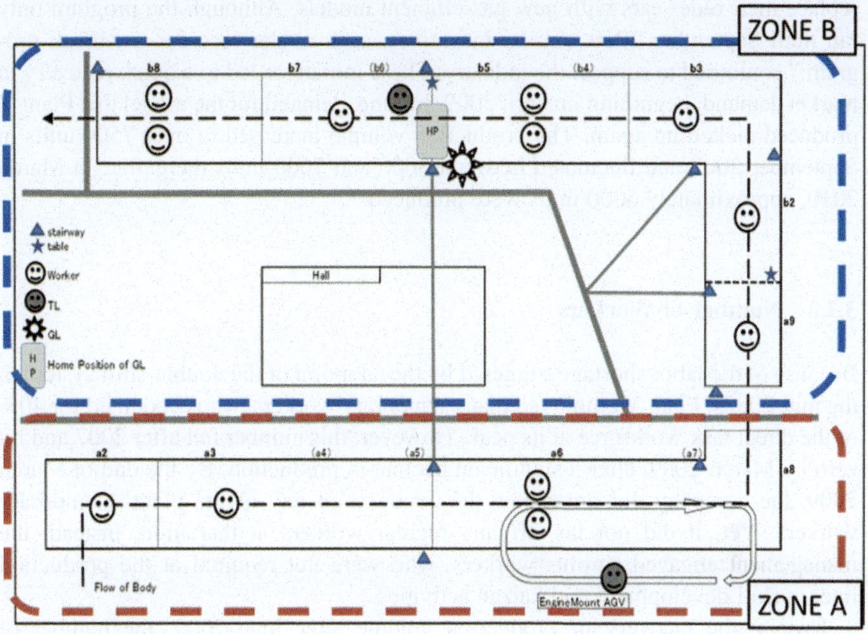

Fig. 3 Layout of Group α

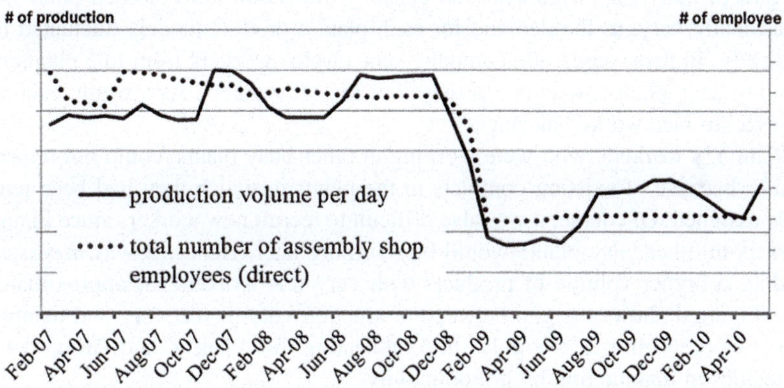

Fig. 4 Gap between production volume recovery and manpower

went from a single-shift to a double-shift system. After November 2008, the production volume dropped significantly because of the global financial crisis. Monthly production fell from 14,000 units in October 2008 to 7000 units in January 2009 and finally to 3000 units in February and March of the same year (see Fig. 4). This drastic decrease caused production to revert to a single-shift system.

In April 2009, the Japanese government introduced an eco-car subsidy as a measure to deal with the economic crisis. This policy encouraged consumers to replace their older cars with new gas-efficient models. Although the program only ran until September 2010, similar incentives, such as the "eco-car tax break program," continued to support the industry. These initiatives led to a quick recovery in market demand, beginning in April 2009. In June, demand for the model that Plant Y produced picked up again. The production volume increased to over 7500 units in September 2009 and fluctuated between 5000 and 7000 units thereafter. In March 2010, approximately 6000 units were produced.

3.2.2 Number of Workers

Because of the labor shortage triggered by the adoption of the double-shift system in the mid-2000s, Plant Y employed many temporary workers, who accounted for 40% of the direct task workforce at its peak. However, this number fell after 2007 and hit zero by March 2009, after a significant decline in production. By the end of March 2009, the company did not renew the contracts of any of the plant's temporary workers. Yet, it did not lay off any regular workers at that time. Instead, the management engaged surplus workers, who were not required at the production line, in skill development and kaizen activities.

Despite the recovery of production volume after June 2009, the number of workers, including backup and regular workers, steadily decreased. This is because the management had sent any surplus workers to assist in the company's other plants in April of that year. In general, the capacity utilization ratio of each plant would occasionally vary, as the demand for each plant's product models fluctuated independently. In these cases, the company sent surplus workers from idle plants in its group to busy plants, so as not to recruit or lay off workers. As a result, each shop floor had to face worker shortage.

Plant Y's workers, who were working in other busy plants, could not return for months because of existing contracts in the plants to which they had been temporarily assigned. Of course, it was also difficult to recruit new workers since Plant Y's workers in other busy plants would be returning later. Hence, it was necessary to produce a higher volume of products with very few workers for approximately a year. Figure 4 shows the gap between production volume recovery and manpower (direct workers) in the assembly shop, indicating that Plant Y and Group α faced difficulties in maintaining high productivity.

3.2.3 Man-Hours per Vehicle

Figure 5 shows the target, actual, and planned man-hours per vehicle from August 2008 to May 2010. It indicates the plant's vehicle production efficiency. As seen in this figure, the actual and target man-hours decreased more steeply after April 2009. This means that Plant Y and Group α had the strength and capability to maintain high productivity even in the face of a tough work environment.

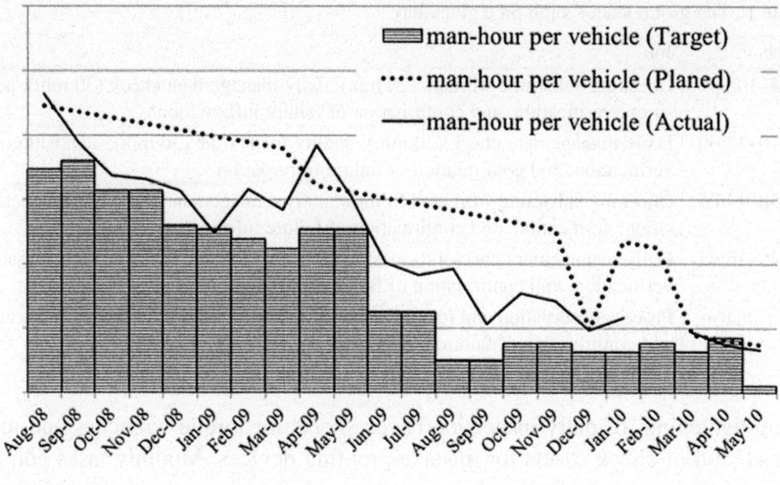

Fig. 5 Man-hours per vehicle

It is certain that the line speed picked up somewhat to respond to the increase in production volume. Yet, productivity rate (man-hours/vehicle) is affected by several other factors in a rather complex way. When the workers' skills are constant, in order to increase line speed, it is necessary to increase the number of workers and not the working hours. Therefore, productivity rate (man-hours/vehicle) will not be enhanced by increasing the line speed only. If the number of workers does not increase, frequent work delays and mistakes occur in each station, which leads to an increase in the number of working hours. As a result, the productivity rate (man-hours/vehicle) will not improve. To increase productivity, it is necessary to develop workers' skills to enable each worker to perform more tasks within the same number of working hours. By doing so, productivity will rise with the same number of better skilled workers and higher line speed (shorter working hours). However, workers' skills cannot be developed overnight. In short, it takes time to improve productivity. The question then is "What happened in Plant Y of Company X?" This will be shown by the time study concerning the GLs described in the following sections.

3.3 *Formal Jobs of GLs*

3.3.1 Basic Job of GLs

The basic job of a GL is divided into daily, weekly, and monthly tasks. Daily tasks consist in preparing a table of malfunction trends and a table of actual line stoppages, conducting analyses on the number of defect process cases, and managing workers' attendance. Weekly tasks include the preparation of an X-R control chart to monitor

Table 1 The group leader's job on a given day

Time	Job
7:55–10:00	Checking substitute workers (20 min); daily management check (30 min); acquisition, verification, and confirmation of failure information
10:10–12:00	Daily management check (20 min); quality focus time (30 min); acquisition, verification, and confirmation of failure information
12:50–14:25	Checking substitute workers (20 min); daily management check (20 min); acquisition, verification, and confirmation of failure information
14:45–16:45	Daily management check (20 min); quality focus time (30 min); acquisition, verification, and confirmation of failure information
Overtime	Process observation and maintenance of standards, etc., acquisition, verification, and confirmation of failure information

screw tightening, of daily inspection check sheets for torque wrenches and torque impact, and of check sheets for mistake-proofing devices. Monthly tasks comprise revision of the standard operating procedure, skill management of workers, and consolidation of kaizen proposals.

On any given day, the GLs' duties are listed on their respective groups' whiteboards, so that the members of each group and the supervisors can see what kinds of tasks the GLs perform at which time. Table 1 shows an example of this procedure.

Checking Substitute Workers When a substitute worker is needed, the GL decides who he/she should be and confirms the status of his or her work, while the line is in operation. A substitute worker is appointed when a worker is on leave or goes on a business trip. In addition, when a substitute worker's level of proficiency is not yet 100% for all the line operation tasks, the GL decides who should perform the operations that the substitute worker does not do proficiently or who should provide support.

Daily Management Check This involves standard tasks such as the filling out, management and analysis of data on tools and mistake-proofing devices, as well as checking malfunction occurrence data using dedicated check sheets.

Acquisition, Verification, and Confirmation of Failure Information When the production line stops because of a failure, by filling out a dedicated sheet, the GL records and analyzes data such as shutdown time, process in which the failure occurred, failure details, and measures taken. Acquisition refers to the GL's activity of information gathering by interviewing workers on specific failures occurred.

Quality Focus Time Quality focus time refers to the time needed to make the concentrated efforts required to improve quality. Quality improvement is conducted through a cycle of measures that include careful examination of standards (e.g., standard operating procedure and work instructions), observation of the process (e.g., work observation, production tools checks), danger prediction activities (e.g., presenting kaizen cases, quality circles, improvement of tasks that are difficult to perform), confirmation of effectiveness (confirmation of malfunction information, checking quality assurance items), and creation of new standards (e.g., standardization or, more specifically, revision of the standard operating procedure or work instructions).

Process Observation and Maintenance of Standards Standards refer to the following seven documents: standard operating procedure, work instructions, work procedure, standardized work combination sheets, new worker education manual, QA assessment, and changing point management. Process observation and maintenance of standards consist in confirming the details of these documents as well as improving them. Improvement of the work environment is another responsibility of the GL and involves, for example, improvement of the shelf where the standard operating procedure is stored, oil injection to the carriage wheels, organizing parts boxes and parts shelves, and replenishing parts.

Within Company X, the GLs could not use their discretion to adjust the line speed (task time). Rather, the line speed was decided at the upper level (e.g., by the plant manager), and GLs were required to maintain productivity and quality under the set line speed (task time). Moreover, an important management index in the plant was the rate of operational availability (time at which the equipment could actually run/time at which the equipment is required to run). Each group was required to achieve a certain level of operational availability.

3.3.2 Assuming the Job of Production Line Manager

A team leader's job is generally to manage a line. However, there may be an unfilled position because of a worker on leave, on a business trip, or supporting another section. In such a case, if there is no excess manpower, the team leader will substitute for the worker and join the line. Since the team leader operating as a worker cannot respond to a malfunction that occurs at other stations, the GL takes on the job of managing the line while performing his/her usual job. When the GL does not need to do this, he/she can concentrate on basic duties.

The person responsible for managing a line on a given day is called "(today's) line chief" on the shop floor. As noted above, this role is usually performed by TLs but sometimes by GLs. The key task of a line chief is to respond to malfunctions, but he/she must perform many other tasks on the line in order to keep production flowing.

Responding to Malfunctions A malfunction at Plant Y is defined as (1) a defect that does not usually occur, (2) the breakdown of a machine before the end of its operating life, (3) any single defect entering at a later stage in the process, or (4) a situation when a worker sees that something is wrong or unusual. In principle, when malfunctions occur, the workers' expected course of action is to (1) stop working, (2) notify the line chief of the malfunction, and (3) wait for the line chief's instructions. If there is a malfunction during operation, the person directly working at that station must turn on the call lamp to notify his/her supervisors.

On the other hand, the line chief's general course of action involves (1) confirming the call lamp, (2) confirming the details of the call, (3) dealing with the task error or (4) dealing with the delayed operation, and (5) confirming the details of the measures taken when the call lamp is on. Precautions to be followed are set for each step:

(1) *Confirming the call lamp*: The line chief must give priority to the current measure. Hence, even when he/she receives other calls, the line chief will not respond until the current measure is completed. In case of an emergency call, the line chief must inform the supervisors and continue to complete the current measure. In addition, as a precaution, the line chief must turn on the lamp when he/she receives an oral notification of the occurrence of a malfunction, so as not to forget it while working on another measure.
(2) *Confirming the details of the call*: When the line chief arrives at the area from where the call was made, he or she will interview the workers to verify the details of the call and then will reconfirm the facts in case something is unclear. The line chief will not turn off the lamp until he/she confirms the details of the malfunction.
(3) *Dealing with the task error*: In the case of a task error, unless there is a delay outside the area of the malfunction, the line chief must ensure that the worker continues to complete the task. The line chief will execute the correctional task.
(4) *Dealing with the delayed operation*: In the case of a delayed operation, the line chief will move the worker and ask him/her to perform the tasks for the next vehicle in order to ensure that the latter can begin during the appropriate time slot. The line chief will confirm all tasks and ensure that, after he/she has substituted a worker, the tasks related to that vehicle are performed and that all tasks are completed from the point when the delay occurred.
(5) *Confirming the details of the measures taken*: Using a relief note, the line chief will record the details of the measures taken, including the order and frame number of the vehicle. The line chief will keep the relief note record and refer to if something else happens.

3.4 Empirical Results

3.4.1 Skills Map Analysis

Table 2 shows Group α's skills map. The data indicate that 8 of its 12 workers were single-station handling, and only the TLs mastered 5 or 6 stations in their own zones. According to an interview with a manager, the GLs and TLs were considered to be the most skilled of all the group members.

Our examination of standard work sheets reveals that there were actually larger numbers of elemental tasks in each station. In addition, task time ranged from 99 to 102 s, which was somewhat shorter than the average for the plants within the

Table 2 Skill map of Group α

	a2-2	a3-2	a6-1	a6-2	a9-2	b2-1	b3-1	b5-2	b7-1	b8-1	b8-2	b5-1
RW1	100%											
RW2		100%	60%									
RW3			100%									
RW4			80%	100%								
RW5			80%		100%							
RW6						100%						
RW7							100%					
RW8										70%	100%	100%
RW9								100%	100%	100%		
RW10									100%			
RW11												100%
RW12										80%	100%	
TL1	100%	100%	100%	100%	80%							100%
TL2					100%	80%	100%	100%	100%	80%	100%	

Note: Station names such as "a2–2" are in the first row. Workers' and team leaders' names, such as RW1 and TL1, are in the first column. The percentage in each cell shows the skill level of each worker and team leader at each station. One hundred percent means that the worker was able to perform all tasks without delay. Eighty percent means that he/she was able to perform tasks but with a certain rate of delay. If the percentage is less than 80%, he/she was unable to carry out tasks by him-/herself company's group. The tasks in each process were considered to be difficult at that time. As a result, the single-station handling workers of Group α were, in fact, multiskilled.

3.4.2 Work Shifts Analysis

Table 3 presents Group α's work shifts for three typical days during our study. On 10 of the 12 days considered, the GLs worked as line chiefs in zone B. The GLs also performed line management in zone B. Moreover, almost every day, the TLs worked on the production line because workers had taken time off and there was no backup. The GLs are usually required to perform administrative activities outside the production line, but, in our study, they frequently had to engage in production line management, which is the TLs' standard job. Meanwhile, the TLs frequently had to work in the production process, which is each worker's standard job.

3.4.3 Time Study of GLs as Line Chiefs

For a close observation of the behavior of the GLs as line chiefs in zone B, we videotaped them twice a day for an hour each time and subsequently analyzed five videos made on the following dates: (A) March 17 morning, (B) March 17 afternoon, (C) April 2 afternoon, (D) April 16 morning, and (E) April 16 afternoon.

For the GLs time study, we classified their behavior as follows: (1) responding to a malfunction, (2) direct tasks (work on the production line), (3) indirect tasks (work away from the production line), (4) communication, (5) confirmation, (6) instruction, and (7) desk work (see Table 4). Note that direct and indirect tasks are performed when there are no emergency calls from workers (tasks involving an emergency call are classified into "responding to a malfunction"). Indirect tasks do not involve activities directly performed at the production line, but rather the effort required to achieve smooth production line flow (e.g., development of tools, parts modifications, and equipment repairs). Desk work refers to GL duties performed at a desk located near the production line. Communication typically includes the conversations between a GL and other members of his/her group, such as workers, team leaders, and others. Conversely, confirmation indicates a GL's monitoring activity through observation of the conditions of torque, line flow, and automobile assembly work in progress, and not conversations with assembly team members. The line chief's tasks can be essentially summarized as (1) responding to malfunctions, (2) direct tasks, and (3) indirect tasks. Communication, confirmation, instruction, and, especially, desk work are a GL's intrinsically required roles. We further divided the 7 major categories above into 27 detailed categories.

Figure 6 shows the ratio of time spent working on the seven major categories during the five videotaped sessions: (1) March 17 morning, (2) March 17 afternoon, (3) April 2 afternoon, (4) April 16 morning, and (5) April 16 afternoon. As the figure indicates, compared to March 17, the time spent on direct tasks increased drastically

Table 3 Line chiefs and work shifts

	Feb 23	March 1	March 9	March 17	April 2	April 9	April 16	April 26	April 27	May 11	May 21	June 14
Line chief (A)	TL1	TL1	TL1	TL1	TL1	TL1	GL1	TL1	TL1	TL3	TL3	GL2
Line chief (B)	GL1	GL1	GL1	GL1	GL2	GL2	GL2	TL2	GL2	GL2	GL2	STL1
a2-2	RW1	RW1	RW1	RW1	TW1	TW1	TW1	RW1	TW1	TW1	TW1	TL1
a3-2	RW2	RW2	RW2	RW2	RW2	RW2	RW2	RW2	RW2	RW2	RW2	RW2
a6-1	RW3	RW3	RW3	RW3	RW3	RW3	Tl1	RW3	RW3	RW3	RW3	RW3
a6-2	RW4	RW4	RW4	RW4	RW4	RW4	RW4	RW4	RW4	RW4	RW4	RW4
a9-2	RW5	RW5	RW5	RW5	TW2	TW2	TW2	TW2	TW2	TW2	TW2	TW2
b2-1	RW6	RW6	RW6	RW6	RW6	RW6	RW6	RW6	RW6	RW6	RW6	RW6
b3-1	RW7	RW7	RW7	RW7	RW7	RW7	RW7	RW7	RW7	RW7	RW7	RW7
b5-1	RW8	RW8	RW8	RW8	RW8	RW16	RW14·	RW14·	RW14·	RW14	RW14·	TW4
							STL1	STL1	TL2		RW9	
b5-2	RW9	RW9	RW9	RW9	RW9	STL1	RW13·	RW9	RW13·	RW13·	RW13	RW13
							RW9		STL1	RW9		
b7-1	RW10	RW10	RW10	RW10	TW3	TW3	TW3	TW3	TW3	TW3	TW3	TW3·
												RW14
b8-1	RW11	TL2	RW11	RW11	TL2	TL2	TL2	RW15	RW15	RW15	RW15	RW15
b8-2	RW12	RW12	TL2	RW12	RW12	RW9	RW12	RW12	RW12	RW12	RW12	RW12

Table 4 Twenty-seven behavioral categories of group leaders as line chiefs

	Major categories	Detailed categories
Line chief's primary roles	Responding to a malfunction	Responding to a malfunction
	Direct tasks	Direct task (engine mount)
		Direct task (center muffler)
		Direct task (undercover)
		Direct task (routine)
		Direct task (others)
		Confirmation after completing direct tasks
	Indirect tasks	Acquisition of and search for necessary parts and tools
		Correction of necessary parts
		Correction of facility and equipment
		Correction of parts conveyor
		Writing of memos
		Managing *kanban*
Group leader's primary roles	Communication	Telephone
		Verbal communication with those in managerial positions
		Verbal communication with workers
		Verbal communication with others
	Confirmation	Confirmation of installation and equipment
		Confirmation of vehicles
		Confirmation of workers' tasks
		Confirmation of the status of the line
		Extracting torque values
	Instruction	Instruction
	Deskwork, etc.	Desk work
		Acquisition of materials for deskwork
		Waiting times
		Others

Note: Time is also movement time

on April 2 and 16, although the time spent on responding to malfunctions did not vary greatly across the five time periods. Relatively less time was spent on desk work on April 2 and 16, while on the morning of March 17, this activity took up around 50% of the total time.

Figure 7 shows the detailed categories for direct tasks, to account for changes in the composition ratio shown in Fig. 6. The increase in time spent on direct tasks was attributed to "direct task (routine)" on April 2 and "direct task (engine mount)" on April 16. The definition "direct task (routine)" refers to the following situation: when a worker's skills are insufficient, a work delay occurs at a certain rate (e.g., every five units); the line chief must thus determine in advance the type and timing of relief

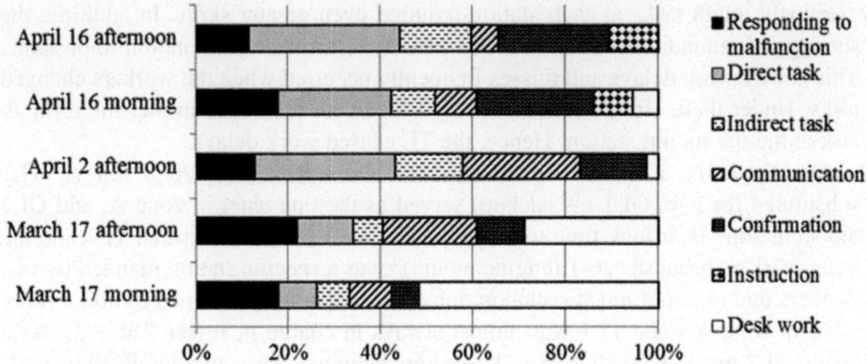

Fig. 6 Time ratios for the seven major categories

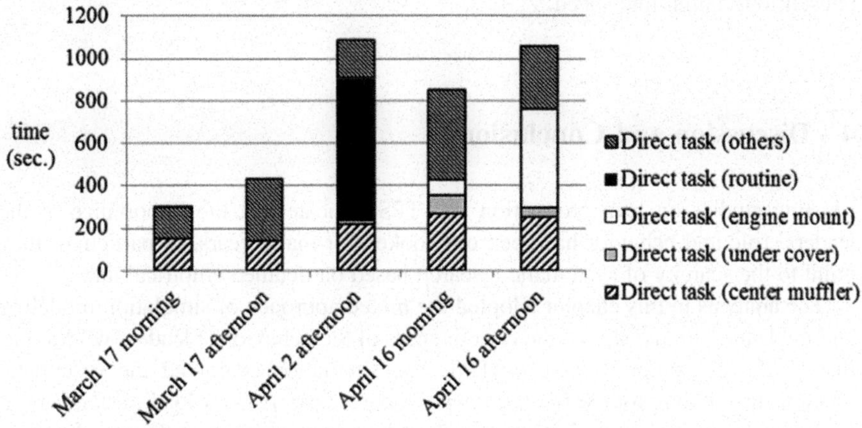

Fig. 7 Breakdown of direct tasks

work required and also perform routine relief work periodically without an emergency call. "Direct task (engine mount)" occurs at the engine mount stations ("a6–1" and "a6–2").

Our observations clarified why few direct tasks occurred on March 17 and why "direct task (routine)" on April 2 and "direct task (engine mount)" on April 16 increased. On March 17, proficient full-time employees performed all processes (see Table 3), and the workers' skill levels were all at 100% proficiency (see Table 2). The GL acting as line chief spent much of his time doing desk work as there were few human-caused errors.

On April 2, a worker in "b8–1" took a day off, and TL2 substituted for him at his station. Accordingly, GL2 served as the line chief in zone B (see Table 3). However, TL2's skill level for "b8–1" was 80%, which caused work delays every five or six units (see Table 2). As a result, GL2 had to periodically help TL2 in "b-8," thus increasing his "direct task (routine)." Even though the TLs were originally skilled, extremely tough tasks at each station required even greater skills. In addition, the shortage of redundant workers and the difficult tasks made job rotation impossible. This is why work delays and misses frequently occurred when the workers changed tasks. Under these circumstances, it was difficult for the TL to master the difficult task at the tire mount station. Hence, the TL caused work delays.

On April 16, a worker in "a6–1" was absent due to sudden illness. TL1 substituted for him, GL1 (as backup) served as the line chief in zone A, and GL2 did so in zone B. It took time to develop effective teamwork between TL1 and the other workers because "a6-1 (engine mount)" was a specific station manned by two workers that required much collaboration. GL1 was not accustomed to doing relief work in zone A since TL1 was almost always in charge of it (see Table 3). As a result, GL2 had to help GL1 and TL1 despite being in charge of zone B. Even a TL who is supposed to be skilled can cause work delays unless he/she works at a given station every day. This is ultimately because it takes a long time to acclimatize oneself to the high line speed.

4 Discussion and Conclusions

Existing studies on lean production and TPS indicate that the importance of the leaders' role and behavior has been overlooked for many years. In particular, they point to the scarcity of systematic research based on detailed empirical data.

The analysis in this chapter adopted the mixed approach of simulation modeling and field study to investigate the effectiveness of the behavior of leaders working as line chiefs in lean production or TPS. The simulation examined the systematic relationships among four variables: (1) production flow, (2) task load (task density), (3) problem occurrence, and (4) frequency of help from line chiefs. The results show that there is a tipping point at which help from a line chief drastically improves production flow when workload is moderately high and problem occurrence is rare. The efficient production flow thus achieved, however, is vulnerable to slight changes in problem occurrence. This tipping point would not have been discovered without computer simulation modeling.

Our field study supported the simulation results. The workers and leaders in Plant Y were forced to do heavy work due to a wide gap between production recovery and manpower decline. In such a severe environment (high line speed and many tasks for each station), even the TLs, who are usually skilled workers, made mistakes and caused delays. When the TLs were not required to substitute for absent workers along the production line, the subject group achieved extremely efficient production flow with very minor help from the GL. However, when the TLs substituted for workers who were away, production flow became drastically unstable, and the GL had to spend much time helping the TLs.

We can conclude that the help behavior of line chiefs (acting as leaders in lean production or TPS) is much more effective when task load is relatively high and problem occurrence is extremely rare. However, there is still a weakness in the efficiency of production flow.

This study did not focus on leader-led kaizen activities. In the future, on the basis of this study's simulation model, we will develop a model where the problem rate, or the number of cells, changes through learning mechanisms. Of course, we need to conduct a more detailed field research on kaizen activities and the leaders' other roles and behavior. This study just begins to shed light on the actual behavior and effectiveness of leaders in lean production or TPS.

References

Appelbaum E, Batt R (1994) The new American workplace: transforming work systems in the United States. ILR Press, Ithaca

Benders J, Van Hootegem G (1999) Teams and their context: moving the team discussion beyond existing dichotomies. J Manag Stud 36(5):609–628

Berggren C (1992) Alternatives to lean production. ILR Press, Ithaca

Berggren C (1993) Lean production: the end of history? Work Employ Soc 7(2):163–188

Cutcher-Gershenfeld J et al (1994) Japanese team-based work systems in North America: explaining the diversity. Calif Manag Rev 37(1):42–64

Dankbaar B (1997) Lean production: denial, confirmation or extension of sociotechnical systems design? Hum Relat 50(5):567–584

Davis JP et al (2007) Developing theory through simulation methods. Acad Manag Rev 32 (2):480–499

Delbridge R et al (2000) Shop floor responsibilities under lean teamworking. Hum Relat 53 (11):1459–1479

Fujimoto T (1999) The evolution of a manufacturing system at Toyota. Oxford University Press, New York

Harrison R et al (2007) Simulation modeling in organizational and management research. Acad Manag Rev 32(4):1229–1245

Inamizu N et al (2014) Group leaders and teamwork in the over-lean production system. J Organ Chang Manag 27(2):188–205

Koike K (2005) Shigotono Keizai Gaku, 3rd ed. (The economics of work in Japan). Toyo Keizai Shinpousha, Tokyo (in Japanese)

MacDuffie JP (1995) Human resource bundles and manufacturing performance-organizational logic and flexible production systems in the world auto industry. Ind Labor Relat Rev 48 (2):197–221

MacDuffie JP (2002) Leaning toward teams. In: Kochan TA, Lipsky DB (eds) Negotiations and change: from the workplace to society. Cornell University Press, Ithaca, pp 94–116

MacDuffie JP et al (1996) Product variety and manufacturing performance: evidence from the international automotive assembly plant study. Manag Sci 42(3):350–369

Mishina K (1995) Toyota motor manufacturing, USA, Inc. HBS Case Number 9-693-019. Harvard Business School, Boston

Monden Y (2012) Toyota production system: an integrated approach to just-in-time, 4th edn. CRC Press, Boca Raton

Morita M (2008) Chimu sagyouhoushiki no tenkai (The development of teamwork). Chikura Shobo, Tokyo (in Japanese)

Nishinari K (2006) Jutai Gaku (Jamology). Sinchosha, Tokyo (in Japanese)

Ohno T (1988) Toyota production system: beyond large-scale production. Productivity Press, Cambridge
Schiff JL (2008) Cellular automata: a discrete view of the world. Wiley, Hoboken
Shah R, Ward PT (2007) Defining and developing measures of lean production. J Oper Manag 25 (2):785–805
Shimizu K (2005) Genbakanrisha ga kataru Toyota no genbakanri: Genbakanrisha no koujutsu kiroku (Shop management at Toyota: Explained by shop managers). Okayama Daigaku Keizaigakkai Zasshi 36(4):203–222 (in Japanese)
Womack JP et al (1990) The machine that changed the world: based on the Massachusetts institute of technology 5-million dollar 5-year study on the future of the automobile. Rawson Associates, New York

The Diversity and Reality of Kaizen in Toyota

Shumpei Iwao

Abstract In the existing literature, *kaizen* (continuous improvement) has often been conceived of as the accumulation of similarly small, mutually independent, incremental process innovations that are conducted by workers, work teams, and their leaders. This chapter attempts to observe continuous improvement in a certain factory for a certain period and to show the diversity and reality of kaizen in Toyota. Through longitudinal observations, seven case studies are examined, showing that (1) kaizen consists of a series of innovations with various scales, such as number of stakeholders, amount of investment, and economic outcomes (e.g., cost reduction effect); (2) kaizen sometimes induces small changes in product design and affects the organizational activities of production design as a small-scale product innovation; and (3) kaizen activities sometimes influence other kaizen activities. With regard to these characteristics of kaizen, this study implies that (4) kaizen management needs organizational design. For example, in Toyota's case, not only work teams but also product/process design engineers contribute to kaizen, and shop floor engineers play a vital role in coordinating between shop floors and engineering departments on the basis of the *staff-in-line structure* of organizations.

1 Introduction

Previous chapters have discussed the dynamic capabilities/abilities of Toyota and other manufacturing firms in Japan. For example, chapter "The Effectiveness of Group Leaders in the Lean Production System: Time Study and Agent-Based Model of Leaders' Behavior" deals with small-scale kaizen by work teams and discusses the basic levels of kaizen abilities. This chapter intends to show the variety and reality of

---

This chapter originally appeared as an article in *Evolutionary and Institutional Economics Review (EIER)* Vol. 14, No. 1 (Iwao 2017).

S. Iwao (✉)
Faculty of Economics, Meiji Gakuin University, Tokyo, Japan
e-mail: iwao@eco.meijigakuin.ac.jp

T. Fujimoto, F. Ikuine (eds.), *Industrial Competitiveness and Design Evolution*,
Evolutionary Economics and Social Complexity Science 12,
https://doi.org/10.1007/978-4-431-55145-4_9

continuous improvements (*kaizen*[1]) through a detailed field study of a leading factory. Particularly, by means of case studies at Toyota's Takaoka plant, we aim to reexamine the conventional notion of kaizen as a series of mutually independent, incremental process innovations driven by operators or work teams. In addition, we explore organizational design for promoting kaizen outcomes based on an innovation perspective.

Until now, kaizen has been regarded as one of the crucial factors in the pursuit of industrial competitiveness indices, such as productivity, manufacturing quality, lead time, and flexibility, in the automobile industry, as well as other industries (Imai 1986; Winter 2003; Anand et al. 2009; Fujimoto 2014; Iwao 2015; Iwao and Marinov 2018).

The existing literature presents a roughly sketched notion of kaizen. It is sometimes recognized as including small but numerous changes in routines written in standard operating procedures (SOP) and small design improvements to artifacts for production, such as production equipment, tools, and software, or time/motion studies (Imai 1986). Kaizen is also regarded as relying on process innovations that improve task efficiency (Anand et al. 2009) or mutually independent activities repeatedly leading to incremental progress, which are sometimes called *continuous innovation*[2] (Boer and Gertsen 2003). Researchers further indicate that innovative activities such as kaizen involve organizational decision-making and problem-solving processes (Thompson 1965; Myers and Marquis 1969; Fujimoto 1999). In addition, because kaizen consists of numerous decision-making/problem-solving activities implemented by workers and their leaders, it is depicted as a different type of innovation than that of research and development divisions (Imai 1986; Bessant and Caffyn 1997). In particular, some studies suggest that the main contributors to kaizen as incremental process innovations are workers/operators and their leaders (Imai 1986; Lindberg and Berger 1997; Bhuiyan and Baghel 2005). In summary, past academic research describes kaizen as consisting of numerous small incremental innovations that (1) have little variability in scale/size, (2) change the way in which products are made and are categorized as process innovations, (3) are mutually independent and have no interaction with other kaizen activities, and (4) are implemented mainly by workers, work teams, and work team/group leaders.

However, some issues still need to be explored: Does all kaizen in a firm only consist of small-scale activities? Are kaizen activities related to each other? Are workers and their leaders the only key players in kaizen? The present chapter will try to answer these questions.

---

[1]This article uses the word *kaizen* as an uncountable noun.

[2]Boer and Gertsen (2003) define continuous innovation as process and product innovation, but they also think of continuous innovation as relatively small and numerous changes.

2 Methods

As mentioned above, previous studies do not seem to provide a full analysis of the conventional notion of kaizen, and, from an empirical point of view, it might be possible to gain more insights by describing and investigating the reality of kaizen in certain firms.

Hence, to observe the detailed interactions of kaizen activities and reveal the reality of this type of innovation, a fact-finding field survey is necessary (Barratt et al. 2011; Eisenhardt 1989). Moreover, our focus is on research questions such as "Are there interactions between small- and large-scale innovations?" and "How can we manage them?", and the best way to approach them is through the case study methodology, as it is suited to answering "why" and "how" questions (Yin 1994). Additionally, we intend to build some new hypotheses (Eisenhardt 1989).

For this purpose, the author participated in kaizen activities at the Toyota Motor factory in Takaoka from August 27, 2013, to September 21, 2013. Seven cases of kaizen were recorded through direct observation and by interviewing participants. The choice fell on Toyota because many studies consider it the leading company in kaizen (Shingo 1981; Ohno 1988; Womack et al. 1990), and Takaoka[3] is a standard factory comprising five units: press and welding, painting, molding, assembly and testing, and quality control. Moreover, the author conducted unstructured 2 h follow-up interviews with factory floor engineers from the body shop at the Takaoka plant on January 31, 2014, February 13, 2014, August 29, 2014, and June 24, 2016, as well as a 2 h interview with Katsuaki Watanabe, Toyota's former president. After these interviews, the collected data were checked and confirmed by the interviewees.

The data cover each kaizen case from its start, when the need for improvement was detected, until its conclusion, when improvement effects were measured.[4] The cases are arranged in ascending order according to the number of man-hours required for coordination, while the scopes of coordination were calculated based on sections and man-hours needed for coordination, and in-house development hours were eliminated.

3 Analytical Framework

From the literature survey presented above, we see that kaizen is described as "accumulation of small incremental process innovations that have little variability in scale" and "mutually independent activities that are contributed to mainly by workers/operators and their leaders in job shops." To further analyze the field data,

[3]Womack et al. (1990) also studies Takaoka.

[4]Following D'Adderio (2011), an organizational routine change was deemed to have occurred when an explicit operating procedure was altered.

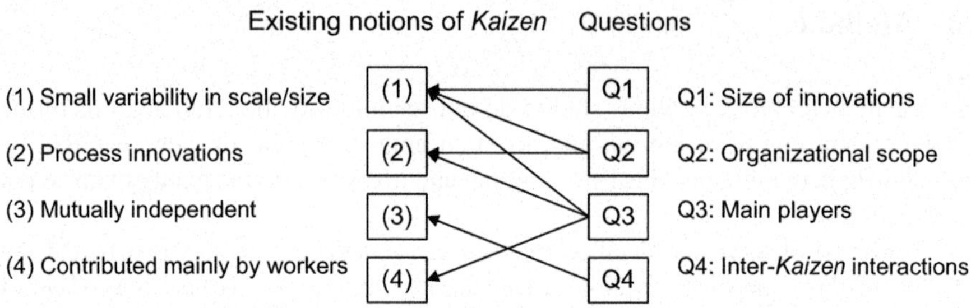

Fig. 1 Relationship between the four questions and existing notions of kaizen

the key research question (Are these existing notions true at all times?) was divided into four secondary questions:

Q1: What is the size of the kaizen case in question?
Q2: How many organizational areas[5] contribute to the kaizen case in question?
Q3: Who are the main contributors to the kaizen case in question?
Q4: Does the kaizen case in question have coordinating interactions with other kaizen cases? If so, which one(s)?

As described in Fig. 1, these four questions are logically linked to the testing of the existing notions of kaizen.

Yet, Q1 remains problematic: How can we estimate the scale of innovation? Of course, the scale of investments and outcomes is one aspect of the scale of innovation (Boer and Gertsen 2003; Imai 1986). However, since innovative activities, such as those of kaizen, always require coordination (Thompson 1965), the scale of coordination must also be measured, and it is adopted here to estimate the scale of each kaizen activity. Furthermore, decision-making efforts and problem-solving through coordination with various stakeholders are needed, and this is why we concentrate on the coordination required to deal with kaizen activities.

Thus, we introduce the concept of "scope of coordination" in addition to the existing measurements of innovation scales, such as the scales of investment and outcomes. Specifically, to measure the scale of each kaizen activity as an innovation, determined by the scope of innovation, the scopes of coordination are measured by the number of stakeholders affected by each kaizen activity, which we call "simple scope of coordination," and by the total man-hours required for coordination, called "weighted scope of coordination." In other words, simple scopes of coordination are calculated as the number of sections involved in kaizen, and weighted scopes of coordination are the man-hours consumed in coordination efforts, which allows us to differentiate between light and heavy coordination efforts. Of course, if those

[5]For example, operational improvement, process/production engineering, product/design engineering, etc. could be included in kaizen projects as innovations.

participating in kaizen spend the same amount of time in each of the seven cases, these two different aspects of the scope of coordination have no meaning. If coordination efforts vary across the seven cases, then the weighted scope of coordination can prove useful to estimate coordination efforts required to carry out kaizen. Coordination effort is also a part of the innovation scale of investment, as firms have to pay wages/salaries for the time consumed in kaizen.

Using specific examples, the scope of coordination is calculated on the basis of organizational sections, such as press and welding, painting, molding, assembly and testing, quality control, product design engineering, production equipment engineering, purchasing, other factories, and other firms. If two operators in welding spend two man-hours on planning and three man-hours on negotiating with each other, the simple scope of coordination is one section, and the weighted scope of coordination is five man-hours. Conversely, if two welding operators, three molding operators, one production engineer, two product design engineers, and four engineers from an equipment supplier meet for half an hour and kaizen is achieved, the simple scope of coordination is five sections (welding, molding, production engineering, product/design engineering, and the equipment supplier), and the weighted scope of coordination is six man-hours.[6]

These two concepts of the scope of coordination make it possible to answer Q1, Q2, and Q3. We also count coordination points, such as the number of meetings, and plot the sum of the two scopes of coordination in each case. After calculating the scale of kaizen as innovation through the scopes of coordination, investments, and cost reduction effects, we will use our analytical framework (Table 1) to summarize and analyze the seven cases in the Discussion section (Table 11).

4 Results of the Comparative Case Study

This study concerns kaizen in competitive factories. Hence, it uses data collected during a field study of the Toyota Takaoka plant, which comprises five units: press and welding, painting, molding, assembly and testing, and quality control. Its organizational chart is presented in Fig. 2.

Our case studies assess the scope of coordination required to complete seven kaizen activities while also considering the extent of changes to artifacts, the scale of investment, and cost reduction effects. Kaizen participants include operators; team leaders; group leaders; shop floor engineers; engineers from other areas, such as design, purchasing, and production equipment; as well as other firm engineers. Operators are designated as A, B, or C. Operators A perform standard tasks and sometimes offer the team or group leaders ideas for kaizen initiatives. Operators B are chiefly responsible for kaizen. Operators C perform preventive maintenance. Kaizen may also require coordination outside the factory, for example, with design

[6]This is calculated as 12 times $0.5 = 6$.

Table 1 An image of the analytical framework summary

Case	Q1	Q2	Q3	Q4
Case 1	–	–	–	–
Case 2	–	–	–	–
Case 3	–	–	–	–
Case 4	–	–	–	–
Case 5	–	–	–	–
Case 6	–	–	–	
Case 7	–	–	–	–

Q1: What is the size of the kaizen case in question?
Q2: How many organizational areas contribute to the kaizen case in question?
Q3: Who are the main contributors to the kaizen case in question?
Q4: Does the kaizen case in question have coordinating interactions with other kaizen cases? If so, which one(s)?

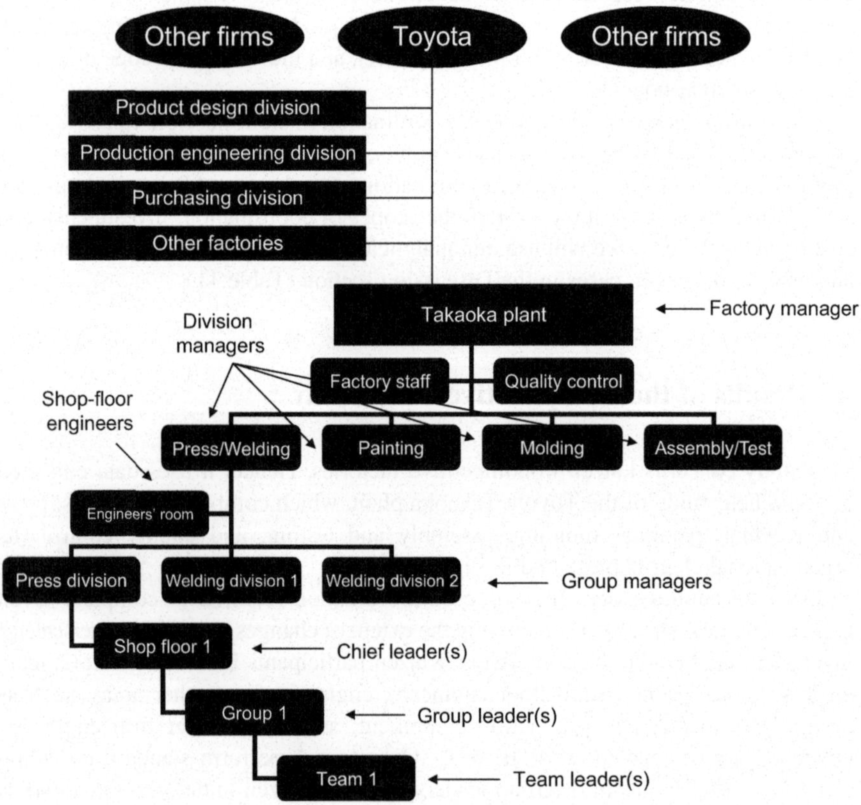

Fig. 2 Organizational chart of Toyota (and other companies)

and production equipment engineers at Toyota's headquarters (rectangles in Fig. 2). These engineers develop new products and equipment and make minor or major alterations to existing products and equipment. Factory managers occasionally suggest kaizen initiatives, as do Operators A and B.

Among these various figures, shop floor engineers have a unique role. They are technical staff located close to the shop floor and not managerial staff. They walk around the shop floor advising operators on technical issues and, if necessary, coordinate and negotiate with multiple actors to realize kaizen. Most of them have master's degrees in factory engineering, and all provide support for changes that affect the shop floor. Shop floor engineers have a wide range of responsibilities that include quality, cost, manufacturing lead time, and flexibility. In other words, they have both specialist and coordinator roles.

4.1 *Case 1: Door Assembly 1*

During the assembly of car doors in the body shop, the required parts are transported from storage to the assembly line, and a button is pressed to indicate that they are ready for assembly. The SOP details each step in this process and its sequence of execution. The need for improvements was first detected (August 27, 2013) when the line operators noticed a 1-second delay between the button being pressed and the ready light switching on. The line operators told the supervisor that this was a wasted time, and together they considered how to reduce it (1 man-hour). The line supervisor concluded that changing the position of the sensor would eliminate the delay. After confirming with the operators that the repositioning posed no safety problems, the supervisor asked the plant maintenance section to move the sensor (0.5 man-hours). He also made necessary changes to the process specification list, thus changing the operation standards. This improvement entailed no expense and reduced manufacturing costs by 140,000 yen annually (August 30, 2013).[7] This kaizen initiative took up about 1 man-hour for coordination.

In the Door Assembly 1 case, the total expenditure was 0 yen, the simple scope of coordination was one section (i.e., only welding), the weighted scope of coordination was 1.5 man-hours, the starting point for kaizen was the line operator, and the finishing point was the line supervisor. Although this kaizen activity was completed, further changes were required by Toyota's headquarters due to production start-up for the Hybrid Harrier (Case 2). Two measurements of the scope of coordination transition are plotted in Fig. 3. The horizontal axis shows the points of coordination, where the scope of coordination was calculated. Table 2 displays a brief summary of this case.

[7] 100 yen was equivalent to approximately US$1 in 2015.

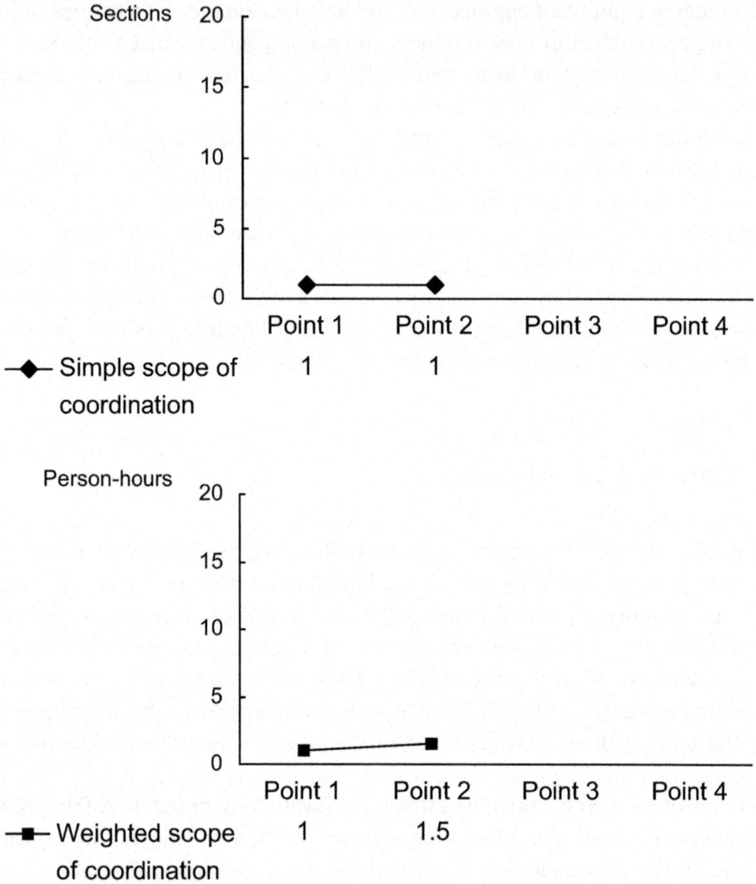

Fig. 3 Transition of the scope of coordination in Case 1

Table 2 Summary of Case 1 based on the analytical framework

Case	Q1	Q2	Q3	Q4
Case 1	Using 1.5 man-hours and 0 yen, 140,000 yen cost reduction	1 section at the Takaoka factory	Workers (line operators) and line chiefs	Leads to Case 2

4.2 *Case 2: Door Assembly 2*

While improvement in the door assembly process was underway, the plant was preparing to launch the Hybrid Harrier. This meant that its production volume was set to increase, which created pressure to reduce takt[8] time further. Since the number of assembled models would also increase, a larger amount of parts would have to be kept in the storage area. Hence, two targets had to be achieved: reduced takt time and improved flexibility.

The need for kaizen was communicated by the group manager to all line chiefs (chief leaders), group leaders, and team leaders (September 2, 2013). The line supervisors decided that their knowledge was insufficient for improvements on that scale, so they talked to other shop floor employees who had special responsibility for improvement. The discussion revealed that after placing the parts on the assembly line, the operators had to take several steps to press the button indicating that the parts were ready for assembly. These steps were wasted motion, and the line supervisors decided that the problem could be eliminated by repositioning the button nearer the parts storage area. This discussion took up about 2 man-hours. The plant maintenance team, however, noted that the repositioning would require electrical rewiring, which entailed expenses. This prompted the involvement of shop floor engineers, who concluded that costs could be reduced significantly if the button and rope switch used to change between the on and off positions were connected via the ceiling (0.5 man-hours). The line operators agreed, and, besides reducing takt time, manufacturing costs went down by 290,000 yen annually after spending 30,000 yen (September 10, 2013).

In this case, the total expenditure was 30,000 yen, and the simple scope of coordination was 2 sections (welding and shop floor engineers' room), the weighted scope of coordination was 2.5 man-hours, and the starting point for kaizen was the group manager, followed by a meeting among line supervisors and a revision of the draft plan for kaizen by shop floor engineers. The finishing point was the line operators. The kaizen activity succeeded and production of the Hybrid Harrier began. This success highlights Takaoka's manufacturing strategic direction of "multiple products on fewer lines" (Case 3). Two measurements of the scope of coordination transition are plotted in Fig. 4, and Table 3 displays a brief summary of this case.

4.3 *Case 3: Door Assembly 3*

Although the Takaoka plant successfully began production of the Hybrid Harrier, the plant management still faced problems. Headquarters had decided that the facility needed to increase its flexibility and implement a production plan for multiple product processes with many operational constraints. This plan called for further improvements in the manufacturing process and provided impetus for kaizen. Shop floor engineers concluded that two major problems had to be solved: the parts

[8]Takt is a Japanese-English term used in factories. Takt time means the time needed to assemble one unit.

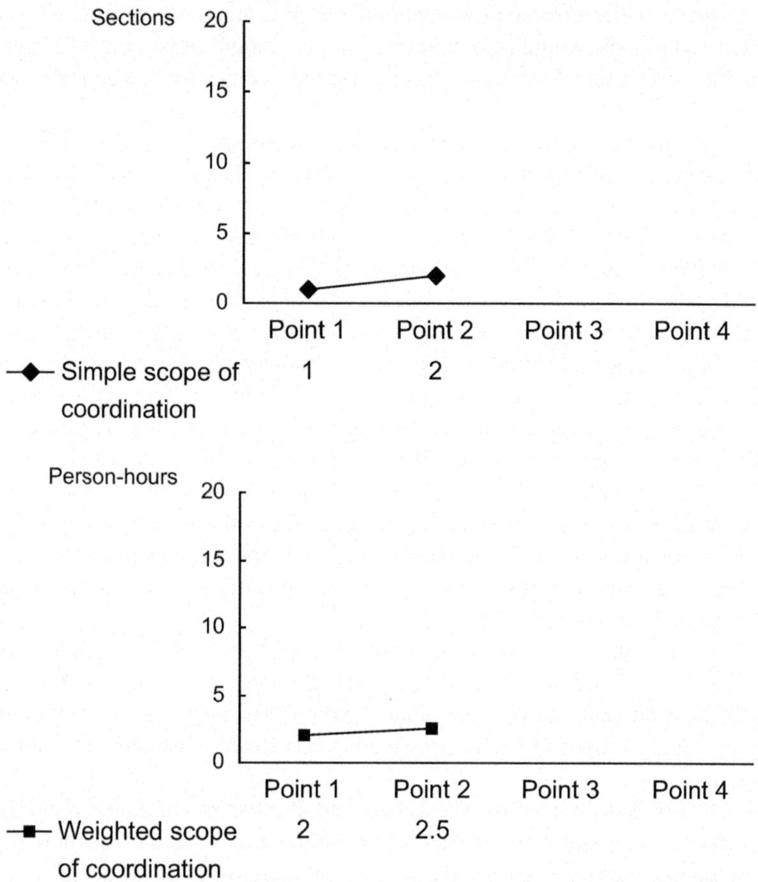

Fig. 4 Transition of the scope of coordination in Case 2

Table 3 Summary of Case 2 based on the analytical framework

Case	Q1	Q2	Q3	Q4
Case 2	Using 2.5 man-hours and 30,000 yen, 290,000 yen cost reduction	2 sections at the Takaoka factory	Workers and shop floor engineers	Leads to Case 3

storage area would be insufficient for the greater numbers of model parts, and the operations required to transport parts from the storage area would increase (September 11, 2013). At the time, shop floor engineers from another plant were working on a project at the Takaoka plant, and they were entrusted with solving these problems. Weekly meetings took approximately 4 man-hours.

In these meetings, insufficient storage space was identified as the most urgent problem, and a solution was soon proposed, i.e., to increase the factory's parts transportation capacity. However, shop floor engineers calculated that this solution would raise manufacturing costs by 1,730,000 yen. Then, line Operators A devised a plan that envisaged changes in how parts were transported. Building upon this suggestion (0.5 man-hours in coordination), the shop floor engineers proposed a radical reorganization of the storage area, with parts arranged by frequency of usage and larger component storage boxes, which they said would eliminate the need for greater transportation capacity. Computer-aided design (CAD) systems were also used in the planning. Negotiations with equipment suppliers (1 man-hour) resulted in manufacturing costs falling by 2,160,000 yen annually, after an expenditure of 300,000 yen (September 20, 2013). In this case, the amounts of investment and component change were moderate.

In the Door Assembly 3 case, the total expenditure was 300,000 yen, the simple scope of coordination was three sections (welding, shop floor engineers' room, and equipment supplier), and the weighted scope of coordination was 5.5 man-hours. The starting point for kaizen was Toyota's headquarters, and the shop floor engineers' room comprised the project team. A draft plan made by the shop floor engineers was revised by the operators, and the finishing point was the introduction of the new equipment. Two measurements of the scope of coordination transition are plotted in Fig. 5, and Table 4 displays a brief summary of this case.

4.4 Case 4: Kaizen Involving Component Changes

After the door subassembly is complete, the doors must be attached to the car body, an operation that requires bolts, nuts, and nail guns. With the changes introduced at Takaoka, the number of bolts and nuts had risen to five different types for each car model. The operators became concerned about the increased likelihood of picking up the wrong bolt or nut. They informed the shop floor engineers of the need for improvement (August, 2013), and the engineers agreed that picking up the wrong bolt or nut would diminish product quality. They suggested reducing the number of bolt and nut types. This conversation consumed 1 man-hour. Achieving commonality of parts, however, meant that major characteristics of the bolts and nuts—such as strength, length, and width—had to be changed and the shop floor engineers had to obtain the consent of product engineers. Long negotiations (4 man-hours) with the product design department led to the number of bolt and nut types being reduced to two. After further negotiations with the purchasing department (2 man-hours), it was agreed that buying large quantities of common parts would reduce total costs (September 30, 2013). The line operators accepted the proposed improvement, which was extended to two additional assembly processes.

Besides improving quality, this zero-investment kaizen activity reduced manufacturing costs by 3,300,000 yen annually. In this case involving component changes, the total expenditure was 0 yen, the simple scope of coordination was four sections (welding, shop floor engineers' room, product design division, and purchasing division), and the weighted scope of coordination was 7 man-hours. The starting point for kaizen was a conversation between a line operator and a shop floor

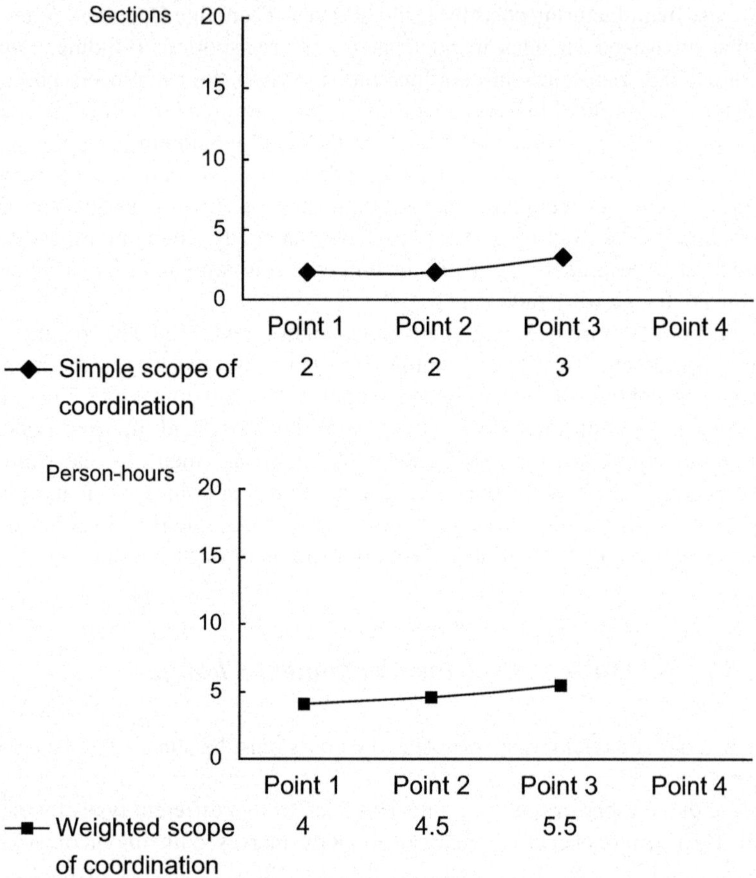

Fig. 5 Transition of the scope of coordination in Case 3

Table 4 Summary of Case 3 based on the analytical framework

Case	Q1	Q2	Q3	Q4
Case 3	Using 5.5 man-hours and 300,000 yen, 2,160,000 yen cost reduction	3 sections at the Takaoka factory and an equipment supplier	Workers, shop floor engineers, and an equipment supplier	Leads to Case 6

engineer, and the shop floor engineers then coordinated with the product design and purchasing divisions. The finishing point was acceptance by the line operators. Two measurements of the scope of coordination transition are plotted in Fig. 6, and Table 5 displays a brief summary of this case.

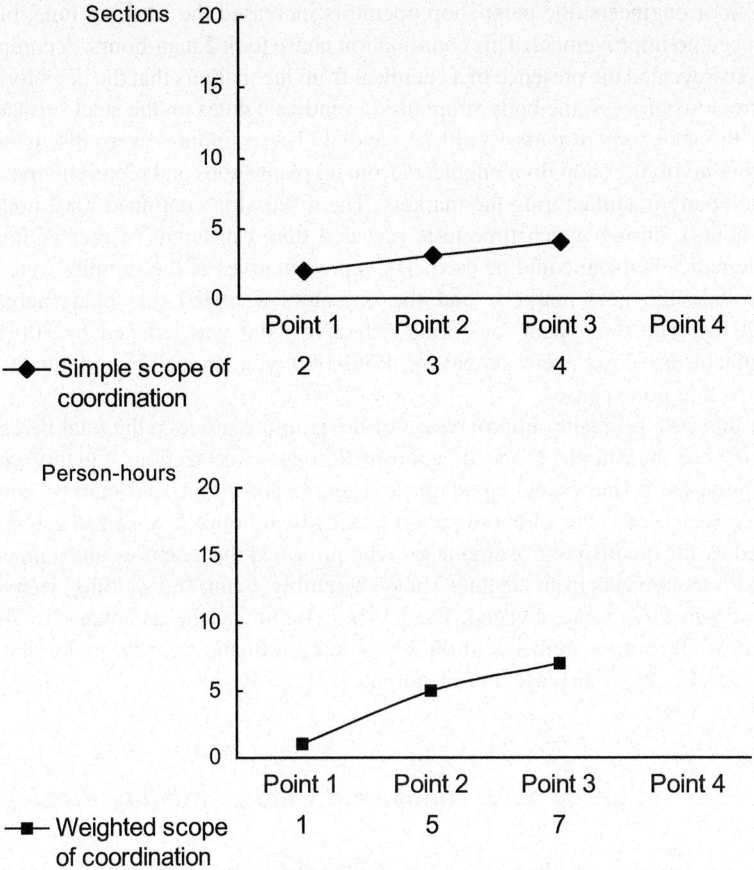

Fig. 6 Transition of the scope of coordination in Case 4

Table 5 Summary of Case 4 based on the analytical framework

Case	Q1	Q2	Q3	Q4
Case 4	Using 7 man-hours and 0 yen, 3,300,000 yen cost reduction	4 sections at the Takaoka factory including design eng.	Workers, shop floor engineers, production engineers, and purchasing division	–

4.5 *Case 5: Quality Improvement of the Painting Process*

Quality control at the Takaoka plant revealed that 1 in 100 units had painting defects, which prompted shop floor engineers to conduct a chemical analysis (2 man-hours). Its results showed that the defective paint contained foreign matter from imperfect washing of the surface before the paint was applied (August, 2013). The cause of the problem was communicated to the assembly and paint shops. At the request of the shop floor engineers, the paint shop operators increased the washing time, but that produced no improvement. This coordination phase took 2 man-hours. A component analysis revealed the presence of a chemical from the markers that the operators from the previous process, the body shop, used to indicate dates on the steel surface or to mark the spots where parts would be welded. These findings were discussed in a meeting involving shop floor engineers from all plant shops and representatives from the company manufacturing the markers. The discussion continued for 2 hours (10 man-hours), during which time tests revealed that a different marker without the problematic chemical could be used. The representatives of the supplier were asked to provide the new markers, and the operators accepted this plan (September 20, 2013). The time spent reworking defective paint was reduced by 300 hours, manufacturing costs went down by 1,560,000 yen annually, and quality was improved at no expense.

In this case of quality improvement of the painting process, the total investment was 0 yen, the simple scope of coordination was six sections (quality control, assembly, paint shop, welding, shop floor engineers' room, and marker supplier), and the weighted scope of coordination was 14 man-hours. The kaizen activity was started by the quality control engineers, who informed the assembly and paint shops. Shop floor engineers from the three shops (assembly, paint, and welding) were called on, and a meeting was convened. The finishing point was the acceptance by the line operators. Two measurements of the scope of coordination transition are plotted in Fig. 7, and Table 6 displays a brief summary of this case.

4.6 Case 6: Large-Scale Equipment Change (Sliding Puzzle)

During body assembly, the hood[9] is attached to the car body. Normally, there are ten storage areas for hood components, from which two operators carry interior and exterior components to the assembly line (Fig. 8). Since the hood components are large, only a limited amount of parts can be stored in the same area, which would create problems if the number of assembled car models per assembly line went up. Yet, at Takaoka, an increase in this number was inevitable because headquarters had required more flexibility on the production line. Hence, the manager of the body shop instructed the shop floor engineers to find ways to accommodate more parts (April, 2013).

While the line supervisors were planning improvements to the storage area (1 man-hour), the shop floor engineers heard that a similar problem had been solved at Toyota's Tahara plant and went there to learn the details (6 man-hours). At Tahara, they discovered that using automated guided vehicles (AGVs) solves the issue of limited storage space in plant management. However, the Takaoka shop floor engineers thought that the AGVs were used inefficiently at the Tahara plant. Upon

[9]The hood is the hinged cover over the engine.

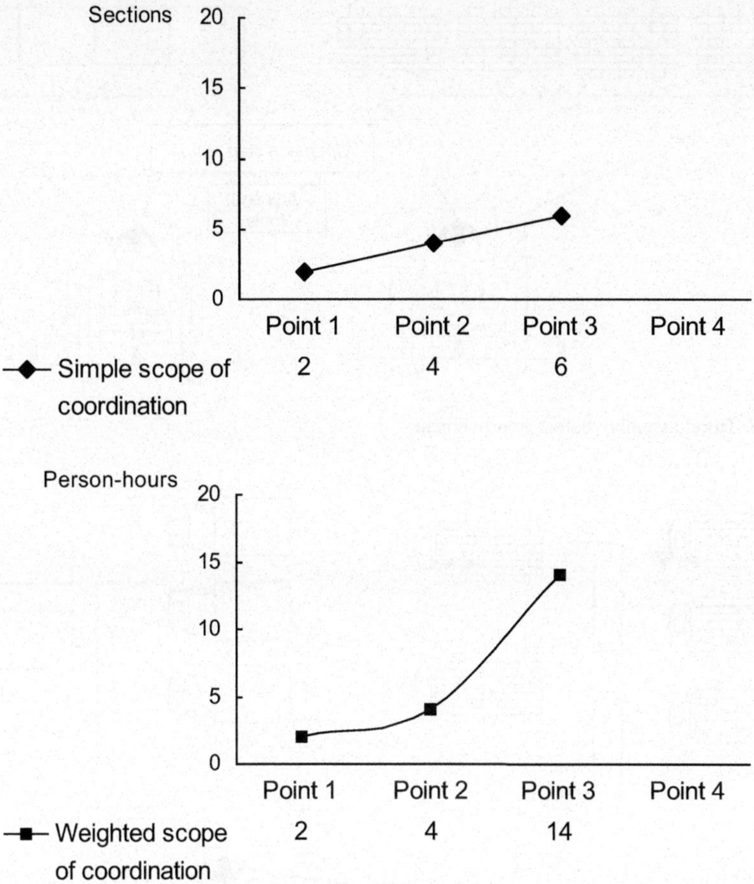

Fig. 7 Transition of the scope of coordination in Case 5

Table 6 Summary of Case 5 based on the analytical framework

Case	Q1	Q2	Q3	Q4
Case 5	Using 14 man-hours and 0 yen, 1,560,000 yen cost reduction	6 sections at the Takaoka factory and a marker producer	Workers, shop floor engineers, and marker producer	–

returning to their plant, they discussed the problem with the line supervisors (5 man-hours). The meeting yielded a plan, proposed by the line supervisors and expanded upon by the shop floor engineers, whereby AGVs could be used efficiently if they moved along routes that resembled a sliding puzzle (Fig. 9). Nonetheless, opposition arose among the operators because they would regularly need to change the AGV batteries. The Takaoka shop floor engineers negotiated with the AGV manufacturer (3 man-hours), who agreed to develop a model with a battery that

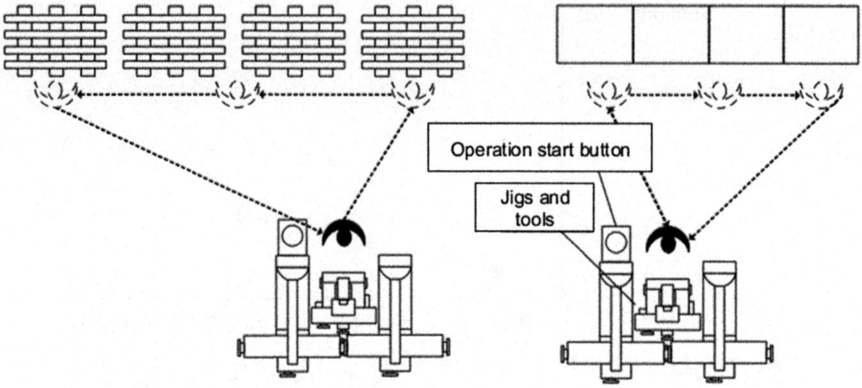

Fig. 8 Hood assembly before improvement

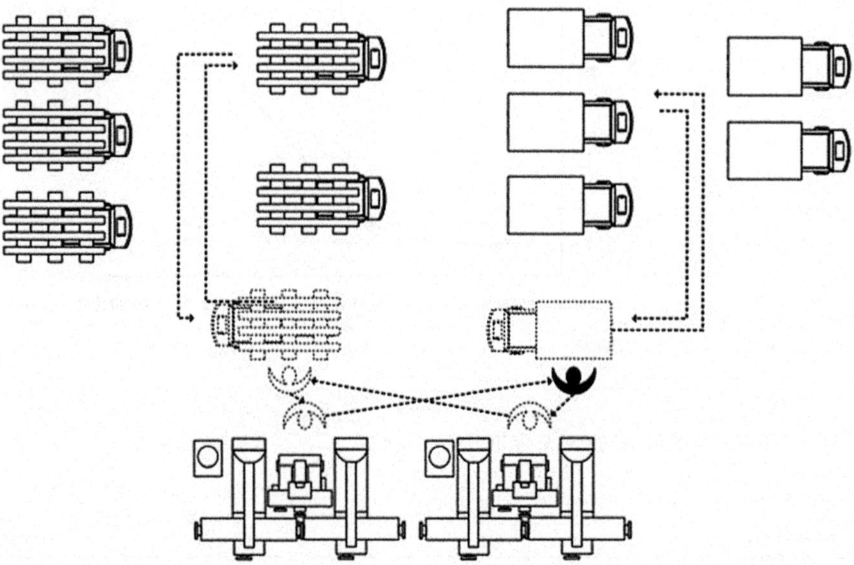

Fig. 9 Hood assembly after improvement

recharged automatically. The investment in equipment was 20 million yen, but it was now possible (November, 2013) to assemble eight different car models on the same line, compared with two car models before the change. The facility's annual manufacturing costs were reduced by 20,000,000 yen.

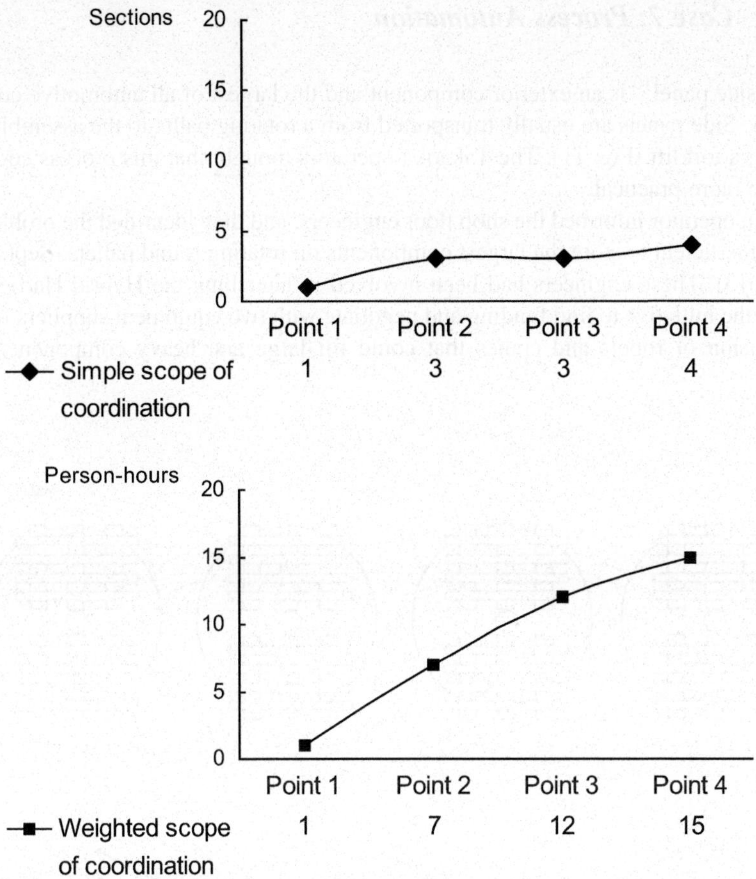

Fig. 10 Transition of the scope of coordination in Case 6

Table 7 Summary of Case 6 based on the analytical framework

Case	Q1	Q2	Q3	Q4
Case 6	Using 15 man-hours and 20,000,000 yen, 20,000,000 yen cost reduction	4 sections in Toyota factories and an AGV manufacturer	Workers, shop floor engineers, and AGV manufacturer	Leads to Case 7

In the case of the sliding puzzle, the total expenditure was 20,000,000 yen, the simple scope of coordination was four sections (welding, shop floor engineers' room, Tahara plant, and AGV manufacturer), and the weighted scope of coordination was 15 man-hours. The starting point for kaizen was Toyota's headquarters. The line supervisors were unable to make improvements at first, but, after visiting the Tahara plant for benchmarking, the shop floor engineers modified Tahara's original idea and even coordinated with the AGV manufacturer and line operators. The finishing point was the acceptance of the new process by the line operators. Two measurements of the scope of coordination transition are plotted in Fig. 10, and Table 7 displays a brief summary of this case.

4.7 Case 7: Process Automation

The side panel[10] is an exterior component and the largest of all automotive components. Side panels are usually transported from a rotating pallet to the assembly line using a forklift (Fig. 11). The Takaoka operators thought that this process could be made more practical.

An operator informed the shop floor engineers, and they identified the problem: it was inefficient to store the largest components on rotating round pallets (September 4, 2013). These engineers had been involved in launching the Hybrid Harrier and took the initiative to find funding and negotiate with two equipment suppliers for the provision of robots and cranes that could lift large and heavy components. The

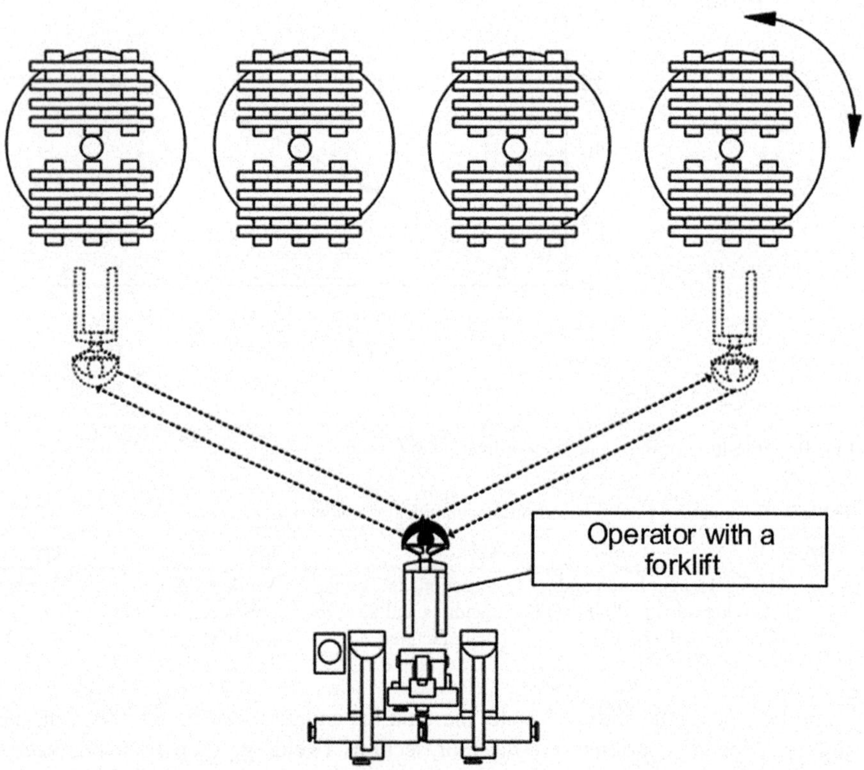

Fig. 11 Side members assembly before automation

[10]The side panel is also called the "side member."

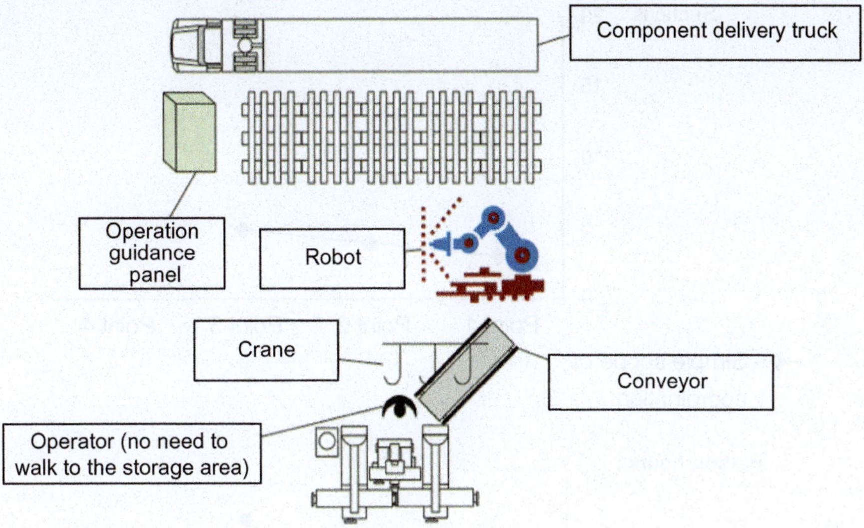

Fig. 12 Side members assembly after automation

robots would eliminate the need for rotating round pallets and provide the opportunity to use rectangular pallets, shortening the distance that the operators had to walk. However, heavy components had to be lifted during line assembly, for which it was proposed to suspend an inexpensive crane from the ceiling. This project required 10 man-hours. Responding to the (stamping press) logistics section's concerns that scheduling component deliveries would become overly complicated, the shop floor engineers created an operation guidance board that displayed all relevant information (1 man-hour). In order to acquire the 52 million yen needed to implement this kaizen, one of the shop floor engineers negotiated with the production engineering department (6 man-hours). Together, these improvements (Fig. 12) reduced yearly manufacturing costs by 20 million yen, with improved flexibility, reduced inventory, and shorter manufacturing lead time (January 2014).

For this kaizen activity, a budget of 52,000,000 yen was needed, the simple scope of coordination was six sections (welding, shop floor engineers' room, logistics section of the stamping press division, production engineering division, and two equipment manufacturers), and the weighted scope of coordination was 17 man-hours. The starting point for kaizen was a line operator, who had a talk with a shop floor engineer. The shop floor engineers coordinated with the equipment manufacturers and the production engineering division. The finishing point was the introduction of the new equipment. Two measurements of the scope of coordination transition are plotted in Fig. 13, and Table 8 displays a brief summary of this case.

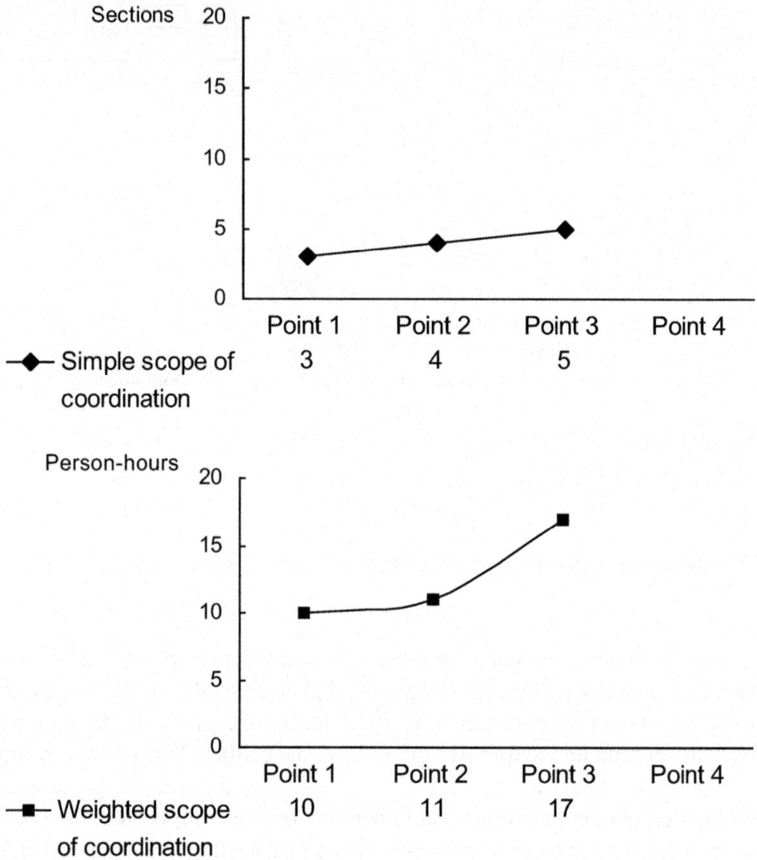

Fig. 13 Transition of the scope of coordination in Case 7

Table 8 Summary of Case 7 based on the analytical framework

Case	Q1	Q2	Q3	Q4
Case 7	Using 17 man-hours and 52,000,000 yen, 20,000,000 yen cost reduction	6 sections, Toyota and equipment manufacturers	Workers, shop floor engineers, production/equipm. engineers, and equipment makers	–

5 Discussion

The cases described in the previous section point to great differentiation in the parameters of innovations through kaizen. For example, investments ranged from 0 yen (Cases 1, 4, and 5) to well over 1 million yen (Cases 6 and 7). Kaizen included both small and large incremental innovations in investments, cost reduction effects,

Table 9 Brief summary of the scales of innovation in the seven cases

Case	Simple scope of coordination	Weighted scope of coordination	Investment	Cost reduction
Case 1	1	1.5	¥ 0	¥ 140,000
Case 2	2	2.5	¥ 30,000	¥ 290,000
Case 3	3	5.5	¥ 300,000	¥ 2,160,000
Case 4	4	7	¥ 0	¥ 3,300,000
Case 5	6	14	¥ 0	¥ 1,560,000
Case 6	4	15	¥ 20,000,000	¥ 20,000,000
Case 7	6	17	¥ 52,000,000	¥ 20,000,000

Table 10 Spearman's rank correlation coefficient of the Toyota cases

	Simple scope of coordination	Weighted scope of coordination	Investment	Cost reduction
Simple scope of coordination	1			
Weighted scope of coordination	0.909 *	1		
Investment	0.283	0.593	1	
Cost reduction	0.679	0.883 *	0.692	1

N = 7 * $p < 0.05$

and scopes of coordination. A summary of these scales is presented in Table 9, with significant variance in investments and cost reduction across the seven cases. Although Cases 1, 4, and 5 required no actual investment,[11] there are a tenfold gap in the scale of investment between Cases 2 and 3, a 66.7 times gap between Cases 3 and 6, and a 2.6 times gap between Cases 6 and 7. Cost reduction effects have a tenfold gap between Cases 1 and 2 and Cases 3, 4, and 5 and even more than tenfold gaps from Cases 3 to 5 and Cases 6 and 7.[12] Moreover, according to Spearman's rank correlation coefficient, these Toyota cases (Table 10) display a strong correlation between the two scopes of coordination and the weighted scope of coordination and performance improvements, such as cost reduction, but not between investment and the scopes of coordination (Table 10). We can infer from these findings that the scope of coordination is a useful measure of the scale of innovation through kaizen.

[11]Of course, a factor not to be overlooked is that the weighted scope of coordination is a kind of investment.

[12]Based on these cases, we can say that kaizen consists of a variable scale of innovations, and we can measure this by the scope of coordination, in addition to the amount of investment and cost reduction effects.

Moreover, many actors and activities interact with one another, and simple/weighted scopes of coordination continue to change because of this on a case-by-case basis. For example, some cases indicate that the scopes of coordination remain small (Cases 1 and 2), while Cases 3 and 4 show that the scopes of coordination can sometimes increase gradually. In Case 6, the simple scope of coordination remains moderate, but the weighted scope of coordination escalates. In addition, Case 4 shows that small kaizen activities can occasionally become large at some point, whereas Case 7 suggests that the opposite can also be true. Furthermore, kaizen can be changed and at times even eliminated through coordination (as seen in Case 5).

Next, kaizen may sometimes require product design changes (Case 4), as well as the involvement not only of operators but also of engineers (Cases 3 to 7). Additionally, some kaizen activities are in line with the policy of "multiple products on fewer lines" (Cases 1, 2, 3, 6, and 7), and others are not (Cases 4 and 5). Based on the analytical framework (Table 1), we can observe these variances in kaizen (Table 11) and draw some conclusions. Indeed, the existing notion of kaizen—as an accumulation of incremental innovations that (1) have little variability in scales, (2) are categorized only as process innovations, (3) are mutually independent innovative activities, and (4) are driven only by workers/operators, work teams, and their

Table 11 Summary of the seven cases based on the analytical framework

Case	Q1	Q2	Q3	Q4
Case 1	Using 1.5 man-hours and 0 yen, 140,000 yen cost reduction	1 section at the Takaoka factory	Workers (line operators) and line supervisor	Leads to Case 2
Case 2	Using 2.5 man-hours and 30,000 yen, 290,000 yen cost reduction	2 sections at the Takaoka factory	Workers and shop floor engineers	Leads to Case 3
Case 3	Using 5.5 man-hours and 300,000 yen, 2,160,000 yen cost reduction	3 sections at the Takaoka factory and an equipment supplier	Workers, shop floor engineers, and an equipment supplier	Leads to Case 6
Case 4	Using 7 man-hours and 0 yen, 3,300,000 yen cost reduction	4 sections at the Takaoka factory including design eng.	Workers, shop floor engineers, production engineers, and purchasing division	–
Case 5	Using 14 man-hours and 0 yen, 1,560,000 yen cost reduction	6 sections at the Takaoka factory and a marker producer	Workers, shop floor engineers, and marker producer	–
Case 6	Using 15 man-hours and 20,000,000 yen, 20,000,000 yen cost reduction	4 sections in Toyota factories and an AGV manufacturer	Workers, shop floor engineers, and AGV manufacturer	Leads to Case 7
Case 7	Using 17 man-hours and 52,000,000 yen, 20,000,000 yen cost reduction	6 sections, Toyota and equipment manufacturers	Workers, shop floor engineers, production/equipm. engineers, and equipment makers	–

leaders—is not completely adequate to describe the kaizen activities in Toyota (Table 11).

Furthermore, due to the emergent nature of kaizen, shop floor engineers are sometimes required to follow up on kaizen activities. For instance, if kaizen affects only one operation and involves small improvements and low technical complexity, as in Case 1, then the authority to initiate and implement the changes is delegated to the shop floor operators and group leaders, who can reasonably predict what will be affected (e.g., safety) and coordinate with the interested parties on a small scale. Conversely, as seen in Cases 2 and 3, which involve equipment changes, line operators and supervisors must rely on shop floor engineers, who possess the required technical knowledge and can drive improvements. In the last four cases, changes involve different functional departments or sections in the plant as well as external firms. In these instances, shop floor engineers are responsible for coordinating with all parties. Depending on the circumstances, shop floor engineers are both improvement experts and coordinators.

In sum, one issue in kaizen management and continuous improvement from the innovation perspective is determining the extent of coordination required to implement kaizen activities. Indeed, Toyota seems consistently able to assess the extent of a kaizen initiative when it is introduced.[13] Moreover, after deciding what scope of coordination is sufficient, Toyota's shop floor engineers take over coordination, switching to the weighted scope of coordination in the seven cases. Decision-making and coordination activities can also be seen in the interactions among the seven kaizen cases (Fig. 14). Particularly, Case 1 was concluded with a change in SOP, but further changes were required to start production on the Hybrid Harrier,

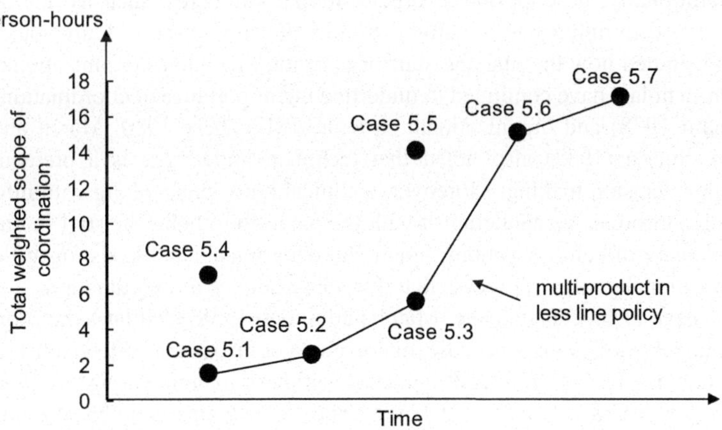

Fig. 14 Interactions among the seven cases

[13]Of course, it is doubtful whether Toyota can always determine the exact extent of coordination needed for kaizen.

and a shop floor engineer served the purpose of coordinating this change (Case 2). Thanks to successful production start-up, the Takaoka plant implemented a manufacturing strategy of "multiple products on fewer lines," with contributions from several actors, including shop floor engineers (Cases 3, 6, and 7) (Fig. 14).

The above evidence leads to the next question: Why are shop floor engineers able to make these decisions? In all of the Takaoka plant shops, shop floor engineers work near the floor and act as both specialists and coordinators, which is why this organizational form is often referred to as *staff in line.*[14] This differs from the traditional *staff and line* organizational structure, in which the staff and line organization mainly advises the line's upper-level managers (Mintzberg 1980).[15] On the contrary, staff in line advises lower levels of the organization, such as workers/operators, and coordinates with other participants depending on their abilities and responsibilities. There are two main characteristics of the staff-in-line organization at Toyota. First, engineers are physically and organizationally near the shop floor/production line and normally act as its technical advisers. Second, they possess the technical knowledge needed to coordinate with a range of actors when the impetus for kaizen is large enough to require coordination. These two characteristics enable smooth coordination among different functions, which is often required when an organizational routine is changed through process or product improvement. The staff-in-line form of organization minimizes the number and intensity of internal conflicts that often accompany change. Since staff-in-line members are organizationally and physically near the plant floor, they naturally maintain contact and communication with line operators and team leaders, who call upon their technical knowledge. Their knowledge and experience enable them to understand how and where changes in organizational routines occur.

Traditionally, descriptions of organizational structures, such as line and staff, were rooted in military terminology (Simon 1947). Ever since Gulick and Urwick (1937) studied how the line and staff organization facilitates organizational coordination, scholars have continued to underline the importance of coordination patterns (Barnard 1938) and organizational structure (Mintzberg 1980; March and Simon 1958). Simon (1947) also noted that technical knowledge is a prerequisite for complex decision-making. Moreover, technical knowledge, along with information collected through constant contact with the plant floor, helps the staff in line reduce the intensity of conflicts among functional departments and leads toward common organizational goals—a process that this study defines as coordination.

A shared technical language (March and Simon 1958) facilitates the coordination function of staff in line because major changes in routine affect other technical departments, such as product/design engineering or manufacturing/equipment engineering (e.g., Cases 4 and 6). At Toyota, staff in line apparently has the authority to

[14]The term "line" here means both organizational positions: line and production line.

[15]Mintzberg (1980) calls an organization resembling staff in line "the Divisionalized Form organization" (p 335). However, here the staff in line is nearer the plant floor than in Mintzberg's model and is unique in serving two roles.

implement changes suggested by other horizontal functional departments or by upper levels of the vertical organization (e.g., Cases 3, 5, and 6). These changes accommodate the realities of the plant floor, which is why the operators accept them with little opposition. Thus, staff in line can coordinate horizontally and vertically within the organization. Changes suggested on the plant shop floor are adjusted during coordination to reflect shared organizational goals. These modifications are accepted on the shop floor because it is understood that they originate from experts. In other words, the shop floor has authorized the staff in line to make decisions related to improvements, which eliminates opposition to these decisions (e.g., Cases 2, 4, 5, and 6).

Thus, the effectiveness of coordination provided by the staff-in-line organization is determined by the technical knowledge of the shop floor engineers as well as by their easy access to shop floors, and, at Toyota, it has four characteristics:

- Collecting information to facilitate potential kaizen improvements
- Assessing which elements of the organization are affected by the proposed changes in routines
- Communicating among technical specialists
- Adjusting the changes proposed by other parties without generating conflict

Yet, a staff-in-line organization does not always guarantee sufficient coordination for kaizen, as coordinators need to be trusted by the operators, team leaders, and group leaders. Toyota's former president, Katsuaki Watanabe, emphasized this while being interviewed: "To realize kaizen, engineers have to be nice, kind, and respect shop floor people. Additionally, staff in line could help people on the plant floor with knowledge and wisdom."[16] The evidence from our case studies suggests that cultivating the beginnings of kaizen entails a certain degree of coordination, which changes with the circumstances. Indeed, estimating this degree and bringing together the necessary participants are crucial requirements for kaizen. Hence, a staff-in-line organizational structure is effective in managing kaizen as innovation at Toyota.

Here, we explored actual cases of continuous improvement (kaizen) and found that it sometimes deviates from its current definition. Small changes in organizations become radical and episodic when positive feedback exists in complex systems (Plowman et al. 2007; Weick and Quinn 1999), and kaizen activities also have that kind of chain effect inside and between the cases. Some kaizen activities were initiated by the operators, and, after completing one kaizen activity, another actor, such as Toyota's headquarters, requested additional changes (e.g., Cases 1 and 3) or vice versa (e.g., piloting preparation for the production of the Hybrid Harrier). In summary, this chapter explains the nature of kaizen as innovation and also partly discusses how to manage it.

[16]Interview conducted on February 13, 2014.

If changes are able to bring about further changes, coordination among these changes is needed. Fujimoto (2007) notes that Toyota frequently generates both large and small changes, such as R&D and continuous improvement in operations, and that even small changes require coordination among vertical hierarchies in manufacturing or horizontal functional departments. According to Fujimoto (1999), Toyota's manufacturing processes based on static organizational routines have repeatedly changed, thanks to kaizen and Toyota's ability to generate, select, and implement new and improved organizational routines at the macro level of the organization. However, it remains unclear how this ability is supported at the micro level of actors in the organization. In other words, the principles underlying Toyota's ability to innovate by changing routines have not been investigated completely. Our analysis clarifies that the coordination systems at Toyota might be a part of the mechanisms that underlie Toyota's kaizen ability.

6 Conclusions and Next Steps

Existing research views kaizen (continuous improvement) as a set of small, mutually independent, and incremental process innovations made repeatedly by workers/operators and their leaders. However, the question "Does the conventional notion of kaizen always correspond with the actual characteristics of kaizen in firms?" has not been sufficiently investigated through empirical research. To answer the said research question, we conducted a series of in-depth field studies at one of the main factories of the Toyota Motor Company, the Takaoka plant. Our results suggest that previous views/notions of kaizen do not entirely reflect the reality of actual kaizen activities. For instance, we found that kaizen sometimes requires small changes in product design. The adequate scale of coordination depends on each case, and the extent of coordination among those participating in kaizen changes with each situation. Additionally, the ability to determine the extent of coordination and bring together the relevant participants is important for the success of innovations through kaizen.[17] Some coordination activities cannot be handled only by the operators and their leaders, and the shop floor engineers play a crucial role in integrating kaizen activities of different sizes, such as coordination of scopes, into one process. Furthermore, shop floor engineers have a staff-in-line organizational structure serving a twofold role, i.e., technical experts and coordinators for kaizen. In other words, effective kaizen, as a series of incremental innovations of various sizes and characteristics, needs an organizational design that facilitates adequate coordination among individual kaizen activities.

Yet, some issues remain open for future research. Investigating the reality of kaizen in industries other than the automotive is an important avenue to explore. In

[17]The scale of innovations (and variability thereof) is also a function of the scope of innovations because innovative activities in this context are organizational in nature and require coordination.

this regard, chapter "The Nature of Surviving Japanese Factories in the Global Competition: An Empirical Analysis of Electrical and Electronics Factories", which analyzes various firms in several industries, suitably complements the discussion offered in this chapter.

References

Anand G et al (2009) Dynamic capabilities through continuous improvement infrastructure. J Oper Manag 27(6):444–461

Barnard CI (1938) The functions of the executive. Harvard University Press, Boston

Barratt M et al (2011) Qualitative case studies in operations management: trends research outcomes and future research implications. J Oper Manag 29(4):329–342

Bessant J, Caffyn S (1997) High-involvement innovation through continuous improvement. Int J Technol Manag 14(1):7–28

Bhuiyan N, Baghel A (2005) An overview of continuous improvement: from the past to the present. Manage Decis 43(5):761–771

Boer H, Gertsen F (2003) From continuous improvement to continuous innovation: a (retro) (per) spective. Int J Technol Manag 26(8):805–827

D'Adderio L (2011) Artifacts at the centre of routines: performing the material turn in routines theory. J Inst Econ 7(2):197–123

Eisenhardt KM (1989) Building theories from case study research. Acad Manag Rev 14 (4):532–550

Fujimoto T (1999) The evolution of a manufacturing system at Toyota. Oxford University Press, New York

Fujimoto T (2007) Competing to be really really good: the behind-the-scenes drama of capability-building competition in the automobile industry. International House of Japan, Tokyo

Fujimoto T (2014) The long tail of the auto industry life cycle. J Prod Innov Manag 31(1):8–16

Gulick L, Urwick L (eds) (1937) Papers on the science of administration. Institute of Public Administration, New York

Imai M (1986) Kaizen: the key to Japan's competitive success. Random House Business Division, New York

Iwao S (2015) Organizational routine and coordinated imitation. Ann Bus Adm Sci 14(5):279–291

Iwao S (2017) Revisiting the existing notion of continuous improvement (*Kaizen*): literature review and field research of Toyota from a perspective of innovation. Evol Inst Econ Rev 14(1):29–59

Iwao S, Marinov M (2018) Linking continuous improvement to manufacturing performance. Benchmark Int J 25(5):1319–1332

Lindberg P, Berger A (1997) Continuous improvement: design organization and management. Int J Technol Manag 14(1):86–101

March JG, Simon HA (1958) Organizations. Blackwell, Cambridge, MA

Mintzberg H (1980) Structure in 5's: a synthesis of the research on organization design. Manag Sci 26(3):322–341

Myers S, Marquis DG (1969) Successful industrial innovations (NSF 69-17). National Science Foundation, Washington, DC

Ohno T (1988) Toyota production system: beyond large-scale production. Productivity Press, New York

Plowman DA et al (2007) Radical change accidentally: the emergence and amplification of small change. Acad Manag J 50(3):515–543

Shingo S (1981) A study of the Toyota production system from an industrial engineering viewpoint. Japan Management Association, Tokyo

Simon HA (1947) (1997) Administrative behavior: a study of decision-making processes in administrative organizations, 4th edn. The Free Press, New York
Thompson VA (1965) Bureaucracy and innovation. Adm Sci Q 10(1):1–20
Weick KE, Quinn RE (1999) Organizational change and development. Ann Rev Psychol 50 (1):361–386
Winter SG (2003) Understanding dynamic capabilities. Strateg Manag J 24(10):991–995
Womack JP, Jones T, Roos D (1990) The machine that changed the world. Rawson Associates, New York
Yin RK (1994) Case study research: design and methods, 2nd edn. Sage, New York

Balancing Standardization and Integration in ICT Systems: An Architectural Perspective

Young Won Park and Takahiro Fujimoto

Abstract In this chapter, we first review the history of Japanese IT system implementation and define various IT requirements that serve both advanced market standards and emerging market needs. Then, we introduce some examples of Korean firms using a global standard IT system (GSIS). The essence of this effective ambidextrous strategy is to maximize strengths and complement weaknesses, which requires the synergistic combination of an integral architecture IT system (i.e., integrated manufacturing IT system (IMIS)) and an open modular architecture IT system (global standard IT system (GSIS)). We therefore present the IT system evolution of Japanese firms from a product architecture perspective. By adopting a dialectic approach, we provide a framework that identifies the dynamic relationships among integrated manufacturing IT system (IMIS), global standard IT system (GSIS), and global integrated manufacturing IT system (GIMIS). Thus, a global integrated manufacturing IT system (GIMIS) is the synergistic combination of IMIS and GSIS. We then investigate further through case studies of Japanese global firms and illustrate how to implement the said global integrated manufacturing IT system (GIMIS).

1 Introduction

Japan's *integrated manufacturing system* has proven its usefulness in numerous integral types of product architecture groups, including manufacturing and export-driven product development projects, since the post-World War II period, and it still does in the current global competition contexts. In practice, various high-performance automotive products, industrial machinery equipment, and electrical

Y. W. Park (✉)
Graduate School of Humanities and Social Sciences, Saitama University, Saitama, Japan
e-mail: ywparkjp@gmail.com

T. Fujimoto
Graduate School of Economics, The University of Tokyo, Bunkyo-Ku, Tokyo, Japan

T. Fujimoto, F. Ikuine (eds.), *Industrial Competitiveness and Design Evolution*,
Evolutionary Economics and Social Complexity Science 12,
https://doi.org/10.1007/978-4-431-55145-4_10

machinery component parts have maintained their relative competitive advantage in the global markets. Behind the said sustainable performance edge, there have been effective teamwork by cross-discipline engineers and serious efforts to build information technology capabilities ensuring effective flows of design information across diverse organizational units. This is what we call *integrated manufacturing IT system* (*IMIS*), which has supported the abovementioned integrated manufacturing system (Fujimoto and Park 2015).

In this chapter, we first review the history of Japanese IT system implementation and define various IT requirements that serve both advanced market standards and emerging market needs. Then, we introduce some examples of Korean firms using a *global standard IT system* (*GSIS*). The essence of this effective ambidextrous strategy is to maximize strengths and complement weaknesses, which requires the synergistic combination of an integral architecture IT system (i.e., integrated manufacturing IT system (IMIS)) and an open modular architecture IT system (global standard IT system (GSIS)). We therefore present the IT system evolution of Japanese firms from a product architecture perspective. By adopting a dialectic approach, we provide a framework that identifies the dynamic relationships among integrated manufacturing IT system (IMIS), global standard IT system (GSIS), and *global integrated manufacturing IT system* (*GIMIS*). Thus, a global integrated manufacturing IT system (GIMIS) is the synergistic combination of IMIS and GSIS. Based on case studies of Japanese global firms, we illustrate how to implement said global integrated manufacturing IT system (GIMIS).

2 Japanese Firms and Integrated Manufacturing IT System (IMIS)

Generally, there are three distinct periods in IT evolution, in terms of IT usage patterns, prior to the information and communication technology (ICT) age: (1) digital revolution period (1980s), (2) network revolution period (1995–2005), and (3) user revolution period (2006–present) (Ministry of Internal Affairs and Communications 2013). Before the network revolution period (1995–2005), most Japanese firms used to develop their own IT systems internally. As a result, integrated manufacturing IT system (IMIS) has supported Japan's Integrated Manufacturing (Park 2009).

The Japanese information service industry first developed in the early 1960s (Japan's Information Service Industry Association 2015). Back then, computers were rare and expensive, and information service firms, with their huge computer systems, offered computing services to selected users (e.g., large banks and manufacturing firms).

In the 1980s, with the rapid increase of software development, these firms focused on creating software rather than providing computing services. Moreover, from the mid-1980s, large global firms established information service subsidiary divisions to make the most of computer application technologies and experiences.

In the 1990s, as business networks and software information processing requirements grew more complex, many large Japanese firms were no longer able to develop effective information systems using their internal capabilities alone. Hence, they outsourced their complex information processing system needs and diverse business solutions to information service firms with extensive technology development capabilities. These information service firms became known as system integrators.

In North America and Europe, system integrators continued to build best practices for different industries on a global scale and took on the role of global standard-setting IT leaders. However, in response to specific user requests, Japanese system integrators focused on building optimal, individually customized systems. They did not go beyond the best practices of Japanese firms and thus failed to develop global industry standards, unlike other global IT systems vendors.

In addition, from the mid-1980s, most Japanese firms discontinued their information processing divisions to reduce the high fixed costs related to IT system services. By the 1990s, due to extremely weak internal IT development capabilities, they increasingly relied on external information system (IS) vendors and consultants for their customized needs. Naturally, the numbers of competent internal information system (IS) personnel remained very small. According to a recent information service system trend report, the average outsourcing ratio in 2014 was 61% (JISA 2015), with projects focusing mostly on the integration between firm-specific legacy systems and global standard IT systems.

In discussing the Japanese manufacturing system, it is important to understand the concept of *monozukuri*, which can be defined in both broad and narrow terms (Fujimoto 2001, 2003). This chapter is based on three broad aspects of manufacturing (monozukuri in Japanese): (1) an integrated manufacturing IT system that connects cross-functional tasks, including strategic management, R&D, design engineering, manufacturing activities, marketing, sales, maintenance, and services; (2) a broad information system focusing on design information, which is embedded in all business processes; (3) and the purpose of the system is sustainable competitiveness based on customer value. Combining these three aspects, we define monozukuri as integrative communication processes of design information through diverse media mechanisms (Fujimoto 2003).

The adoption of IMIS has contributed to the effective utilization of IT systems integrating all functional processes, including product concept, system and detail design, purchasing, production, marketing, logistics, and services. IMIS has mainly dealt with design information for developing and manufacturing new products, specifically in relation to quality assurance, cost reduction, effective design, production efficiency, standardization, data management, and product information management. However, it has not been able to appropriately respond to the fast changes in global monozukuri.

For a long period, Japanese manufacturing firms have been making intense efforts to defend their competitive market position in integrative monozukuri products and meet the challenges posed by open modular electronic products from emerging competitors in Korea, Taiwan, and China. In such turbulent competing environments, Japanese firms have evolved their IT systems into what we call an integrated manufacturing IT system (IMIS). Its drawback is that IMIS is context specific and oriented toward local application, which is excellent for products targeting the Japanese domestic market and advanced markets in North America and the European Union. Yet, its application complexity does not fit with the features of a global standard IT system (GSIS), which is mostly implemented for open modular product types targeting emerging markets.

3 Korean Global Firms and Global Standard IT System (GSIS)

3.1 Global Standard IT System (GSIS)

Japanese firms struggling to balance their internal technology with global standards may find help close at hand. Korea has shown that a global standard IT system (GSIS) is feasible. Samsung Electronics's key strategy has been to use reverse engineering in order to understand the technologies and functions that Western and Japanese producers have launched. The firm then revises and redefines those functions based on the market price for each region. Furthermore, it markets its differentiated products using sports advertising and other marketing tools. Samsung's emphasis on groundbreaking innovation can be seen in the firm's internal structure. Instead of the usual R&D division, Samsung has an R&BD (research and business development) division. By integrating the technology and business development areas, Samsung has focused on applied research, with design and marketing that take local needs into account. This implies that Korean firms have absorbed the basic IT research processes of Japanese and Western firms and used Western patent policies to focus on acquiring new technologies as well as recruiting highly talented employees.

This emphasis on localization is hard to find in Japanese electronics firms, which struggle largely due to a lack of local responsiveness. Indeed, Samsung has shown that simply customizing products is not sufficient. It has created a "local expert system" that dispatches employees to foreign countries with a single task: to explore the region and absorb the local lifestyle. Based on the insights that they gain from this different perspective, the local experts then create products that fulfill the needs of the area's actual lifestyle.

Besides factors such as a drop in the value of the won (the Korean currency) until 2010, a top-down corporate structure, and government support, the success of Korean firms can also be attributed to leveraging product development systems that fit the global markets and a global standard IT system (GSIS). To survive in the household electronics industry, characterized by open modular architectures, the ability to balance an open innovation strategy, which uses external technology and management resources, with integral core technology, to differentiate the products manufactured, is more valuable than ever (Park et al. 2012; Hong and Park 2014).

3.2 *Examples of Korean Global Firms*

Korean firms use cross-functional teams to create products that meet customer needs. While collaboration across departments can be frequently seen in Japanese firms (Clark and Fujimoto 1991), Korean cross-functional teams are further supported by a global standard IT system (GSIS) (Park 2017).

3.2.1 Samsung Electronics's VIP Center

Samsung's investment in the company's information interface began in 1991, when it created an in-house announcement and email system (Park 2016). It then sets up systems, such as SINGLE and TOPIC and an ERP system in 2001. Through an internal system concentrating all of the management information in one place, Samsung has helped its executives identify problems faster. According to the firm's CIO, not only has this enabled them to solve issues at an earlier stage, it has also decreased spans of control. Supported by these systems, Samsung has succeeded in integrating its entire global production process, exemplified by the selling of China-made products in Russia (Park et al. 2012).

All of these successes stem from the creation of an E-CIM Center and VIP Center. After its new management declaration in 1993, Samsung set up a taskforce of 60 people to conduct product development analysis. Based on this information, the company carried out benchmarking and prediction activities to identify its position in the global industry. In addition, Samsung introduced the *E-CIM Master Plan*, which set the goal to increase product development capacity to three times the level of that time, and systematized all the product and component documents of the conglomerate's four groups into one data analysis system. One of its many activities was to adapt a new development process that used CAD data in order to comply with the requirements of concurrent engineering (CE). Through these revolutionary steps—such as implementing EVT, DVT, Rapid Prototype Interface, Product Data Management, and BOM—referred to as *Engineering Computer Integrated Manufacturing*, Samsung secured immense improvements, among which reducing production lead times from 2 years to 4 months.

Before achieving these successes, however, Samsung Electronics faced the problem of discrete IT systems that could only be used within individual departments. In order to overcome communication barriers across departments, Samsung developed PDM, a horizontal interdepartmental interface for the control of BOM information by the central management. PDM helped solve a number of internal issues, but it also enhanced the company's connection with its suppliers and resulted in outstanding sales of its Bordeaux TVs, largely due to decreased stocks and shorter delivery times. This example clearly shows that Samsung succeeded in increasing its market share by using its IT systems to match firm-level goals and customer demands (Chang 2008). Since its products require extremely fast production, the company set up a VIP center focusing on the accomplishment of horizontal innovation and concurrent engineering.

3.2.2 LG's TDR Center

LG was founded in 1958 thanks to financial loans and technology from Germany and Japan. Aided by the government's export-driven policy of the 1970s, LG developed large-scale mass production systems and expanded into the developed nations of the West. However, as exports to the USA and Western Europe became difficult in the 1970s, due to a wave of New Protectionism, LG shifted its strategy to opening factories in the USA, Germany, and the UK to produce and sell locally. In the 1990s, LG began to implement a global competitiveness strategy that involved entering emerging markets, changing its brand name from Goldstar to LG, and investing in sports marketing. While most of its production sites are located China and India, 70% of its sales are made in the USA and Europe.

LG's cross-functional team, which began as a revolutionary management initiative in 1996, was first called Tear-Down Redesign (TDR) and was then renamed Tear-Down Reengineering by Mr. Kim, the CEO of LGEIL (LG India). The work of this team allowed LG to develop unique products, among which a color TV that incorporated a cricket mode and multi-language system, closely based on local needs. Supported by its TDR, LG has continued to release products such as refrigerators with door-lock function, air conditioners that can cleanse the air, washing machines that keep mice away, and microwave ovens that can be used with unstable voltages.

Such innovations have been supported by a global standard IT system, which has played a significant role in facilitating the relay of know-how from LG Electronics' parent factory to its foreign plants. This global information channel is especially pertinent to LG's TV business, as quick communication between LG Philips LCD and other suppliers is critical. To achieve enhanced coordination, LG also implemented the ERP and SCM systems. Supply chain integration was accomplished through design modeling, and a function was made available for monthly, weekly, and daily planning of capacity allocations. Thanks to these efforts to integrate the supply chain through process, system, and organizational revamping and master planning lead times, which had previously taken 23 days and went down to 3 days, and customer response times were shortened from 10 to 3 days. Hence, greater integration resulted not only in increased visibility but also in increased customer responsiveness.

The above case studies of two Korean firms illustrate how business success can be achieved by responding to global needs thanks to the implementation of changes in various aspects of the organization. Moreover, shifting to open modular products leads firms to compete in markets with relatively short product lifecycles, inevitably causing the speed of production to become a key determining factor of competitiveness. To succeed in this battle, an efficient and coordinated product development process spanning several departments is necessary. What can Japanese firms learn from the examples of these Korean companies?

4 Proposal of a Global Integrated Manufacturing IT System (GIMIS)

Japanese integrated manufacturing (monozukuri) is based on factory-level embedded system knowledge (Fujimoto and Park 2015), and Japanese monozukuri capabilities in terms of technological depth and quality processes are well documented. Japanese software system providers—once disregarded due to their perceived weak competitiveness—are comparable with their US counterparts in selected areas (Cusumano 2004; Cole and Nakata 2014). Yet, the existing IT systems are not quite suited to the global expansion of Japanese monozukuri system capabilities, due to their being characterized by large amounts of embedded knowledge at the factory level. The said embedded knowledge should not be limited to the firm level, but it should flow according to the concept of FOA (flow-oriented approach) (Tomita et al. 1999; Oku et al. 2010; Park et al. 2011; Fujimoto and Park 2015). An integrated manufacturing IT system (IMIS) aims to respond to specific customer requests point by point, support product development design processes, and maintain field-level routine requirements within the factories. Therefore, an IMIS is mostly a user-initiated IT system, not a vendor-designed IT system.

By contrast, the implementation of a global standard IT system (e.g., ERP) allows firms to immediately adopt the best business processes of top global firms. However, due to the rapid development of IT technologies, all IT systems, without exceptions, keep upgrading their internal capabilities. A brief review of recent developments indicates continuous progress in all IT system areas, such as MRP (material requirements planning), ERP (enterprise resource planning), CRM (customer relationship management), SCM (supply chain management) IT solutions, CC (cloud computing), SOA (service-oriented architecture), IoT (Internet of Things), Industry 4.0, and AI (artificial intelligence).

In view of such breathtaking speed in technological change, it is unreasonable to overlook global IT standards. Naturally, Japanese firms are more likely to adopt global standard IT systems (e.g., ERP and SCM packages) that go beyond firm-specific IT system development.

However, the adoption of external IT systems often implies that firm-specific contexts and organizational identity are neglected. In addition, even the best systems become outdated and rigid over time, hence unable to respond flexibly and quickly to dynamic and ever-changing needs. The only way to remedy these shortcomings is to consider user initiatives and develop a unique system that reflects firm-specific

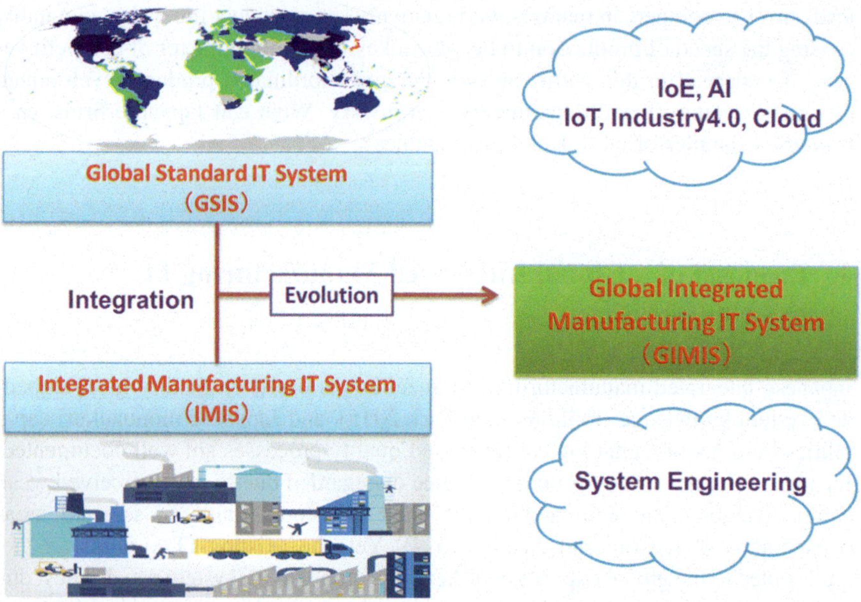

Fig. 1 Global integrated manufacturing IT system (GIMIS). (Adapted from Fujimoto and Park 2015)

identity-based requirements. For the sustainable delivery of outstanding products that exceed customer requirements, it is crucial to build an IT system that ensures integration of product development processes and organizational capabilities. The essence of this winning strategy is a firm's ambidextrousness, which highlights strengths and complements weaknesses. This new range of organizational capabilities thrives on integral architecture for the integrated manufacturing of complex products (e.g., automobiles and medical equipment), which is a typical trait of outstanding Japanese manufacturing firms. Yet, it is also capable of adopting an open modular architecture for consumer products (e.g., electronics), which requires a large number of suppliers with limited manufacturing capabilities. In this way, it is possible to attain long-term global competitiveness by penetrating both emerging and advanced markets. Said ambidextrous strategy uses both integrated manufacturing IT for integral architecture products and global standard IT for global modular products. In other words, we argue in favor of a GIMIS (global integrated manufacturing IT system) that integrates IMIS (integrated manufacturing IT system) and GSIS (global standard IT system). Figure 1 illustrates an IT strategy for Japan's monozukuri industry that combines both (1) IMIS (integrated manufacturing IT system) and (2) GSIS (global standard IT system).

As global competition intensifies also in emerging economies within a framework of overall free trade, in spite of strong national opposition, Japanese firms pursue "global long-term optimum management," which requires both absolute domestic advantage (based on operational productivity at the factory field level) and relative

Table 1 Comparison between IMIS and GSIS

	IMIS (integrated manufacturing IT system)	GSIS (global standard IT system)
Purpose	Knowledge transfer of outstanding Japanese productivity performance	Knowledge transfer of outstanding best practices of non-Japanese global firms
Strengths	Japanese domestic IMIS-based productivity performance	Global standards of Integrated IT practices (e.g., MRP, ERP, SCM Packages)
Basis	Complex production processes, efficient teamwork methods, and cross-discipline workforce	Global standards of leading firms in advanced markets that also impact on the methods adopted in emerging economies
IT system location	Domestic production facilities	Global production facilities

global advantage (based on total cost competitiveness). This double-edged competitive sword combines IMIS (integrated manufacturing IT system) and GSIS (global standard IT system), concentrating on neither IMIS nor GSIS.

In the 2010–2012 period, global firms in emerging economies (including China) experienced increasing pressure from rising labor costs. Naturally, outstanding factory-level productivity performance through the monozukuri system is receiving growing attention as a practical way to maintain production networks in emerging economies as global export bases. Japanese firms adopt global long-term optimum management by (1) maintaining highly productive manufacturing facilities in Japan, as domestic knowledge transfer cores, and (2) linking emerging and advanced economies through global production centers. The most appropriate IT systems that support global long-term optimum management are IMIS (integrated manufacturing IT system) for domestic production facilities and GSIS (global standard IT system) for global production centers.

Table 1 summarizes the features of IMIS and GSIS, the two crucial branches of GIMIS. Although so far no Japanese firm has successfully merged IMIS and GSIS, there are signs of a shift toward GIMIS (global integrated manufacturing IT system).

For example, as a practical step toward the integration of IMIS and GSIS, Komatsu, the Japanese global construction equipment firm, uses the global standard software ERP, which facilitates daily collection of various cost performance reports regarding global production centers and standardizes both CAD design information and BOM production requirements.

Figure 2 shows a BOM-based architectural analysis framework, which is an extension of the Komatsu-style GIMIS (global integrated manufacturing IT system). In general, a bill of materials (BOM) is a record of component codes and assembly sequences. Manufacturing firms use BOMs to monitor their complex flows from order receipt to order fulfillment, detailing the various steps of product planning, design, procurement, production, and maintenance activities. Depending on the nature of their business operations, firms may use the so-called design engineering BOM (E-BOM), manufacturing BOM (M-BOM), services BOM (S-BOM), and

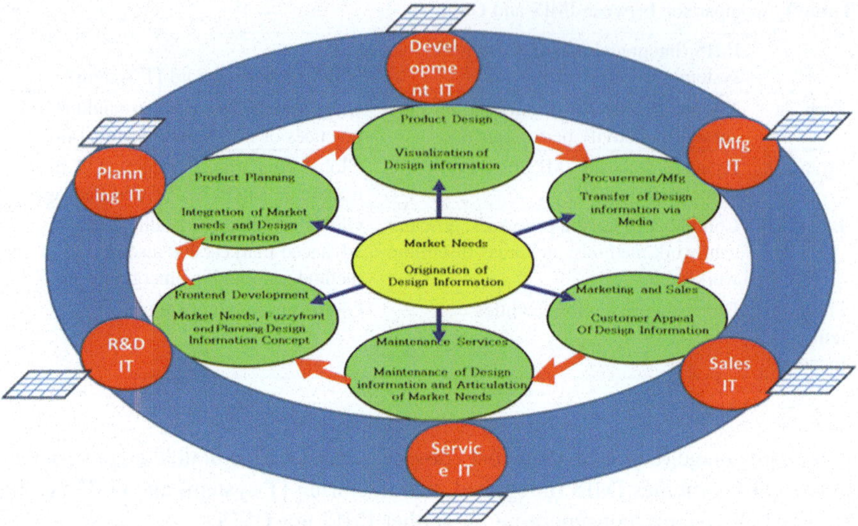

Fig. 2 Model of architectural analysis based on BOM of design information. (Adapted from Fujimoto and Park 2015)

master BOM (MA-BOM). The master BOM is a massive database that covers business processes including design, manufacturing, procurement, and maintenance services.

Komatsu adopted KOMTRAX (in 2001) and CMS (Cash Management System) (in 2007) in its construction equipment to manage operational efficiency, energy consumption, and IT utilization rates in real time from distant locations. In 2009, Komatsu also implemented a standardized BOM system (including both design features and components codes) in all its production facilities around the world. Such standardization efforts ensure flexible transfer of parts among production facilities and simultaneous production in factory work units across the globe. Prior to the standardization of components codes, Komatsu strengthened its network infrastructure for data exchange from any parts of the world. In October 2008, it delegated its network operations management to the US-based firm Verizon to achieve effective worldwide connectivity.

Furthermore, Komatsu's internally developed BOM system, called G-DMS (Global Data Management System), unified data communication between R&D, CAD design divisions, and factory ERP (Infor ERP LN). G-DMS is a single global system in which each upstream local design unit uses general CAD systems and each downstream local factory unit operates an MRP system installed with Infor ERP LN. Design data flow from the CAD system via G-DMS to ERP. The internal development teams of Komatsu also created interface mechanisms between CAD and ERP. In this way, Komatsu's G-DMS uses multiple global hubs to connect all local systems with standardized CAD and ERP systems.

The broad operational scope of G-DMS includes CAD-based component parts data (BOM data), BOM editing functions, CAD-applied electronic design blueprints, CAD data release functions, paperless electronic design blueprint approval, automatic notices and communication functions for global design changes, M-BOM (manufacturing BOM) editing functions, direct data transfer of Infor ERP LN, REACH (Registration, Evaluation, Authorization and Restriction of Chemicals), and automatic evaluation of materials fitness tests according to export regulations. The concept of G-DMS was first introduced in 2004, actual development took place in 2005, and by 2006 the system had already been implemented in the Chattanooga facility (TN, USA). In December 2006, Komatsu began large-scale production of D52-22, a midsize bulldozer jointly developed by Japanese and American design teams. In 2007, G-DMS was installed in domestic facilities in Japan, and then it was introduced in the company's facilities in Europe (2008), Brazil (2008), and China and Russia (2011). While extending the scope of G-DMS implementation, Komatsu also expanded its database and kept improving overall system functions, also adding target cost functions for all new equipment. By 2009, G-DMS included a cost BOM in its database, which enabled the management to monitor factory-level production efficiency and productivity data, thus significantly reducing the number of manufacturing processes involving extensive human activities. Komatsu's GIMIS (global integrated manufacturing IT system) is an example of how to combine IMIS (integrated manufacturing IT system), to tap the full potential of Japanese manufacturing capabilities, and GSIS (global standard IT system), to address fast-changing global market requirements.

5 Conclusion

In Japan, during the post-World War II period, a historical context of "shared destiny" among people led firms as a whole to own value-added flows in the form of integrated manufacturing work environments. The Japanese style of IT support for IMIS needs reflected this emphasis on IMIS-centered IT systems and enabled Japanese factories to attain outstanding field-level productivity and flexibility. However, the focus on IMIS did not promote standards that could be shared by both corporate strategic divisions and factory field operations, thus creating relatively weak global system linkages.

On the other hand, the concept of GSIS (global standard IT system), with its strong emphasis on the needs of specialized functional segments, is not necessarily compatible with an approach based on IMIS (integrated manufacturing IT system). Japanese manufacturing firms (with their monozukuri orientation) focus on sustainable competitiveness. Therefore, implementing GSIS by sacrificing field-level productivity performance is tantamount to seeking globalization for globalization's sake.

This chapter highlights the strengths of IMIS (integrated manufacturing IT system) and points out the need for GSIS (global standard IT system). In other

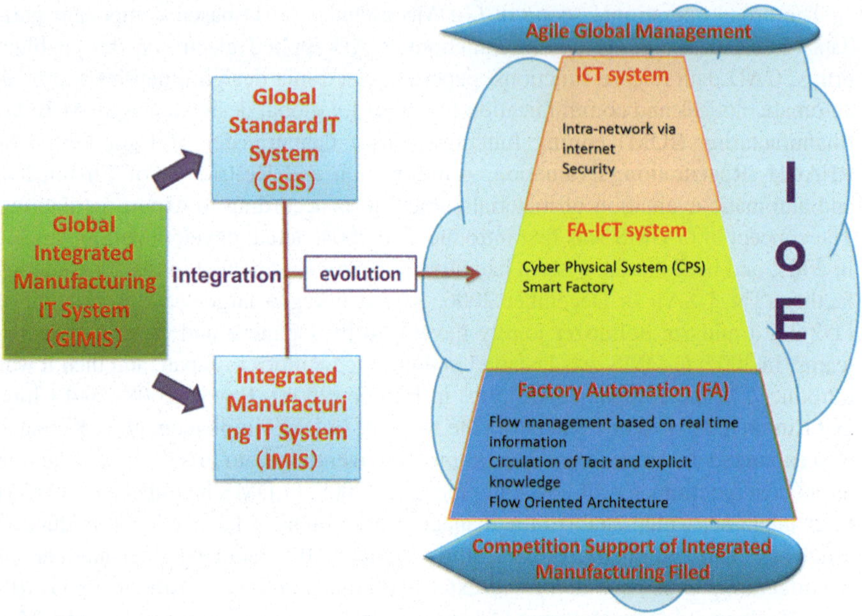

Fig. 3 Ideal global integrated IT system linked with field level

words, we argue in favor of a GIMIS (global integrated manufacturing IT system) that integrates both IMIS and GSIS. In conclusion, Fig. 3 shows an ideal GIMIS (global integrated manufacturing IT system), which might be a way to fulfill the dynamic requirements of the emerging IoT and Industry 4.0 and, at the same time, to achieve high levels of intelligent system specifications for IMIS needs sensing, complex control mechanisms, and intra- and interfirm network connections. Future research may examine related issues further to provide useful guidelines for global firms, caught between the needs of unique firm contexts and global market requirements.

References

Chang SJ (2008) Sony vs. Samsung. John Wiley & Sons, Tokyo

Clark K, Fujimoto T (1991) Product development performance: strategy, organization and management in the world auto industry. Harvard Business Review Press, Boston

Cole RE, Nakata Y (2014) The Japanese software industry: what went wrong and what can we learn from IT? Calif Manag Rev 57(1):16–43

Cusumano MA (2004) The business of software: what every manager, programmer, and entrepreneur must know in good times and bad. Free Press/Simon & Schuster, New York

Fujimoto T (2001) Industry argument on architecture. In: Fujimoto T et al (eds) Business architecture. Yuhikaku, Tokyo, pp 3–26

Fujimoto T (2003) Competing to be really good. International House of Japan, Tokyo

Fujimoto T, Park YW (2015) Case study: Monozukuri utilizing IT. Nikkei BP, Tokyo
Hong P, Park YW (2014) Building network capabilities in turbulent competitive environments: business success stories from the BRICs. CRC Press, Boca Raton
JISA (2015) Japan's information service industry in 2014. Japan Information Technology Services Industry Association, Tokyo
Ministry of Internal Affairs and Communications (2013) White Paper 2013 (in Japanese)
Oku M et al (2010) Integrated manufacturing information system case series (1): an example of management (FOA Concept) by IMIS principles. MMRC Discussion Paper 306:1–18 (in Japanese)
Park YW (2009) Core competence and IT strategy. Waseda, Tokyo (in Japanese)
Park YW (2016) Adding to the way of reformation. In: YDC (eds) Why don't Japanese industries succeed in a solution business? Nikkan Kogyo Shimbun, pp 180–196 (in Japanese)
Park YW (2017) Business architecture strategy and platform-based ecosystems. Springer, Singapore
Park YW et al (2011) Core competence and architecture strategy: framework for product strategy analysis. MMRC Discussion Paper 376:1–42 (in Japanese)
Park YW et al (2012) Implementation of product strategy with differentiated standards. Int J Technol Manag 57(1/2/3):166–184
Tomita N et al (1999) Flow oriented approach for human-centered agile manufacturing systems. ISADS99 (Fourth international symposium on autonomous decentralized systems) IEEE Computer Society Press, pp 1–9

Part III
Architectural Evolution and Demand Creation in the Digital Industries

Let us now turn to the second main aspect regarding the evolution of industries and firms during the period after the 1990s, i.e., architectural evolution and demand creation in the digital industries, many of which are characterized both by open-modular architectures with industry-standard interfaces linking mutually complementary products and components and by competition among rival platforms.

One of the main features of the platform-driven digital industry is that demand for goods that are complementary within a platform tends to have a mutually reinforcing effect. That is, increased demand for a core product/component usually brings about increased demand for its complementary products/components, which results in a cumulative effect of network externality. This cumulative effect sometimes leads to the creation of major platform-leading firms (e.g., Apple, Google, Amazon, Facebook, Microsoft, Intel, etc.) that, within a relatively short time, take over certain segments of the digital industry and generate huge revenues and profits from the extremely fast-growing platforms that they dominate. Yet, the question whether this sort of mutual demand creation from complementary relationships might continue endlessly remains unanswered at this point. Moreover, although the platform-leading giants just mentioned are heavily concentrated in the US as of the 2010s, some prototypical cases of platform-leading firms may also be found outside America—Nokia (Finland) and NTT Docomo (Japan) in the case of traditional cell phones, as well as Nintendo and Sony in the case of video games.

Another important finding is that the abovementioned cumulative demand creation effect between complementary goods can be found at the level of both products and components. This is particularly true when low-cost firms in emerging countries, with relatively low technological integration capabilities, are involved in such cumulative process in global competition. That is, when a platform and its core products are architecturally open-modular, with industry-standard interfaces established not only among products but also among their components, low-cost firms in the emerging nations are soon able to develop core products. They do this by combining the design information that is available in the open domain of the digital platform in question.

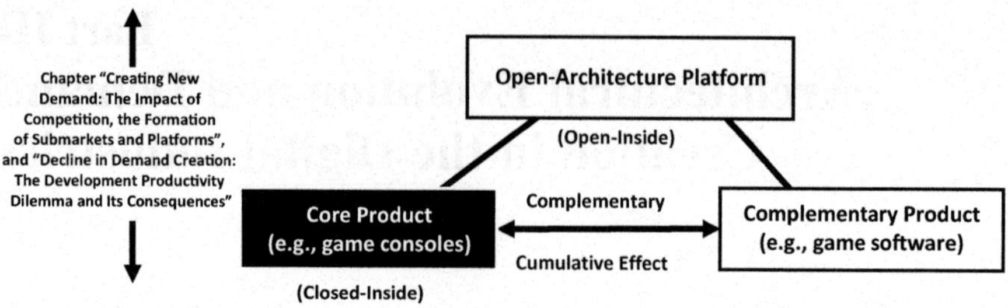

Fig. 1 Open architecture platform with complementary goods, a simple case

Based on the above characterization of the open-architecture sectors, we start Part III with the case of the video game industry in Japan. Chapter "Creating New Demand: The Impact of Competition, the Formation of Submarkets and Platforms" illustrates the basic nature of the video game industry, where mutual demand creation between game hardware and software did occur, as also seen for Internet-driven digital products, such as smart phones and their application software. Chapter "Decline in Demand Creation: The Development Productivity Dilemma and Its Consequences" shows the mechanism of diminishing effects of cumulative demand creation as the video game industry lifecycle progresses.

It is worth noting here that the video game industry is a relatively simple prototypical case of platform-based industry, which can be represented by a single-layer structure (see Fig. 1) with core products (i.e., video game consoles as hardware) and complementary goods (i.e., game software). The core products themselves, which are architecturally closed-inside, are developed and produced by technologically capable firms, such as Nintendo and Sony, in advanced nations, such as Japan. Apple's smartphones (iPhone), with internally developed operating system and hardware, may be described in much the same way.

Chapter "Investigating the Creation and Diffusion of Knowledge for Demand Creation: The Case of the Telecommunications Industry" and Chapter "The Impact of Platform Providers' Knowledge on Interfirm Division of Labor: The Case of the Mobile Phone Industry", on the other hand, focus on the more recent and complicated mobile phone segment (conventional cellular phones and Google/Qualcomm-based smart phones), characterized by a multi-layer structure for mutual demand creation with complementary products/components within an open-architecture platform/product. It should be noted that there are at least two layers of open-architecture platform/products in the case of mobile phones of this type.

As illustrated in Fig. 2, the core products themselves (e.g., conventional cellular phones and smart phones) are architecturally open-modular-inside. Thus, even with relatively limited technological capabilities, *brand firms* in emerging countries can produce such core products by utilizing design information and production knowledge in the open domain developed by *core component firms* (e.g., Qualcomm as platform provider) and/or *contract manufacturers*.

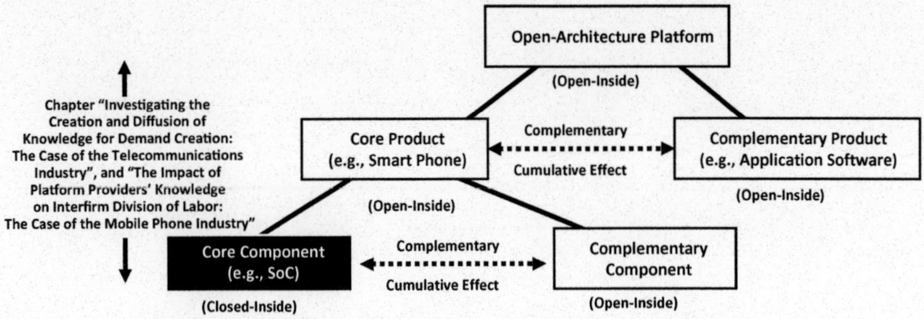

Fig. 2 Open architecture platform with complementary goods, a multi-layer case

Let us now look at each of the above cases in greater detail in Chapters from "Creating New Demand: The Impact of Competition, the Formation of Submarkets and Platforms", "Decline in Demand Creation: The Development Productivity Dilemma and Its Consequences", "Investigating the Creation and Diffusion of Knowledge for Demand Creation: The Case of the Telecommunications Industry", and "The Impact of Platform Providers' Knowledge on Interfirm Division of Labor: The Case of the Mobile Phone Industry".

Creating New Demand: The Competition, Formation of Submarkets, and Platforms

Fumihiko Ikuine

Abstract How do companies go about creating demand? This chapter presents the case of console games (home video games) in Japan. This market was formed as a completely new market in the 1980s. The early days of software for consoles (game software) are described, and the process of creating demand for this market is considered. Emphasis is placed on the impact of corporate competition on genres, the formation of submarkets, and platforms. Especially, we discuss the contribution of *Space Invaders* as a successful predecessor and the role of the *Family Computer* as a platform. Game software relies on an open architecture with paired (complementary) products. Since these markets are open, it is difficult to define them in a simple and consistent manner. Rather, they are more like agglomerations of multiple submarkets, within which interactions among companies take place. Diversity both within and across these submarkets attracts large numbers of consumers. Therefore, multiple submarkets coexist within the video game market and create huge demand as a whole. The combination of a platform and a successful predecessor enabled the coexistence of submarkets, brought about competition among companies, and was a driving force that stimulated vigorous demand for game software.

1 Introduction

Improvements in workplace (genba) capability are rewarded by consumers, who purchase products and services, and workplace evolution and demand creation work closely together (Nakazwa et al. 2016). This chapter explores the process of demand creation along with increased product diversity. The case of console games (home

F. Ikuine (✉)
Faculty of Engineering, Information and Systems, University of Tsukuba, Tsukuba, Ibaraki, Japan
e-mail: fikuine@sk.tsukuba.ac.jp
URL: http://www.trios.tsukuba.ac.jp/en/researcher/0000003030

T. Fujimoto, F. Ikuine (eds.), *Industrial Competitiveness and Design Evolution*, Evolutionary Economics and Social Complexity Science 12, https://doi.org/10.1007/978-4-431-55145-4_11

video games)[1] in Japan is an excellent research subject to shed light on demand creation. The reason for this is that a situation in which no previous demand existed completely changed in the early 1980s.

The business of software for consoles (game software) in Japan was first established in the 1980s, and by the late 1990s, it had grown into a 500 billion yen market, with many titles also being exported overseas. Several products launched in rapid succession helped this expansion. Indeed, from 1983 to 1999, 7162 products were released by 472 companies,[2] which means that on average, 421 products were made available per year or more than 1 new product every day.

Game software is a product that provides enjoyment. Players manipulate the character (avatar) and interact with the computer, and this is combined with changes in images and sounds. Three key characteristics derive from this process. Firstly, the situation, which lets the player have fun through representation and manipulation, varies from user to user. Since it is hard to know in advance which situations might attract the most users, companies are confronted with a range of uncertain and ambiguous needs. Secondly, users buy new products because they become bored with the process that each individual game provides. Hence, to induce users to make a purchase, companies need to constantly add elements that differ from those in existing products. Thirdly, developers at competing companies can grasp the ideas and representations included in good products by playing the games themselves. Moreover, game software is a wholly digital product. It is thus relatively easy for developers to imitate other products if they have a certain amount of knowledge about game software. Based on these characteristics, companies concentrate their efforts on creating new and interesting products in rapid succession.

This chapter will describe the rise of the console game business in the 1980s to try and shed light on two main questions: How exactly was new demand for game software created? And how were companies able to achieve differentiation among products that were being launched one after the other? Our analysis places particular emphasis on such factors as corporate competition, product categories of video games (genres), and platforms.

[1]Games using computers can be categorized according to the hardware used. Computer games using hardware specifically designed to play games at home are called console games. A console game combines specialized hardware and the software running on that hardware. Other types of computer games are arcade games, using hardware for playing games in public businesses; PC games, using personal computers; handheld console games, using portable gaming hardware; and so on.

[2]These figures exclude games that were merely different versions of existing products but ran on specific hardware (platforms). If these are included, the number of products increases to 9879.

2 Literature Survey

The business of game software in Japan has attracted the attention of researchers inside and outside of the country. Representative research includes works focusing on Nintendo, a key player during the 1980s and 1990s (Kohashi and Kagono 1995; Subramanian et al. 2011; Uemura et al. 2013), as well as studies of game software development and the game software business (Aoyama and Izushi 2003; Ikuine 2012; Izushi and Aoyama 2006; Shintaku et al. 2003; Storz 2008). Most of the existing research asserts that the business of game software in Japan has grown because several entrants have been offering many and diverse products. This chapter supplements the existing research by taking a similar stance.

On the surface, the game software sector resembles the digital camera market discussed by Benner and Tripsas (2012). Initially, the entry of a wide array of companies led to the establishment of this new industry. According to them, both the main features and the definition of the product (i.e., what a digital camera is) were decided through corporate competition. The digital camera market took shape as companies struggled to remain viable. This chapter explains the early days of the game software business through a comparison with Benner and Tripsas (2012).

Nonetheless, the case of game software differs from that of the digital still camera in that the game software market consists of multiple product categories, each of which forms an essentially independent submarket. Each game software submarket attracted different users and developed independently, which led to the growth of the entire market. By focusing on product categories (genres), the evolutional process of game software will be described to explore how interfirm competition brought diversity to the sector, how the diversity of game software drove the formation of submarkets, and how the growth of said submarkets created huge demand. At the end of the chapter, we will discuss the role of a platform and of a successful predecessor as the basis for product differentiation.

3 Case Description and Interpretation: Demand Creation of Game Software

3.1 Early Days of the Video Game Business

Two businesses influenced the establishment of home video games in Japan.[3] One was the computer game business in the USA, and the other was the arcade game[4] business in Japan.

[3]For the history of home video games prior to 1983, see Fujita (1998, 1999a, 1999b, 2006), Akagi (2005), Uemura et al. (2013), Nakamura (2016), and Kondo (2017).

[4]Arcade games are games placed by businesses in public places (bars, shopping malls, amusement arcades). Because arcade games used more expensive hardware than home video game consoles, technological standards remained high until the 1990s. See Akagi (2005) regarding the vicissitudes of the arcade game industry.

There are different stories about the origin of computer games, but most researchers agree on the important role played by Nolan Bushnell, a key predecessor in their development. Nolan Bushnell founded Atari in the USA. In 1972, Atari released *Pong*—a new type of entertainment used in public places—and this drove the company's success and rapid growth. Around the same time, Ralph Baer developed the *Brown Box*, which made it possible to play games at home. As computer performance improved in the 1980s, the business of computer games grew in the USA. Game consoles, in which one could insert various cartridges for playing multiple games (consoles with exchangeable cartridges), arrived on the scene in 1976–1977. Among these, the most representative products were Fairchild's *Video Entertainment System (VES)* and Atari's *Atari 2600*. While some of them garnered a measure of success, stable demand remained elusive in the USA.

In Japan, the situation was similar to that of the USA, even though the computer game business started a little later. Taito released the original arcade game *Space Invaders* in 1978, which proved to be a turning point in the history of the sector. The craze for this game, labeled a "social phenomenon," led to the widespread recognition of computer games and spurred many companies to trade *Space Invaders* or develop similar games. During the *Space Invaders*' boom, these entrants accumulated financial resources, related technology, and business practices.

Just like in the USA, Japanese companies started selling consoles with exchangeable cartridges using their own standards, and over ten types of gaming consoles were on the market by the early 1980s. Only Nintendo survived and established a viable console game business, with the launch of the *Family Computer* in 1983.

3.2 *Start of the New* Family Computer *Market*

Nintendo released 3 pieces of game software in conjunction with the launch of the *Family Computer* and then went on to release 16 additional titles within the next year. Examples of the earliest game software included transplants from such arcade games as *Donkey Kong* and *Mario Bros.*, computerized table games like *Connect Five*[5] and *Mahjong*, and sports-themed games such as *Baseball* and *Tennis*. Early products covered four genres.[6] The appeal of the *Family Computer* console was that it used exchangeable cartridges, so that a variety of games could be played on a single piece of hardware. Nintendo was thus able to offer various games to children and adults alike, providing a new form of entertainment as well as new ways to play old games like Mahjong.

[5]A classic Japanese game resembling tic-tac-toe.

[6]The term "genre," used loosely by both companies and users, is a subjective determination of product classification, rather than an objective indicator of product content. This chapter uses the term "genre" consistently with its usage by Ambit Inc. See Ikuine (2012) for the pros and cons of using "genre" as an index.

Both the *Family Computer* hardware and its compatible software sold well. Seeing this, other companies started to develop game software that could be played on Nintendo's *Family Computer*. In 1984, Hudson and Namco entered this business by closing contracts with Nintendo. Namco was a major producer of arcade games, while Hudson had achieved success with games for personal computers (PC games). These two companies were instrumental in transplanting arcade games and PC games to the *Family Computer* console. The entry of Hudson gave rise to a new genre, i.e., the puzzle game, with such titles as *Nuts and Milk* and *Roadrunner*, while Namco pioneered the shooter genre with titles like *Galaxian* and *Xevious*. *Pac-Man* and other games also enjoyed popularity within genres introduced by Nintendo. At the start of the business, three firms transplanted various games, including original games, arcade-origin games, and PC-origin games, to the *Family Computer* console. This triggered the growth of both individual submarkets and the market as a whole (Fig. 1).

The success achieved by Nintendo, Hudson, and Namco with their *Family Computer* titles encouraged the expansion of the sector. Sixteen companies entered the market in 1985, followed by another 19 in 1986. During the 1980s, at least ten companies entered the market every year.

The number of new products also grew, reaching 67 in 1985 and 114 in 1986. In terms of genres, one new genre was added in 1985, while four new genres first

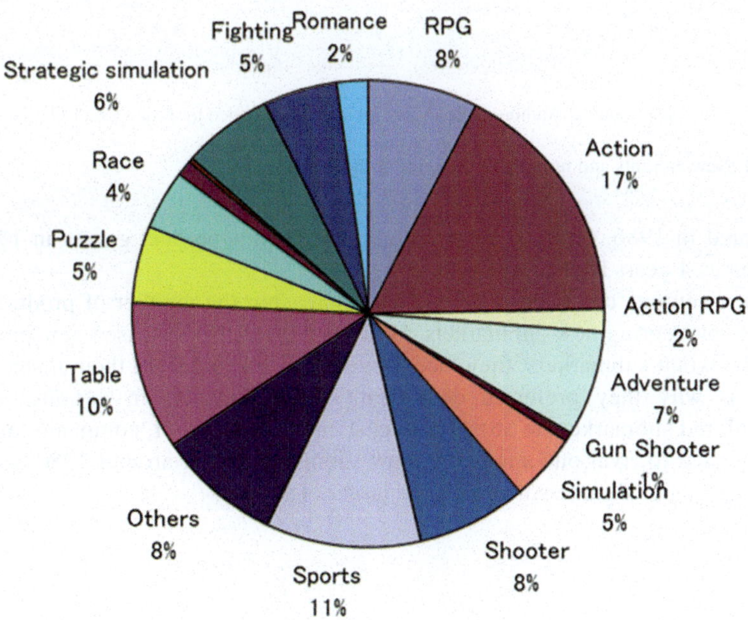

Data source: Author-created based on the data provided by Ambit Co., Ltd.

Fig. 1 Classification of game software by genres

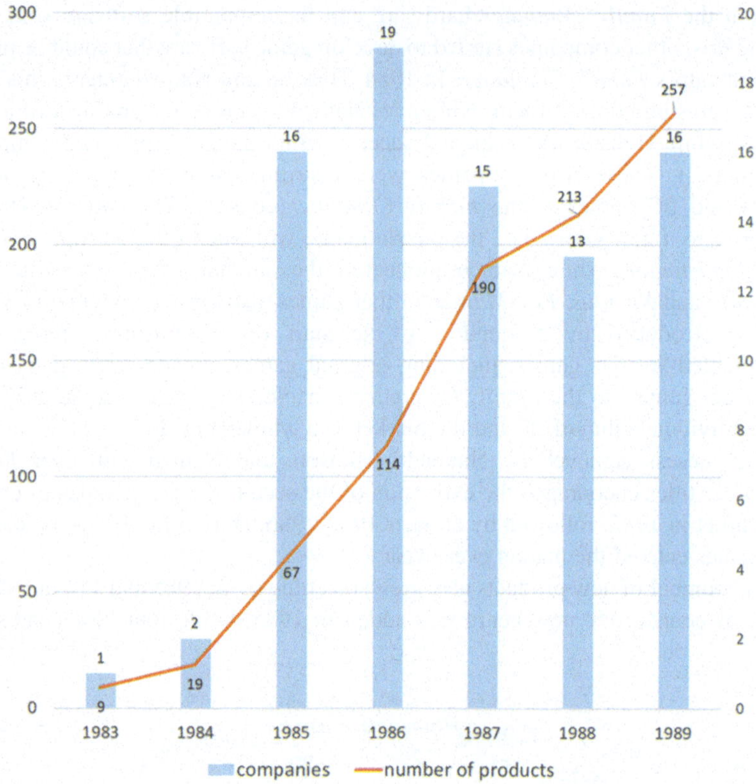

Data source: Author-created based on the data provided by Ambit Co., Ltd.

Fig. 2 New entrants and number of products in the 1980s

appeared in 1986. Overall, the industry went from 4 product genres in 1983 to 12 genres 4 years later (Fig. 2).[7]

New entrants contributed not only to increasing the number of products but also to pioneering new submarkets. However, when they released new products for the extant submarkets, they faced severe competition from their antecessors. This is why they preferred developing novel products in various, so far unexplored submarkets to attract new customers. As a result, companies aggressively tried to rival one another in new submarkets until around 1986, and the product space of game software grew larger (Table 1).

[7]The breakdown of companies that introduced the first product in a genre are Nintendo with six genres, Namco with one genre, Hudson with one genre, Enix with two genres, SoftPro with one genre, and Toshiba EMI with one genre.

Table 1 The first products in each genre

	The year of first product of genre	The year of first million seller (the first product of series)	The year of first million seller
RPG (role-playing game)	1986	1986	1986
Action	1983	1983	1983
Action RPG	1986	1986	1986
Adventure	1985	1996	1996
Gun shooting	1984	x	x
Simulation	1986	1997	1995
Shooting	1984	1984	1984
Sports	1983	1984	1983
Other	1983	1998	1996
Table game	1983	x	1983
Puzzle	1984	1984	1984
Racing	1984	1984	1984
Simulation breeding	1993		x
Action, music-based	1996	1998	1996
Simulation, strategy	1986	x	1997
Action, fighting	1988	1992	1992
Simulation, love story	1987	1994	1994

Data source: Author created based on data provided by Ambit Co., Ltd., Tokyo Gungu Ningyo-Donnya Kyokai "Monthly Toy Journal" (Gekkan Toi Jahnal)

Yet, corporate efforts to form submarkets by introducing titles in new genres began to decline around 1987. Only four new genres were introduced during the period from 1987 to 1999. After 1987, competition focused mostly on differentiating products within the genres introduced until 1986 (Ikuine 2012).

3.3 *Interfirm Relationships in the RPG Submarket*[8]

How did new product releases establish submarkets, and how did competition develop within those submarkets? Let us look at role-playing games (RPGs), the

[8]Fujii (2017) describes the early days of the RPG submarket and analyzes the strategies of Enix and Square, emphasizing the rivalry between the two companies, top management's strategic decision-making, and the process of building a business system, along with organizational capability. This section is based on Fujii's analysis.

most popular game software genre in Japan and unique from a global perspective.[9]

In May 1986, the first RPG, *Dragon Quest* (also known as *Dragon Warrior*), was released by Enix, a new entrant that had already created two titles in 1985. The development of *Dragon Quest* was influenced by table-talk RPGs and RPGs for personal computers, which made it easier to play table-talk RPGs on computers (Wada and Ichikoji 2016). Using features from existing games while also incorporating its own unique aspects, *Dragon Quest* was enjoyed by large numbers of gamers and eventually became a million seller.[10] This game displayed two new key elements (Wada 2017). First, players could reasonably reach the final goal by accomplishing a series of achievable targets for which they were awarded experience points. Second, they could enjoy a fully fledged story and feel as if they were playing the role of a hero/heroine in a virtual world, which led them to keep playing the game to the end without getting bored. Thanks to the addition of new elements departing from traditional RPGs, *Dragon Quest* resulted in a game that many Japanese users wanted play.

Encouraged by the success of Enix's *Dragon Quest*, more and more companies launched products in the RPG genre. In 1986, three products, including *Dragon Quest*, were launched by three producers. This grew to 18 new titles released by 14 companies in 1987. Despite a decline to 11 products by 8 companies in 1988, the number continued to rise in 1989 and thereafter, with 59 games launched in 1994 (Fig. 3).

Many titles by various companies appeared on the market year after year, among which the well-known *Final Fantasy* series by Square. In *Final Fantasy*, Square took the basic elements of the RPG from *Dragon Quest* and added beautiful images, impressive music, and other dramatic effects. As a result, *Final Fantasy 3* also became a million seller. Square succeeded in nurturing its own product series to rival *Dragon Quest* and was furthermore able to release spin-offs, thus serializing the *Final Fantasy* saga. Meanwhile, genre pioneer Enix released additional titles in the *Dragon Quest* series every few years after 1990. Enix's *Dragon Quest* and Square's *Final Fantasy* series became million-seller RPG products in Japan.[11] Enix and Square continued to launch new products in their game series, which became million sellers and established the RPG submarket.

[9]This is also known as "J-RPG" because of its unique content.

[10]*Million sellers* are products that have logged unit sales of 1 million or more. As the penetration of game consoles in Japanese homes at 10 million devices is considered to be the de facto standard, titles that have penetrated 10% of this hardware market are viewed as successful. See Ikuine (2012) for further details.

[11]There were only two million-seller RPGs released by other companies, i.e., one by Nintendo and one by Sony Computer Entertainment (SCE).

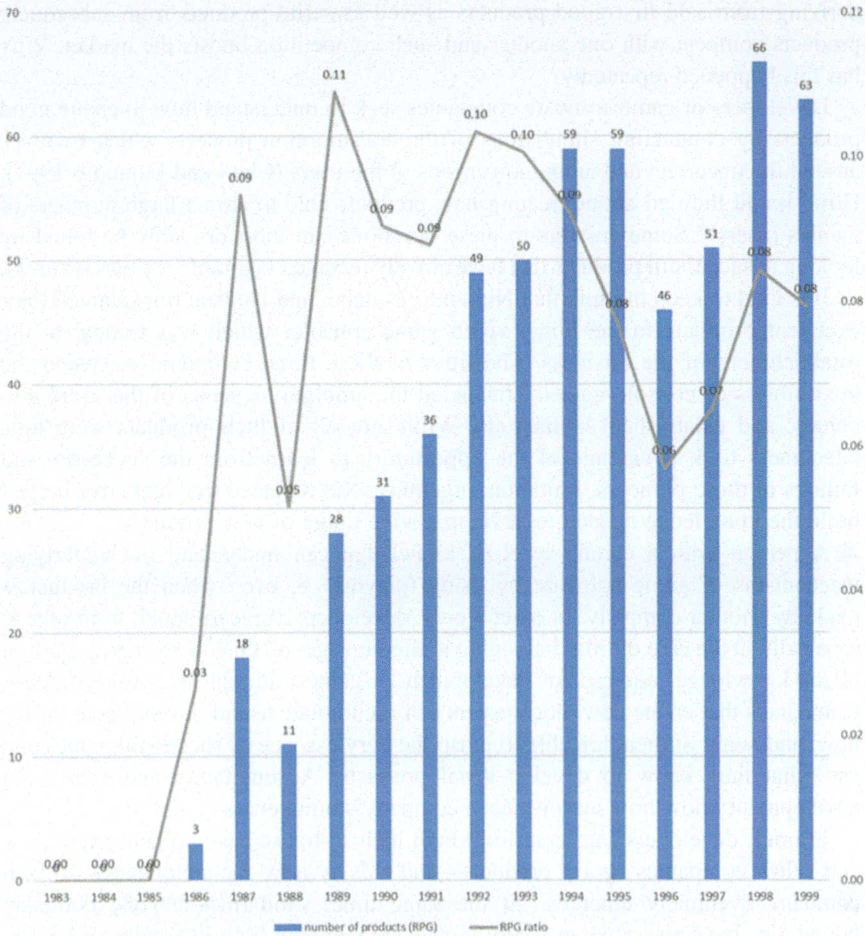

Data source: Author-created based on the data provided by Ambit Co., Ltd.

Fig. 3 Overview of the RPG submarket

3.4 *Similarity in Differentiation: Competition as Interlocution Revisited*

The phenomenon that occurred in the RPG submarket has also been seen in other genres. Once the first outstanding products, like *Dragon Quest*, are released, the new submarket grows, and the genre is established. Then, titles based on the first excellent products but containing some new elements are launched. Serial products deriving from said first good products as well as serial products from subsequent products compete with one another and such competition boosts the market. Why has this happened repeatedly?

Developers of game software companies seek to understand how to create good products by conducting simulations of the consumption process, which revolves around the uncertain and ambiguous needs of the users (Clark and Fujimoto 1991). How should they go about creating new products able to attract large numbers of gamers (users)? Some answers to these questions can most probably be found by looking at successful products that have already received favorable reviews by users.

It is vital to keep in mind that Nintendo, Namco, and Hudson transplanted their excellent products to the home video game console, which was critical to the establishment of the business. The titles of these three companies revealed the possibilities of console games, channeled the ambiguous needs of the users into genres, and established submarkets. While not all of their products were hits, latecomers took advantage of the opportunity to learn from the successes and failures of these pioneers. Imitation, ingenuity, and repeated trial and error helped build the knowledge needed to develop a wide range of new products.

A person with a certain level of knowledge can understand the underlying mechanisms of game software by using (playing) it, even when the product is made by another company. In other words, developers can learn much from others, especially in the case of titles belonging to the same genre. Of course, a great portion of the knowledge required for development is gained through in-house products. Companies that create new blockbusters can accumulate useful development know-how, and some of them are able to grasp the very essence of successful games and use what they know to develop serial products. Accumulating and employing development know-how nurtures each company's uniqueness (taste).

In brief, developers can learn from both their in-house development experience and other companies' good products—and this is how the uniqueness of each company eventually emerges. At the same time, similarities beyond company boundaries have also emerged, and knowledge gathered both internally and externally has been instrumental in the continued development of moderately differentiated products. As a result, product similarities within specific genres as well as similarities within the Japanese market have gradually taken shape. To survive in the highly competitive Japanese game software market, companies tend to behave homogeneously and develop homogeneous products.

Through the above activities, developers have learnt to recognize what constitutes a game, which in turn affects managers, distributors, and users. Thus, a fixed concept, or common perception, of game software has spread uniformly across Japan.

3.5 *Coexistence of Multiple Dominant Designs: In Contrast to Benner and Tripsas (2012)*

With companies pursuing moderate differentiation, each piece of game software was somewhat alike but also somewhat different from the next. More specifically, almost every product used a plus-shaped control button to manipulate the characters, adjust the difficulty level, and calculate and display experience levels and scores. On the other hand, the game rules, character portrayals, and gaming environments enjoyed by the users were rather varied. Commonalities observed amid the many differences were considered attributes of a single genre. Gradually, both commonalities and differences started to be recognized, and this is how the genres of game software were determined. Hence, the definitions of game software, as well as its genres, were loosely formed ex post facto.

Despite ambiguous boundaries, the establishment of a genre made it effective to learn through products, as mentioned above. Learning within the same genre also occurred in the case of RPGs. Produced by outstanding innovators, *Dragon Quest* provided many clues as to how the users' appetite for purchasing could be stimulated within the RPG genre. Indeed, Square was one of the main actors able to take full advantage of such clues. It is not so much that game software in general served as a model to emulate but rather that specific products representative of each genre became the reference templates. *Dragon Quest* was a *dominant design*, serving as model worthy of emulation within the RPG genre.

At the same time, *Dragon Quest* was quite different from Nintendo's first 16 products. It also considerably differed from *Galaxian* and *Xevious* by Namco. Therefore, it was recognized as not belonging to any of the genres existing in 1985. In the same manner, Nintendo's first 16 products were categorized into 4 genres. This means that many users, businesspeople, and developers acknowledged the existence of product genres from the very early days of the industry.

Since then, competition through differentiation has taken place within the established genres, and multiple submarkets have emerged. Said genres and the formulation of product definitions are not akin to the case of digital cameras, which Benner and Tripsas (2012) studied. In game software, several dominant designs coexist within the same market. A submarket having a dominant design at its core is recognized as a genre—and it is these separate genres that attract the users. A definition of game applicable to the entire video game market remains ambiguous as companies continue to come up with various types of gaming experiences. This results in the creation of a product space accommodating many diverse products and generating enormous demand overall.

4 Discussion: Platform Effect on Demand Creation[12]

The platform for the use of game software was the central factor supporting this loosely defined product and the coexistence of multiple genres. In other words, the existence of platforms like the *Family Computer* sets the stage for companies to compete by differentiation.

Platforms affected game software companies in two ways. Firstly, they provided an opportunity to make up for a lack of management resources, particularly in terms of technology. It is safe to say that game software companies lacked sufficient resources to develop both software and hardware before they entered the business. Nevertheless, once hardware and software were separated, thus paving the way for the development of software alone, scarce resources were no longer an issue.[13] Enix and Square were typical examples. The other effect linked to the establishment of platforms was that wannabe entrants were confident that potential demand for game software did indeed exist and, as a result, they were stimulated to enter the business.

On the other hand, the establishment of platforms also had an impact on consumers, since platforms helped generate a certain amount of demand by organizing the needs of gamers. Before the appearance of platforms, the concepts of what and how one wanted to play remained undefined. Consumers, each with their diverse wants and needs, vaguely thought "I want to play video games," without clear preferences as to which games they actually wished to play.

After the appearance of platforms, consumers, who still harbored various and sundry desires, learnt how to satisfy them. As long as they had a platform, they could expect to be entertained by playing some game created for that platform. With the introduction of multiple games, consumers discovered that their latent desires could be met with specific products. Consumers' needs were thus generated, bundled, and organized, which resulted in growing demand. In other words, the demand was not fragmented by the power of platforms.

Such demand, organized through the existence of platforms, enabled companies to engage in development activities to thoroughly meet it. Moreover, the further systemization of organized needs was accomplished through multiple dominant designs, as characterized by *Dragon Quest*. The platform became the conduit for stimulating mass demand by bundling needs and leveraging commonalities among diverse user preferences.

The second effect of platforms was to build up expectations about the future. The experience with successful predecessors, like *Space Invaders*, gave users the feeling that playing video games was worth spending money and time on. Positive memories reinforced their belief that they enjoyed gaming. At the same time, the

[12]Tatsumoto (2017) reviews the effect of platforms on demand creation drawing on the *two-sided market* theory. In other words, he explains how platforms can contribute to demand creation. Based on his work, this section discusses why platforms can contribute to demand creation.

[13]Nevertheless, the so-called platform paradox proposed by Wada et al. (2014) occurred simultaneously.

experience with successful predecessors told users that they would get bored in the near future and led them to look at video games with a mixture of anticipation and dread. In such a situation, platforms integrated their present state, past memories, and future expectations, allowing them to overcome the anxiety of knowing that they would eventually get bored and to expect their desires to be fulfilled by new games in the future. From this point of view, the combination of a platform and a successful predecessor was extremely important.

The third effect of platforms was to boost potential demand. Once users had a platform, they became less reluctant to purchase new game software, simply because they already owned the device, which obviously resulted in additional demand. For their part, companies were in a position to make predictions about such additional demand and aggressively invest to develop products in anticipation of it. Hence, additional demand and aggressive development investments stimulated each other and, through this mechanism, overall demand for platform-based game software grew larger than the fragmented market itself.

As for the Japanese game software market, *Space Invaders*, representing the successful predecessor, boosted potential demand, as did the *Family Computer*, as the reference platform. Users had a very optimistic outlook, ascribable to the combination of platform and successful predecessor, and happily bought both the hardware and the software. Expecting additional demand induced by the platform, companies wasted no time in developing new software. As a result, many diverse games were launched, and the demand for game software grew beyond expectations.

5 Conclusions and Remarks

Japan's console game business has successfully created a huge demand through a sequence of key events. At first, successful predecessors appeared, as typified by *Space Invaders*. The release of the *Family Computer* in 1983 was a turning point, and from 1984 onward, many entrepreneurs and users became involved in console games. Thus, a new market was born.

Multiple dominant designs were generated in this new market because of corporate competition based on differentiation. Due to partitions between genres in the game software product space, companies could easily find ways to differentiate themselves, thereby facilitating product selection by consumers. Indeed, the rapid succession of new products has managed to sustain product demand over a 30-year period continuing to this day.

The existence of such platforms as the *Family Computer* influenced demand creation. Platforms did not merely provide business opportunities to market entrants or represent vehicles for learning. Rather, they kept the market unified and prevented it from fragmenting on the basis of separate genres.

The mechanism of competition through differentiation, occurring in game software and its associated submarkets, is similar to that discussed by Numagami et al. (1992). In the case of game software, however, several companies participated in successive interlocution and interactions that led to a certain amount of homogeneity. Further consideration should be given to the subject of interlocution among many companies.

Moreover, the formulation of product categories and markets should be perceived as the result of multiple processes both inside and outside the organizations (Miyao 2011, 2016). This chapter has merely touched upon these processes, which we would like to investigate further building on existing research and drawing on the abundant findings of video game case studies.

References

Akagi M (2005) It starts with Pong (Sore ha Pong kara hajimatta: Akeido TV ge-mu no naritachi). Amusement Press Inc, Toyo (in Japanese)

Aoyama Y, Izushi H (2003) Hardware gimmick or cultural innovation? Technological, cultural, and social foundations of the Japanese video game industry. Res Policy 32(3):423–444

Benner MJ, Tripsas M (2012) The influence of prior industry affiliation on framing in nascent industries: the evolution of digital cameras. Strateg Manag J 33(3):277–302

Clark KB, Fujimoto T (1991) Product development performance. Harvard Business School Press, Boston, MA

Fujii D (2017) The management of technological innovation (Gijutsu teki inobeshon no manejiment). Chuou keizaisha, Tokyo (in Japanese)

Fujita N (1998) The rise and collapse of American videogame industry -the rise of videogame industry (1)- (Beikoku ni okeru bideo gemu sangyo no keisei to kyugekina houkai). Kyoto Univ Econ Rev 162(5,6):54–71 (in Japanese). https://repository.kulib.kyoto-u.ac.jp/dspace/handle/2433/45249?mode=full

Fujita N (1999a) Japanese videogame industry before "Family Computer" the rise of videogame industry (2)- (Famikon tojo mae no nihon no bideo gemu sangyo). Kyoto Uni Econ Rev 163 (3):59–76 (in Japanese). https://repository.kulib.kyoto-u.ac.jp/dspace/handle/2433/45271

Fujita N (1999b) The development of "Family Computer" and the rise of videogame industry -the rise of videogame industry (3)- (Famikon kaihatsu to bideo gemu sangyo keisei katei no sougouteki kousatsu). Kyoto Uni Eco Rev 163(5,6):69–86 (in Japanese). https://repository.kulib.kyoto-u.ac.jp/dspace/handle/2433/45282

Fujita N (2006) The formation and evolution of the modern video game industry: from the technological history perspective (Gendai bideo ge-mu sangyo no keisei to hatten). Ph.D. Dissertation, Kyoto University (in Japanese). https://repository.kulib.kyoto-u.ac.jp/dspace/handle/2433/143921?mode=full

Ikuine F (2012) Development productivity dilemma: the innovation patterns in the digital age (Kaihatsu seisan-sei no jirenma: Dejitaru-ka jidai no inobeishon Patan). Yuhikaku, Tokyo (in Japanese)

Izushi H, Aoyama Y (2006) Industry evolution and cross-sectoral skill transfers: a comparative analysis of the video game industry in Japan, the United States, and the United Kingdom. Environ Plan A 38(10):1843–1861

Kohashi R, Kagono T (1995) The exchange and development of images: a study of the Japanese video game industry. Paper presented at the Hitotsubashi-organization science conference on Asian research in organizations: emerging paradigms in organization studies. Kunitachi, Tokyo

Kondo H (2017) The formation of global supply chain: the case of home video game consoles in Japan. (gemu sangyo ni okeru sapurai chein no kokusaika: kateiyo gemu sijou no hatten to buhin kosei no henka). In: Abe E (ed) Global companies: the historical perspective of globalization. Bunshindo, Tokyo (in Japanese)

Miyao M (2011) New product development to redefine product categories: a study with the social shaping of technology approach (Seihin kategori wo saiteigi suru shin-seihin kaihatsu: gijutsu no syakaiteki keisei apurochi ni yoru kento). Organ Sci 44(3):120–131 (in Japanese)

Miyao M (2016) New product development and market creation (Seihin kaihatsu to shijo souzou: Gijutsu no shakaiteki keisei apurochi ni yoru tankyu). Hakuto shobo, Tokyo (in Japanese)

Nakamura A (2016) The evolution of the home video game console of the early days in the North America: Atari, Inc. and Nintendo of America Inc. (Hokubei dejitaru gemu sangyo keiseiki ni okeru kateiyou bideo gemu sennyo purattofomu keisei ni kansuru hikaku jirei kenkyu). Ritsumeikan Uni Coll Images Sci J 9:19–43 (in Japanese)

Nakazwa T et al (2016). Counterattack of Monodukuri (Monodukuri kara no hangeki) Chikuma shobo, Tokyo (in Japanese)

Numagami T et al (1992) Competition as interlocution: reinterpretation for competitive behavior in calculator industry (Taiwa toshiteno kyoso: Dentaku sangyo niokeru kyoso kodo no saikaishaku). Organ Sci 26(2):64–79 (in Japanese)

Shintaku J et al (eds) (2003) Economic analysis on game industry (Gemu sangyo no keizai bunseki: Kontentsu sangyo hatten no kozo to senryaku). Toyo Keizai Shimpo-sha, Tokyo (in Japanese)

Storz C (2008) Dynamics in innovation systems: evidence from Japan's game software industry. Res Policy 37(9):1480–1491

Subramanian AM et al (2011) Capability reconfiguration of incumbent firms: Nintendo in the video game industry. Technovation 31(5):228–239

Tatsumoto H (2017) Platform strategy for global market: the strategic use of open standards and the management of business ecosystem. (Prattofohm kigyo no guro-baru senryaku). Yuhikaku, Tokyo (in Jpananese)

Uemura M et al (2013) The life and times of the Nintendo Famicom: the birth of TV games (Famikon to sono jidai). NTT Shuppan, Tokyo (in Japanese)

Wada T (2017) History of Japanese role-playing games. Annals Bus Adm Sci 16:137. https://doi.org/10.7880/abas.0170228a

Wada T, Ichikoji T (2016) Different perception of continuity about innovation: where the MMORPG came from (Gijutsu ninshiki no chigai ni yoru inobeshon no renzokusei ninshiki no sai: MMORPG ha dokokara kitanoka). In: Hosoi K (chair), the second Kyoto game conference, symposium conducted at the Ritsumeikan Center for Game Studies, Ritsumeikan Univeristy, Kyoto

Wada T et al (2014) Platform paradox. Annals Bus Adm Sci 13:91–103. https://doi.org/10.7880/abas.13.91

Data

CESA Games White Paper (2002). Computer Entertainment Suppliers Association, Tokyo

Data provided by Ambit Co., Ltd.

Data provided by Media Create Co., Ltd.

Research for Information and Media Society (1995). Dentsu Innovation Institute, Tokyo

The Toy Journal (1991, 1992, 1993, 1994, 1995, 1996, 1997, 1998, 1999, 2000). Tokyo Toy and Doll Wholesalers Cooperative Association, Tokyo

White Paper on Leisure (1982, 1983, 1984, 1985, 1986, 1987, 1988, 1989, 1990, 1991, 1992, 1993, 1994, 1995, 1996, 1997, 1998, 1999, 2000). Leisure Development Center, Tokyo

Decline in Demand Creation: The Development Productivity Dilemma and Its Consequences

Fumihiko Ikuine

Abstract What are the consequences of demand creation by firms? Specifically, why does the ability to stimulate demand decrease in certain areas, as demonstrated by the life cycle model? When certain industries are in decline, where might that *lost* demand be redirected? In this chapter, we look at the game software business in Japan to illustrate the mechanisms that reduce the ability to spur demand. In the early days of an industry, development activities thrive, so that firms can respond to vigorous demand. As more and more time is spent on continued development activities, an increasingly large amount of development know-how (knowledge) is accumulated. However, this accumulation of knowledge gives a specific direction to said development efforts, and companies are faced with an implicit bias against creating completely new products. This bias inhibits product diversification. Users who witness this phenomenon often shift their buying intentions elsewhere, which in turn reduces demand. In fact, users tend to transition to other areas via platforms, particularly in circumstances where information technology flourishes. Thus, the boundaries of the market as the venue for demand creation become blurred, so that the market fuses with another market and eventually become *extinct*.

1 Introduction

As discussed in the previous chapter, Japan's console game (home video games) market rose to prominence in the early 1980s. While there were several options for the business of games based on computers, in Japan the sector developed thanks to a combination of hardware and software for use at home.[1] Since Nintendo's *Family*

[1]Uemura et al. (2013) provides factual details about the launch of the *Family Computer*.

F. Ikuine (✉)
Faculty of Engineering, Information and Systems, University of Tsukuba, Tsukuba, Ibaraki, Japan
e-mail: fikuine@sk.tsukuba.ac.jp
URL: http://www.trios.tsukuba.ac.jp/en/researcher/0000003030

T. Fujimoto, F. Ikuine (eds.), *Industrial Competitiveness and Design Evolution*, Evolutionary Economics and Social Complexity Science 12,
https://doi.org/10.1007/978-4-431-55145-4_12

Computer platform adopted an exchangeable cartridge system, many other companies entered this business, which soon increased the variety of console game software that could be played on the *Family Computer*, and the industry grew at a rapid pace. Nevertheless, the sector quickly matured. In the 1990s, handheld game consoles were launched on the market. The early twenty-first century saw the advent of online games,[2] as well as games played on mobile phones or social networking services (SNS), while the home video game business matured and then went into decline.

This chapter will explore the product evolution of game software from the perspective of innovation. By interpreting product evolution, we will discuss how product diversification is generated, how it deteriorates, and what type of impact it has on demand creation. In addition, problems related to research focusing on this industry will be considered. The question posed at the end of this chapter is how a set of products and services comes to be perceived as an *industry*.

2 Literature Survey

2.1 Past Researches

Extensive research has analyzed the topic of changes in industries over time. Representative examples are Kotler (2000), which discussed the industry life cycle in the field of marketing, and Abernathy (1978), which investigated models of innovation patterns. Studies of innovation patterns, among which Utterback and Abernathy (1975) and Utterback (1994), looked at the fact that it becomes increasingly difficult to create novel products and technologies over time. They argued that this phenomenon is tied to the rise and fall of both industries and firms.

Following in the footsteps of Abernathy (1978), subsequent works reviewed in greater detail the factors that influence innovation patterns. Key examples included Henderson and Clark (1990), which studied organizational inertia; Shintaku (1994), which discussed corporate strategy; and Christensen (1997), which explored value network transformation.

Based on these studies, in this chapter we will turn our attention to the changes that have occurred in the game software industry over time. In Japan, the business of console games (home video games) went from being a new industry to reaching maturity in less than 20 years. We will also discuss how home video games evolved during this period. Thanks to the availability of information on individual products, we can rely on appropriate evidence to consider the question of how an industry matures after its inception.

[2]Research on online games includes Nojima (2002, 2008).

2.2 *Definition and Categories of Innovation*

In examining the process of product evolution for what concerns game software, this chapter builds on two previous studies of innovation: Abernathy (1978) and Abernathy and Clark (1985). We apply the types of innovation proposed by Abernathy and Clark (1985) to game software and identify innovation patterns similar to those in Abernathy (1978).

The definition of product innovation used in this chapter is "the phenomena leading to the renewal, growth, and maintenance of the home video game industry through novelty and maturation that do not rely on hardware (i.e., game consoles)." In other words, phenomena that induce changes in the market/consumer linkages are recognized as innovations. Based on this definition, we divide product innovation into two categories, depending on the degree of market/consumer linkage.

The first category is creative innovation, which disrupts the existing market/consumer linkages or renders them obsolete. When it came about, this type of innovation multiplied the products known as game software and was embraced by consumers with such sentiments as "I never knew that such fun games existed." Increasing the possibilities for game software products motivated users to purchase them, thus creating demand—that is, reinventing, advancing, and maintaining the industry. In other words, creative innovation is innovation that expands the product space (Hotelling 1929).

The second category is inherent innovation, which maintains or entrenches the existing market/consumer linkages. This type of innovation is limited by the inherent possibilities already pioneered by creative innovation. However, it increases the degree of completeness of products within the range pioneered by creative innovation, thus inducing users to purchase them. In other words, inherent innovation is innovation that creates demand by enhancing the existing product space.

In the next sections, we will describe the evolutionary process of game software along with the occurrence of these two types of innovation. We will then interpret the innovation patterns and factors behind them, discussing their effects on demand creation.

3 Case Description and Interpretation: The Evolution of Game Software and the Development Productivity Dilemma

3.1 *Innovation Patterns of Game Software*

This section explores the evolution of game software from 1983 to 1999, focusing on product innovations. During this time, game software transitioned from a period of creative innovation to a period of inherent innovation. Table 1 summarized the main product innovations.

Table 1 Transition of game software in Japan

					Inheritance period												
	Establishment period				Preservation period								Re-establishment period			Preservation period	
Year	1983	1984	1985	1986	1987	1988	1989	1990	1991	1992	1993	1994	1995	1996	1997	1998	1999
First products of a series	Many				Few								Few			Few	
Number of genres appearing	12				3								1			0	
Number of genres in which million-seller products appeared	8				1								5			0	
Types of million-seller products	First products of a new series/first products of a new genre				Products of existing series/products of existing genres								First products of a new series/products of existing genres			Products of existing series/ products of existing genres	
Main types of innovation occurring	Creative innovation				Inherent innovation								Hardware-dependent innovation			Inherent innovation	
Typical products	Mario Bros. (1983, FC, Nintendo, NG, NS)				Dragon Quest (Warrior) 2 (1987, FC, Enix, EG, ES)				Street Fighter 2 Turbo (1993, SF, Capcom, EG, ES)				Derby Stallion 3 (1995, SF, Ascii, EG, ES)			Resident Evil (Biohazard) 2 (1998, PS, Capcom, EG, ES)	
Product name (first release, hardware, publisher, new or existing genre, new or existing series)	Baseball (1983, FC, Nintendo, NG, other)				Family Stadium '87 (1987, FC, Namco, EG, ES)				Final Fantasy 6 (1994, SF, Square, EG, ES)				Super Donkey Kong 2 (1995, SF, Nintendo, EG, NS)				
	Roadrunner (1984, FC, Hudson, EG, NS)				Dragon Quest (Warrior) 3 (1988, FC, Enix, EG, ES)				Super Donkey Kong (1994, SF, Nintendo, EG, NS)				Virtua Fighter 2 (1995, SS, Sega, EG, ES)			The Legend of Zelda: Ocarina of Time (1998, N64, Nintendo, EG, ES)	
	F1 Race (1984, FC, Nintendo, NG, NS)				Super Mario Bros. 3 (1988, FC, Nintendo, EG, ES)				Virtua Fighter (1994, SS, Sega, EG, NS)				Dragon Quest (Warrior) 6 (1995, SF, Enix, EG, ES)				
	Xevious (1994, FC, Namco, EG, NS)				Family Stadium '88 (1988, FC, Namco, EG, ES)								Resident Evil (Biohazard) (1996, PS, Capcom, EG, NS)				

	Excitebike (1984, FC, Nintendo, EG, NS)	Tetris (1988, FC, ESP, EG, NS)		Super Mario Bros. 64 (1996, N64, Nintendo, EG, ES)	Final Fantasy 8 (1999, PS, Square, EG, ES)
	Super Mario Bros. (1985, FC, Nintendo, EG, NS)	Dragon Quest (Warrior) 4 (1990, FC, Enix, EG, ES)		PaRappa the Rapper (1996, PS, Sony Computer Entertainment, NG, other)	Hot Shots Golf (Everybody's Golf) 2 (1999, PS, Sony Computer Entertainment, EG, ES)
	M.U.S.C.L.E. (1985, FC, Bandai, EG, NS)	Final Fantasy 3 (1990, FC, Square, EG, ES)		Final Fantasy 7 (1997, PS, Square, EG, ES)	Resident Evil (Biohazard) 3 (1999, PS, Capcom, EG, ES)
	TwinBee (1986, FC, Konami, EG, NS)	Super Mario World (1990, SF, Nintendo, EG, ES)		Hot Shots Golf (Everybody's Golf) (1997, PS, Sony Computer Entertainment, EG, NS)	Donkey Kong 64 (1999, N64, Nintendo, EG, ES)
	The Legend of Zelda (1986, FC, Nintendo, NG, NS)	The Legend of Zelda: A Link to the Past (1991, SF, Nintendo, EG, ES)		Gran Turismo (1997, PS, Sony Computer Entertainment, EG, NS)	Gran Turismo 2 (1999, PS, Sony Computer Entertainment, EG, ES)
	Nemesis (Gradius) (1986, FC, Konami, EG, NS)	Street Fighter 2 (1992, SF, Capcom, EG, NS)			
	Dragon Quest (Warrior) (1986, FC, Enix, NG, NS)	Dragon Quest (Warrior) 5 (1992, SF, Enix, EG, ES)			
	Super Mario Bros. 2 (1986, FC, Nintendo, EG, ES)	Final Fantasy 4 (1991, SF, Square, EG, ES)			
	Pro Yakyuu (Professional Baseball) Family Stadium (1986, fc, Namco, EG, NS)	Final Fantasy 5 (1992, SF, Square, EG, ES)			

Data source: Created by the author based on data provided by Ambit Co., Ltd
NS, first product of a new series; ES, product of an existing series
FC, Nintendo Entertainment System (NES); SF, Super Nintendo Entertainment System (SNES); N64, Nintendo 64

The period from 1983 to 1986 was characterized by creative innovation. Titles like *Mario Bros.* and *Super Mario Bros.*, *Dragon Quest* (*Dragon Warrior*), and *The Legend of Zelda* were certainly worthy of the label of "creative innovations." Highly innovative products leading to the establishment of new genres and new series were developed and launched, and, as a result, the game software industry grew rather quickly.

In contrast, the period starting in 1987 concentrated on inherent innovation. The years from 1987 to 1994 witnessed such events as the launch of *Dragon Quest 2* (*Dragon Warrior 2*) and *Super Mario Bros. 3*. Many other products based on genres and series already existing before 1987 were released as well. Most users bought products that were already part of a genre or series.

After 1987, there were phases when this trend was even more pronounced, specifically from 1987 to 1994 and from 1998 to 1999. Even though major changes in hardware occurred during this period, the contents of game software themselves drew on existing genres or series. Nevertheless, a different picture emerged between 1995 and 1997. Sony Computer Entertainment (SCE) launched the hardware named *PlayStation*, and the number of titles compatible with the *PlayStation* increased. The *PlayStation* adopted breakthrough technologies known as 3D computer graphics and the CD-ROM. Proactive utilization of these technologies led to the launch of products that attracted many users, including games belonging to existing series and genres, which in turn expanded the market.

Therefore, while creative innovation peaked about 3–4 years after the game software business was formed, it subsequently stagnated and was followed by a period of inherent innovation. In other words, the expansion of the product space through creative innovation occurred before 1986, while after 1987 the focus shifted toward filling up the previously established product space with inherent innovations.

Now let us look at the trends in terms of scale of the game software market during the periods identified in Table 1. Figure 1 shows that the size of the market grew every year until 1986 and was particularly large in 1986, 1995, and 1997.

The game software business grew by means of creative innovation until about 1986 and then matured rapidly through inherent innovation. In the years 1995–1997, there was a spurt in market growth due to changes not only in game software but also in the hardware on which the games ran. After 2000, online games, games played on mobile phones, and SNS games became popular. The demand for playing computer games was met by a new industry created by new companies, and the business of console games stagnated.

3.2 *The Development Productivity Dilemma: The Factor of Innovation Patterns*

What caused the transitions in the game software industry mentioned above? How did the home video game software industry emerge, grow, and then mature? To try

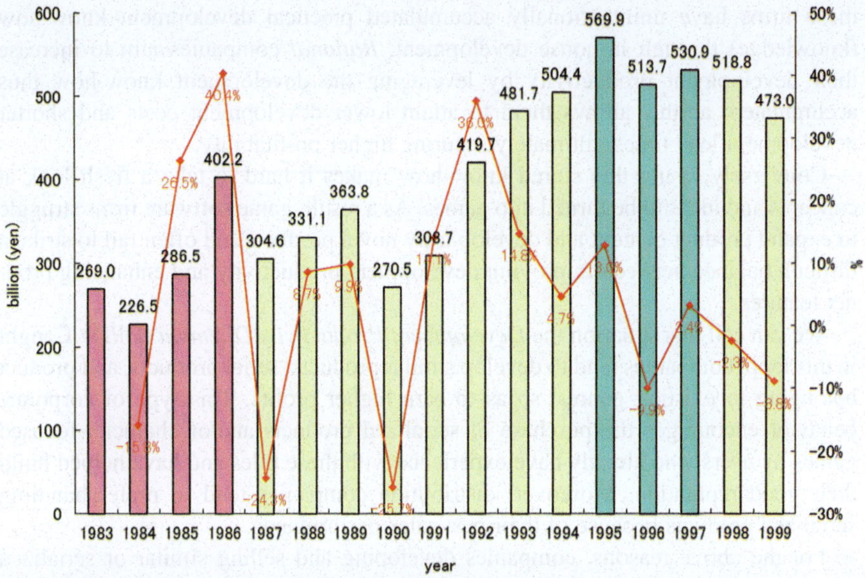

Fig. 1 Trends in game software market scale (1983–2006)
The Toy Journal (1991, 1992, 1993, 1994, 1995, 1996, 1997, 1998, 1999, 2000). Tokyo Toy and Doll Wholesalers Cooperative Association, Tokyo
White Paper on Leisure (1982, 1983, 1984, 1985, 1986, 1987, 1988, 1989, 1990, 1991, 1992, 1993, 1994, 1995, 1996, 1997, 1998, 1999, 2000). Leisure Development Center, Tokyo
Research for Information and Media Society (1995). Dentsu Innovation Institute, Tokyo
Notes: The red line shows the growth rate over the previous year

and answer these questions, let us analyze the behavior of the firms operating in the game software industry.

In developing game software, companies can choose between two different strategies: (1) appealing to users on the basis of novelty or (2) appealing to users with products similar to existing products. The novelty-based product strategy is both high risk and high return. The risk is that consumers may fail to recognize the wholly new product as a game. At the same time, the said wholly new product has no rivals and is thus very likely to capture the consumers' attention. Conversely, the similarity-based strategy of appealing to users by releasing titles similar to existing products is both low risk and low return. The risk is low because consumers can easily recognize familiar features and assess the attractiveness of the product by relying on their experience with similar titles. At the same time, the game will be one among many comparable products and will be confronted with severe competition.

As game software companies remain on the market year after year, they have to deal with an implicit bias against creating completely new products. They naturally tend to prioritize the development of titles similar to those already released, i.e., of products belonging to existing series and/or genres. The reason for such an approach is that, since their corporate activities have been continuing for a long period of time, these firms have unintentionally accumulated practical development know-how (knowledge) through in-house development. *Rational* companies aim to increase their development productivity by leveraging the development know-how thus accumulated, as this allows them to attain lower development costs and shorter development lead times, ultimately securing higher profitability.

Conversely, using this stored know-how makes it hard to take a fresh look at concepts and ideas to be turned into games. As a result, game software firms struggle to expand product content and develop truly novel products and often fail to strike a difficult balance between improving development productivity and enhancing product features.[3]

We can call this situation the *Development Productivity Dilemma* (DPD). Caught in this loop, companies tend to develop similar products, serial products, and product belonging to existing genres, so as to earn higher profits. This type of corporate behavior encourages the purchase of serialized products and/or character-focused games by users who already have experience with these titles and have helped build their good reputation. Moreover, distribution companies tend to prefer handling serialized products because of their past sales performance.

For the above reasons, companies developing and selling similar or serialized products based on their accumulated know-how can improve revenues and overall performance. This, in turn, grants them an advantage in further strengthening their development know-how, human resources, capital, and other areas. Thus, a competitive advantage is gained vis-à-vis other companies, ultimately creating barriers to new entrants. The ensuing situation is that a number of companies already in the business are gradually weeded out, fewer new companies enter the market,[4] while those companies that are still operating successfully become increasingly homogeneous. They make the most of their in-house know-how and prioritize product development in line with previously well-received products (Fig. 2).

In addition, the presence of platforms may have accelerated the Development Productivity Dilemma. As discussed in the previous chapter, the establishment of platforms increased demand creation in the early days of the business but actually reduced it at later stages. Platforms aim to create monopolies or oligopolies so that economies of scale can be achieved. When companies producing monopolistic or oligopolistic platforms try to boost their profits, they raise the platform usage fees charged to companies supplying complementary goods, and, as a consequence, these suppliers become less profitable. When developing such complementary goods as

[3]Companies that do not carry out in-house development are not caught in the Development Productivity Dilemma. They can take a fresh look at concepts and ideas and may create novel products. However, they cannot improve their productivity at the same time and are consequently unable to reduce development lead times and cut development costs. Therefore, they find themselves in a situation of competitive disadvantage compared to companies that develop their products in house and accumulate development know-how.

[4]Shintaku, et al. (2003) clearly observes this phenomenon of fewer new companies entering a market, established companies being weeding out, and revenues being concentrating in a small number of companies.

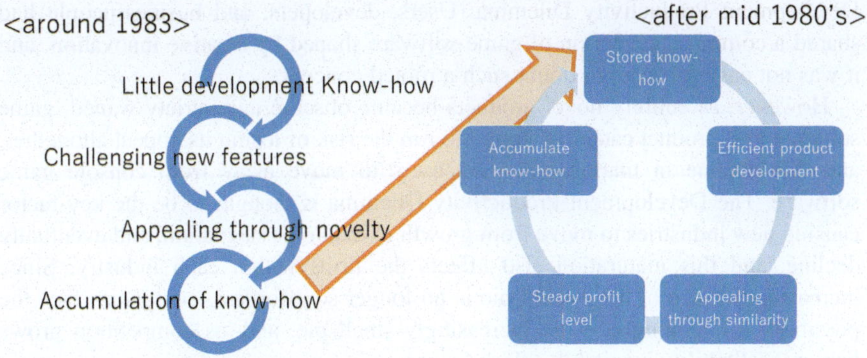

Fig. 2 Changes in development strategy

game software becomes less profitable, companies supplying game software are less inclined to take risks. In such a situation, they choose the product strategy of appealing to users through similarity with existing titles, due to its low risk, triggering a dampening effect on the creativity of their developers. This reinforces their tendency to develop products based on existing genres and series by using their accumulated development know-how, and they are less willing to invest in novel products leading to creative innovation.

As a result of this convergence in how companies behave, the industry as a whole is full of follow-on products that draw on existing genres and series.[5] Moreover, it becomes ever more difficult for novel titles in new genres or series to be developed, as well as to generate large revenues. This type of product shift manifests itself with the passing of time and, in the game software business, it brought about a move from creative innovation to inherent innovation. Product diversification therefore comes to a halt, and the forces of demand creation weaken.

The above changes started to be seen in game software rather soon, around the mid-1980s. After creative innovation diminished in the late 1980s, the market was supported by inherent innovation. From then on, companies and businesspeople who had shared a common idea of what constituted a video game prior to 1986 continued to launch products based on creative innovation. These titles remained popular among users, who continued to buy game software for more than 10 years. After the mid-1980s, inherent innovations occurred one after the other, also attracting users. Owing to either inherent innovations or serialized bestsellers, users continued to look for new titles, the market did not shrink, and the sector as a whole did not suffer from declining profits. Consequently, companies could not detect any difficulties in their development activities, which is precisely the problem with the

[5]Wada (2011) verifies this hypothesis by using game software sales' data. Evidence from the sales patterns of game software confirms that heavier exploitation of development know-how causes less novelty across titles, which changes their sales patterns.

Development Productivity Dilemma. Users, developers, and businesspeople had shared a common perception of game software shaped by creative innovation, and it was not easy for them to doubt such a rooted concept.

However, as entirely novel products became obscure and variety waned, game software as a product category started to run the risk of losing its appeal altogether, and this decline in marketability led users to move away from console game software. The Development Productivity Dilemma is thought to be the key factor causing new industries to move from growth to maturity, stagnation, and eventually decline, and this maturation also affects the firms within each industry. Since increasing the size of the overall pie is no longer so effortless, companies face the possibility of fighting over an increasingly small pie, and, as competition grows fiercer, profitability inevitably suffers.

Businesses try to strengthen their competitive advantage by improving development productivity. However, in environments in which the Development Productivity Dilemma is predominant, the fact that several companies make simultaneous efforts to secure this advantage may cause the market size to shrink. Specifically, firms do not necessarily derive value from their competitive advantage and may even have to face a reduction in its value. The situation is paradoxical: building competitive advantage leads to diminishing the value of competitive advantage, and declining profits are ascribable to that very same competitive advantage.

In other words, when a Development Productivity Dilemma exists, competitive advantage must be built by improving development productivity. Yet, if most companies improve their development productivity, this will not necessarily lead to corporate growth or better competitive advantage. The ultimate irony is that consolidating one's competitive advantage diminishes the fruits (profitability) of the competitive advantage that was built. Therefore, the phenomenon of the so-called Development Productivity Dilemma seems to cause new businesses to move from growth to maturity and decline.

3.3 Where Has Demand Gone?

The previous section described how the video game industry grew, matured, and declined. Where exactly has demand for game software gone? Has it simply disappeared?

Unlike automobiles, home electronics, and even game consoles, game software is a digital good. Hence, there are no physical constraints associated with purchasing it, such as where to put it, how long it can be used for, and what to do when it stops working. It is therefore difficult to think that the decline in demand discussed above has been caused by a transition from new demand to replacement demand. Rather, the reason why demand for game software has fallen is that users have become bored and no attractive new products have been released. As previously discussed, developers of game software have been caught up in a Development Productivity

Dilemma. Therefore, they have been unable to create attractive products, likely causing users to move away from game software.

Although users have actually moved away from game software, this does not necessarily mean that they no longer want to play computer games, as confirmed by the rise in popularity of mobile phone games since 2000. It is not difficult to imagine that users who have stopped buying console games have started to play titles running on different types of computing devices. Hence, our interpretation suggests that people who like playing computer games have shifted their buying preferences to other games, which has resulted in declining demand for console game software.

As explained in the introduction, game software—i.e., software for consoles—is simply one of the existing types of computer games, along with games played on handheld consoles, online games, games on mobile phones, and games on social networks. If we consider only console games, we miss the coevolution of hardware and software, which has generated several related industries. For instance, console-based RPGs have been succeeded by the so-called multiplayer online role-playing games (MORPGs) or massive multiplayer online RPGs (MMORPGs), and puzzle games developed for consoles have influenced smartphone games. An example of a single product is *Pokemon GO*, whose characters as well as game system have their origins in Nintendo's game software.

The above game types are characterized by overlapping customers and overlapping companies. Furthermore, the perception of what constitutes a game has had a considerable influence on the common perception of console games. This situation is similar to the continuity between the horse and buggy industry and the auto industry, which began with the horseless carriage. A similar mixture of multiple industries also exists in computer games.

4 Discussion: Industry as Unit of Analysis

When thinking in such terms, analyses performed at the industry level appear to be quite arbitrary. Yet, regardless of whether they deal with product life cycles, innovation patterns, or design hierarchies, most studies take the existence of a clearly defined industry as a given. In other words, they investigate what occurred *after* the industry's establishment and are therefore unable to answer questions about how the industry was established and exactly what constitutes the industry itself (Shimizu et al. 2016). Until Sect. 3 of this chapter, in looking at its evolution from establishment to maturity, we regarded game software as a single industry. Was this appropriate?

A broad definition of what constitutes game software was not formulated before the industry's establishment, but it gradually came into being later on. Companies performed software development through moderate differentiation from existing products. Consumers experienced the available products and eventually started to associate the salient elements of each title with elements common to game software genres. The definition of what constitutes game software and the elements identified within game software were thus loosely determined as specific products were developed and consumer accumulated experience.

How the game software industry began was retroactively analyzed based on an ex post facto definition of its product. In fact, we usually describe the beginnings of an industry based on what that industry looked like after it took shape. Furthermore, we may recognize as dominant designs those products that more fully embody product categories that are still in existence. If that is the case, both the definition of the industry and the recognition of dominant designs may vary depending on how the observer interprets past facts. Generally speaking, the scope for defining an industry differs according to the chronological perspective of the observer.

Moreover, because of the inability to extricate oneself from an analysis at the industry level, it is more difficult to grasp interactions that transcend the industry. Arbitrary definitions of industries and descriptions restricted by industry boundaries are common problems in industrial research. The reason why these issues have grown so serious is that tie-ups among products and services have become commonplace. The proliferation of information technology (IT) has also been a significant contributing factor.

In the past, consumers developed their own consumption style by combining various products and services that they bought separately. IT has served to make the bundling of each consumer's own products and services explicit, thereby enabling companies to become involved. Indeed, companies can intervene even on products and services that consumers already own, for example, by linking users' smartphones to game consoles, PCs, televisions, home electronic devices, and so on, after these have actually been purchased. Platforms like those proposed by Gawer and Cusumano (2002) have therefore become commonplace in many areas.[6] In other words, what has made industry-level analyses useful in both research and actual practice is not that the products and services are bundled together, but that consumers evaluate products and services on a standalone basis.

However, it is possible to attach another meaning to the concept of industry as a unit of analysis if we recognize that changes have occurred in technologies and markets. An industry as a unit of analysis reflects the relationship among the products and/or services that we use in our everyday life. Even though all products and/or services can be connected through IT, it would be mistaken to accept such a broad relationship covering everything. Our past experience and common sense will help us determine the appropriateness of the relationship among the products and services in question. Such common sense and past experience are essential in order to establish the appropriate boundaries of an industry, which we can use as a unit of analysis. As nowadays all

[6]The sudden increase in research on business ecosystems since 2000, as indicated by the survey in Tatsumoto (2012, 2017), reflects the potential and realities of tie-ups among products and services made possible by information technology.

products and/or services can potentially be connected through IT, the relationship that most people can agree on will mark the industry boundaries.[7]

5 Conclusions and Remarks

This chapter explored the product evolution of game software. Major product innovations occurred during the process of game software industrialization. In the early days of the industry, several creative innovations were generated, the definition of what constitutes a game was established, and genres were created to categorize products. Creative innovation subsequently stagnated, and the focus shifted toward differentiation within existing genres. Stagnation gradually spread to consumers too, and the power to stimulate demand and motivate users to buy products declined. The industry thus matured. The inherent phenomenon of the so-called Development Productivity Dilemma in the creation of new products emerged and became increasingly severe over time (Ikuine 2012, 2013).

The case of Japan's game software as an industry provides some useful hints on the subjects of demand creation and product evolution. At the same time, some issues with industry-based research have become evident. This is because the concept of industry is an arbitrary unit of analysis relying on ex post facto recognition. Basing the question of what constitutes an industry solely on the shaping of said industry makes it easy to overlook innovation that transcends industry boundaries. In an age when IT has enabled connections among countless products and services, overcoming this problem seems to be of critical importance.

One way to tackle this issue is to find out how industry boundaries come into being. In other words, we need to explore how opportunities for forming new industries come about. To achieve this goal, it might be helpful to focus on the people who created the industry, so that we can understand what they saw and what they intended to do. What perceptions and ideas motivated these people under conditions of considerable uncertainty and anxiety, in which they did not know whether an industry would come into being in the first place? Collecting case studies and oral histories might be key to shedding more light on this topic.

References

Abernathy WJ (1978) The productivity dilemma: roadblock to innovation in the automobile industry. Johns Hopkins University Press, Baltimore

[7]Following Aoshima and Kusunoki (2008), innovation that changes the scope of this relationship or that changes companies and other economic entities at the international level can be called *System Redefinition*.

Abernathy WJ, Clark KB (1985) Innovation: mapping the winds of creative destruction. Res Policy 14:3–22

Aoshima Y, Kusunoki K (2008) Innovation as system re-definition (Sisutemu saiteigi tositeno inobesyon). Hitotsubashi Bus Rev 55:58–77 (in Japanese)

Christensen CM (1997) The innovator's dilemma: when the technologies cause great firms to fail (rev ed). Harvard Business School Press, Boston

Clark KB (1985) The interaction of design hierarchies and market concepts in technological evolution. Res Policy 14:235–251. https://doi.org/10.1016/0048-7333(85)90007-1

Gawer A, Cusumano MA (2002) Platform leadership: how Intel, Microsoft, and Cisco Drive Industry Innovation. Harvard Business School Press, Boston

Henderson RM, Clark KB (1990) Architectural innovation: the reconfiguration of existing product technologies and the failure of established firms. Adm Sci Q 35:9–30

Hotelling H (1929) Stability in competition. Econ J 39(153):41–57. https://doi.org/10.2307/2224214

Ikuine F (2012) Development productivity dilemma: the innovation patterns in the digital age (Kaihatsu seisansei no jinemma: dejitaru ka jidai no inobesyon pataan). Yuhikaku, Tokyo (in Japanese)

Ikuine F (2013) The evolution of the home video game software industry in Japan: an empirical study on factors in the Industry's evolution. Evol Inst Econ Rev 9(1):37–50

Kotler P (2000) Marketing management, 10th edn. Prentice Hall, Upper Saddle River

Nojima M (2002) Community and business strategy: case studies in the online game industry (Komyunitei to kigyousenryaku no tekigo moderu: onrain gemu sangyo no jirei). Akamon Management Review 1:527–560. Retrieved from: http://www.gbrc.jp/ (in Japanese)

Nojima M (2008) Hitoha naze katachi no nai mono wo kaunoka: kasousekaino bijinesu moderu. NTT Publishing (in Japanese)

Shimizu M et al (2016) From fun to business: the industrialization of computer games (Asobi kara bizinesu he: nihon to amerika ni okeru gemu no sangyo-ka). 2nd Kyoto game conference, amusement and innovation: toward the history of the game industry, 17th March 2016, Ritsumeikan University, Kyoto (In Japanese)

Shintaku J (1994) Competitive strategies of Japanese firms: De-maturity and competitive behavior in matured industries (Nihon kigyo no kyousou senryaku: datsu seijuku no gijutsu tankan to kigyo kodo). Yuhikaku, Tokyo (in Japanese)

Shintaku J et al (eds) (2003) Economic analysis on game industry (Gemu sangyo no keizai bunseki: Kontentsu sangyo hatten no kozo to senryaku). Toyo Keizai Shimpo-sha, Tokyo (in Japanese)

Tatsumoto H (2012) Competitive strategy of platform business (Puratto fomu kigyo no kyoso senryaku). MMRC discussion paper series no. 396. Retrieved from: http://merc.e.u-tokyo.ac.jp/mmrc/dp/pdf/MMRC396_2012.pdf

Tatsumoto H (2017) Platform strategy for global market: the strategic use of open standards and the management of business ecosystem (Prattofohm kigyo no guro-baru senryaku). Yuhikaku, Tokyo (in Japanese)

Uemura M et al (2013) The life and times of the Nintendo Famicom: the birth of TV games (Famikon to sono jidai). NTT Shuppan, Tokyo (in Japanese)

Utterback JM (1994) Mastering the dynamics of innovation. Harvard Business School Press, Boston

Utterback JM, Abernathy WJ (1975) A dynamic model of process and product innovation. Omega 3:639–656. https://doi.org/10.1016/0305-0483(75)90068-7

Wada T (2011) Exploitation reduces novelty: an empirical analysis of the Japanese video game industry. Annals Bus Adm Sci 10:1–12. https://doi.org/10.7880/abas.10.1

Data

CESA Games white paper (2002). Computer Entertainment Suppliers Association, Tokyo
Data provided by Ambit Co., Ltd.
Data provided by Media Create Co., Ltd.
Research for Information and Media Society (1995). Dentsu innovation institute, Tokyo
The Toy Journal (1991, 1992, 1993, 1994, 1995, 1996, 1997, 1998, 1999, 2000). Tokyo Toy and Doll Wholesalers Cooperative Association, Tokyo
White Paper on Leisure (1982, 1983, 1984, 1985, 1986, 1987, 1988, 1989, 1990, 1991, 1992, 1993, 1994, 1995, 1996, 1997, 1998, 1999, 2000). Leisure Development Center, Tokyo

Investigating the Creation and Diffusion of Knowledge for Demand Creation: The Case of the Telecommunications Industry

Wei Huang, Masanori Yasumoto, and Jing-Ming Shiu

Abstract This chapter attempts to explain how a leading firm contributes to diffusing technologies for implementation by developing and disclosing the technologies of standard-essential patents (SEPs) through system development. Technology diffusion plays a critical role in creating demand. Disclosure of technologies to the public through standardization is reported to accelerate the process of demand creation. However, it is still hard for firms to learn and maintain the knowledge and relevant technologies to develop complex systems: system knowledge. By analyzing a key case in the telecommunications industry, this chapter suggests that demand creation is encouraged by the technological foundations (i.e., system knowledge and relevant technologies) residing in the leading firms' system integration capabilities to implement technology specifications. Such technological foundations are shaped by the strategic decision to provide technologies for implementation rather than total solution systems.

1 Introduction

Technology diffusion is a critical tool to create demand. Versatile technologies (i.e., general-purpose technologies) have a strong economic impact on a variety of fields/industries and create new markets (Helpman 1998; Lipsey et al. 2005). Particularly, technologies such as ICT (information and communications technology) have profoundly influenced the global economy. By offering revolutionary products, services, and operational tools, ICT plays a crucial role in demand creation and market expansion. For instance, firms expect the stream of IoT (Internet of Things) to

---

W. Huang (✉)
University of Tokyo, Tokyo, Japan

M. Yasumoto
Yokohama National University, Yokohama, Japan

J.-M. Shiu
National Cheng Kung University, Tainan, Taiwan

T. Fujimoto, F. Ikuine (eds.), *Industrial Competitiveness and Design Evolution*,
Evolutionary Economics and Social Complexity Science 12,
https://doi.org/10.1007/978-4-431-55145-4_13

generate trillions of dollars worldwide, while mobile telecommunications technologies have so far developed a market worth more than 3 trillion dollars by providing a variety of systems and services including applications (Bezerra et al. 2015).

Furthermore, the integration of ICT into traditional industries has accelerated the flow of knowledge among firms, which, in turn, promotes knowledge accumulation in either quantity or quality. Such knowledge can help multiple firms collectively create demand, and connected communities of firms—business ecosystems (Moore 1993; Iansiti and Levien 2004; Teece 2007)—are critical not only in emerging industries but also in traditional sectors. In order to achieve product innovation and boost market demand, firms need to cooperate with other organizations in relevant areas, even beyond industry boundaries.

The formation of ecosystems requires technologies and knowledge assets to be developed and shared. In particular, standardization of technologies encourages diverse firms to develop their business and collectively shape relevant markets. Technological foundations constitute the basis for demand creation. Therefore, how to build technological foundations across firms has emerged as a significant issue.

For instance, ICT firms disclose their technologies to the public through standardization, in order to encourage a variety of other firms to enter the market. Single dominant firms, such as IC and OS platform leaders, can drive industrial development by leveraging the network externalities of standardized solutions and technologies (e.g., Gawer and Cusumano 2002). Meanwhile, growing numbers of firms attempt to collectively standardize fundamental technologies (i.e., specifications) in order to secure interoperability (e.g., Cargill 1997; Farrell and Saloner 1988; Simcoe 2012; Leiponen 2008; Weiss and Cargill 1992; Xia et al. 2012). In the development of complex product systems (CoPSs), the relationships among different companies tend to become complex. Therefore, standardization—for instance, in the telecommunications industry—based on the consensus of different parties helps market formation by encouraging interfirm transactions across a wide range of sectors. Also, the adoption of IoT in a variety of industries, including manufacturing industries, makes standardization a priority, especially to increase interoperability and compatibility.

However, firms are required to have the appropriate knowledge and relevant technologies to implement standards into products/services (West 2006). Demand creation will not be accelerated until diverse firms can acquire and exploit such knowledge and relevant technologies for implementation. By reflecting on this, we will try to shed light on how a leading firm can combine nurturing knowledge and relevant technologies for implementation with contributing to the diffusion of technologies even at the risk of technology spillovers.

In the next section, we will review the relevant academic research. Due to overconcentration on business aspects, few empirical studies have been conducted from a technological perspective. This work, by contrast, mainly attempts to examine the technological foundation level, which is the basis for ecosystem formation. In Sections 2 and 3, we will review some antecedents and propose our research perspective and framework. In Sect. 4, we will use standard-essential patents

(SEPs) to perform a network analysis on the quantitative data of technology specifications and discuss the position of leading firms. We will then examine the case of Qualcomm (Sect. 5), which is one of the leading firms in the telecommunications industry, to investigate the role of system knowledge and the process of building and deploying it. Finally, we will discuss how firms can maintain their system knowledge advantage and capabilities in the development of their business and products, and we will explore the influence of system knowledge on the process of knowledge spillover.

2 Literature Review

The spread of general-purpose technologies across markets requires cumulative improvements with complementary technologies by multiple firms under the support of institutional settings. Particularly, the diffusion of technologies to create demand is encouraged when relevant specifications (e.g., system architectures and interfaces) are standardized enough to secure interoperability and compatibility among a variety of elements. Standardization not only helps to coordinate knowledge dispersed across firms but also encourages knowledge transfer among them (Steinmuller 2003). Especially in industries/fields characterized by complex systems, standardization through the disclosure of technological information drives technology diffusion, innovation by multiple firms, and market expansion leveraged by network effects (European Commission 2014). Publishing technological information as technological specifications is expected to encourage the participation of diverse firms including complementors, users, and even competitors (David and Greenstein 1990; Katz and Shapiro 1986; West 2003).

Drawing on such technological information, firms can mutually connect their products and services to draw more benefits from them (i.e., network externalities and scale merits), even though each firm has its own sector of specialization (David and Greenstein 1990; Katz and Shapiro 1986). Even those actors which are not the original developers of technologies may exploit them to innovate products and services, and, as a result, firms will collectively shape new markets generating huge demand.

Predominant platform firms, above all platform leaders in the ICT industries, are presumed to create demand by providing standardized platforms to encourage complementors and users to participate in ecosystems (Baldwin and Woodard 2009; Gawer and Cusumano 2002; West 2003). However, nowadays many critical standards are set up through industrial consensus among multiple firms (Blind and Mangelsdorf 2016; Cargill 1997; Simcoe 2012; Leiponen 2008; Xia et al. 2012), and, in the case of complex systems, even leaders can hardly cope with the systemic complexity of interrelated technologies (Cargill 1997; Steinmuller 2003). Therefore, by achieving industrial consensus, firms attempt to collectively shape shared standard technologies.

The basic functionalities and requirements of system goods, such as telecommunications systems (Davies 1996; Davies and Brady 2000; Hobday 1998), are defined by a set of non-exclusive technology specifications prescribed through collaborative activities in standard bodies (e.g., consortiums). Complementors, users, and even competitors can freely refer to the technology specifications of standards and thereby develop products/services, sub-systems, or components (e.g., UNIX from 1980s to early 1990s in mainframe and workstation industries: West and Dedrick 2001; West 2003; IEEE 802.11 WiFi standardization: Garud and Kumaraswamy 1995; Van de Kaa and de Bruijn 2015). Once made open, technology specifications will almost certainly be used by free riders (Kristiansen and Thum 1997). The European Commission (2014, p 28) suggests that information related to standardization is critical for product innovation in many firms and, at the same time, it creates externality and spillover issues.

While cooperating with each other for standardization, firms compete in the implementation of standardized technologies (Simcoe 2006). In order to cope with such a standardization-competition setting, firms contributing to standardization can claim and protect implementation rights on technology specifications through patents (West 2006). They can declare to standard organizations that their proprietary patents are critically related to technology specifications, i.e., SEPs (standard-essential patents) (European Commission 2014). SEPs can help each firm secure its advantages under standardization (Bekkers 2001; Bekkers et al. 2002a; Blind and Thumm 2004; Granstrand 1999). Otherwise, there would be no incentives to develop and improve technologies, and implementation, as well as technological evolution, would prove impossible in the long run.

Yet, it is known that patents also disclose technology information and innovation processes that may cause knowledge spillover problems (Jaffe and Trajtenberg 2002) and SEPs hardly guarantee exclusivity (Blind and Thumm 2004; David and Greenstein 1990). Even with patents used to protect technologies as legal rights, open access to technical information, more specifically citations, increases the likelihood of technological leaks. Indeed, research has shown that the flow of knowledge though the citation of patents related to standards helps new entrants nurture their proprietary technologies (He et al. 2006; Leiponen 2008; Kang and Motohashi 2015).

New entrants can lower their knowledge search costs by using these technology specifications and SEPs while also efficiently acquiring knowledge from the technologies of leading firms. As such, SEPs, as well as technology specifications, encourage innovation processes within and across industries. For example, by studying Motorola patent citations by competitors, He et al. (2006) describe how Nokia, Ericsson, and Samsung managed to catch up with the former. Moreover, Kang and Motohashi (2015) explain that Korean and Chinese firms, new entrants, caught up with incumbent standard setters Nokia, Motorola, and Ericsson by citing many SEPs declared through the telecommunications standardization organizations. New entrants can cite (or refer to) technology specifications included in standardized specifications and/or certain leading firms' SEPs and thereby develop technologies and apply for proprietary patents.

As such, technological information from a few leaders helps various firms assimilate knowledge and relevant technologies to implement standard specifications into products/services. By encouraging innovations, the process of establishing technological foundations, initiated by a few leading firms, will bolster demand creation. In other words, demand creation largely rests on a few leading firms, which can both generate and diffuse technologies relevant to the implementation of standards across various firms. Yet, the risk of technology spillovers from a leading firm can undermine its advantage. Such a dilemma raises a critical question: how can a leading firm maintain the advantage of its knowledge and relevant technologies for implementation while also diffusing critical technologies to create demand?

3 Perspective and Framework

This section illustrates our perspective and framework to understand demand creation by leading firms under collective standardization (Fig. 1). Although technologies are standardized, the specifications for most modern complex systems are incomplete, due to product complexity and technology changes, so that divergent interpretations reduce actual compatibility until they are reconciled (West 2006). For example, omissions from the initial GSM mobile phone standard prevented competing firms from providing interoperable products for years (Funk 2002). Technology specifications for standards do not secure interoperability until implementation is made available to end users.

Moreover, even when SEPs are available, it is still difficult, most of all for new entrants, to exploit standardized technologies, and their implementation (i.e., system developments) can be the most critical bottleneck of technology diffusion for demand creation (West 2006). In order to develop standard-based systems, firms are required to have system knowledge of (1) the behavior of the system (functionalities and requirements) and (2) how the system performs its functions by linking various components together (implementation) (Baldwin 2010; Brusoni et al. 2001; Henderson and Clark 1990; Prencipe 2005).

Hence, it is hardly possible to develop complex systems without system knowledge across a variety of technologies (Brusoni et al. 2001; Davies and Brady 2000;

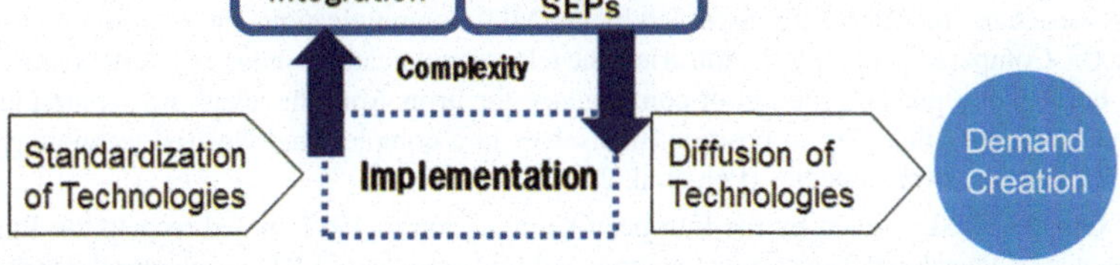

Fig. 1 Framework

Pavitt 2005). The more dispersed the production of knowledge and the more complex the products, the greater the need for firms to make use of explicit system knowledge (Brusoni et al. 2001; Prencipe 2005). For instance, developing telecommunications systems, which are regarded as complex systems, requires the integration of a variety of technologies, such as core networks, base stations, mobile phones, and diverse software (Davies 1996). This type of system knowledge cannot be easily obtained by simply referring to technology specifications or patents, and this is all the more challenging for new entrants. System knowledge that enables the implementation of technologies by defining the relationships among multiple technologies or elements is implicitly held and often concealed as know-how.

The technologies of SEPs that are viewed as necessary for the implementation of standards are developed as part of sets of technologies of the systems concerned. We can thus infer that each of SEP technologies, though developed as an individual element, reflects system knowledge for implementation. When developing the systems concerned and relevant technologies, leading firms nurture, accumulate, and hold system knowledge by appropriating sets of patents, SEPs, necessary for implementation. A set of SEPs, which reflects system knowledge, helps firms preserve their implementation rights, thereby driving the industry and license revenue flows in line with technological progress.

In order to explore the technological foundations of demand creation, we will look at how a leading firm contributing to the diffusion of relevant technologies by disclosing SEPs nurtures such technologies through system engineering, i.e., system integration for implementation. This might shed light on how huge demand is created thanks to the strategies of a handful of leading firms.

4 Data Analysis of Leading Firms in the Telecommunications Industry

We chose the telecommunications industry as the subject of our analysis.[1] Telecommunications systems consist of core networks, base stations, mobile phones, and various hardware and software related sub-systems and are regarded as complex product systems (e.g., Prencipe 2005). With the development of communication technologies, technology specifications are established and disclosed to the public through a number of standardization bodies.[2] This allows firms to easily develop

[1]The issues of the industry were discussed by the MIT Communications Future Program (CFP) in year 2005. Compared to other ICT industries, the telecommunications industry is still in flux. This is partly because it represents a point of convergence for firms from the telephony, computing, and Internet industries, all trying to position themselves in a complex multilayered technology space with different platform strategies (Pon et al. 2014).

[2]For example, ETSI, formed by the European Commission in 1987, aimed to help the European Conference of Postal and Telecommunications Administrations (CEPT) accelerate the technology standardization of 2G GSM from 1988 to 1991. Since 1998, 3GPP, as the central standardization body, has been in charge of coordinating other standard bodies for 3G WCDMA (UMTS) standardization.

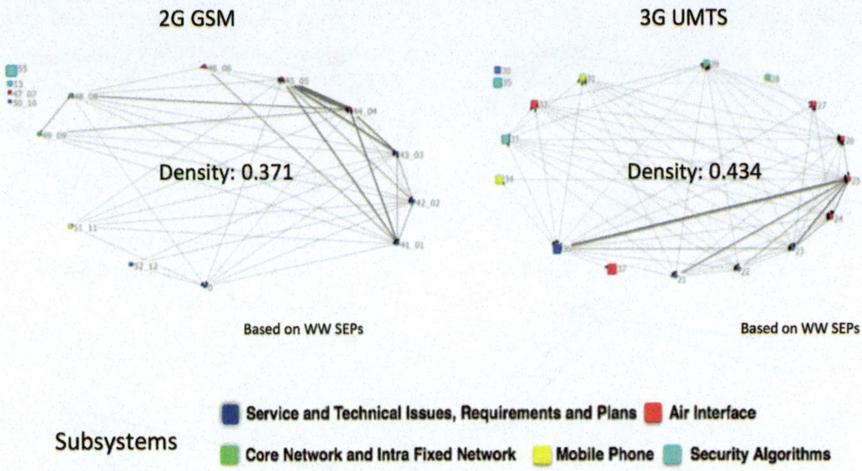

Fig. 2 Complexity of 2G GSM and 3G UMTS

telecommunications systems at low cost and within a short time by adhering to the technology specifications.

However, technology specifications do not secure interoperability and compatibility until implementation is made available to customers (Funk 2002). Different sorts of technology specifications of complex product systems represent interdependent relationships among sub-systems whose different functionalities and requirements should be taken into account. Thus, the way in which the *systematic implementation* of technology specifications into products/services occurs can reflect the firms' know-how capabilities (Shiu and Yasumoto 2017a), which help them to develop superior products to satisfy customer needs in growing markets.

Extant research indicates that firms can influence the direction of technology development by creating technology specifications under standardization (e.g., Funk 2002). Hence, by classifying technology specifications according to the structure of the telecommunications system, we can understand what technological elements are implemented in a systematic manner and how. Furthermore, studying the firms' SEPs related to technology specifications may reveal more about their knowledge of the implementation of those technology specifications (Shiu and Yasumoto 2017b).

As Fig. 2 (Shiu and Yasumoto 2015) shows, there are two different levels of complexity for firms to deal with in developing a telecommunications system. The colors of the nodes represent technology specifications classified according to the different sub-systems, while the lines are the firms' SEPs related to technology specifications. What clearly emerges is that developing a telecommunications system becomes ever more complex. Indeed, compared to 2G GSM, the development of 3G UMTS requires firms to pay greater attention to the technological interdependence among *service and technical issues*, *core network and intra-fixed*

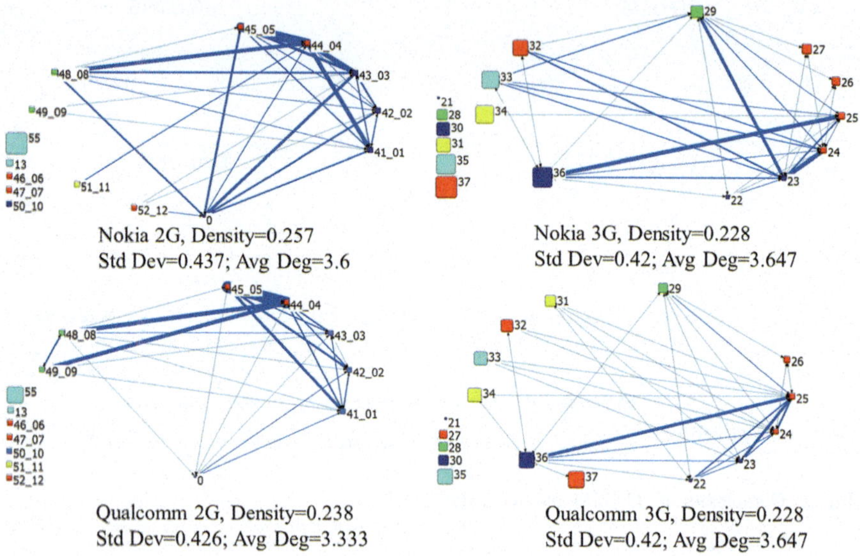

Fig. 3 Systematic implementation rights of Nokia and Qualcomm

network, *air interface*, *mobile phone*, and *security algorithms* (i.e., the densities calculated using UCInet show that 3G UMTS is more complex than 2G GSM).

The densities in Fig. 2, representing the complexity of implementing technology specifications, show that system integration is critical for integrating technologies into products/services to create demand. At the same time, system integration supports the firms' business models to meet market growth. It is generally believed that stronger system integration stems from more vertical integration in a firm's business. For instance, Nokia, Ericsson, and Motorola provided not only mobile devices but also infrastructures, services, and so on to complement their technology innovations (i.e., from 2G GSM to 3G UMTS). Since firms are more likely to deal with technological complexity by enhancing their capability for system integration, vertically integrated firms will have greater competitive advantages and be better equipped to meet customer needs by providing superior system products than non-vertically integrated firms.

When firms implement technology specifications from a systemic perspective, they need to possess the appropriate system knowledge to deal with the complexity of technology developments. Nevertheless, even in that case, they might change their business from vertical integration to vertical disintegration. As we will discuss later, although Qualcomm quit the CDMA infrastructure and mobile business, it further strengthened its R&D capability in those technologies, in addition to its chipset business. As illustrated in Fig. 3 (Shiu and Yasumoto 2017a), Qualcomm possesses degrees of systematic *implementation rights* (i.e., density, calculated using UCInet; Shiu and Yasumoto 2017a, c) on the technology specifications of a

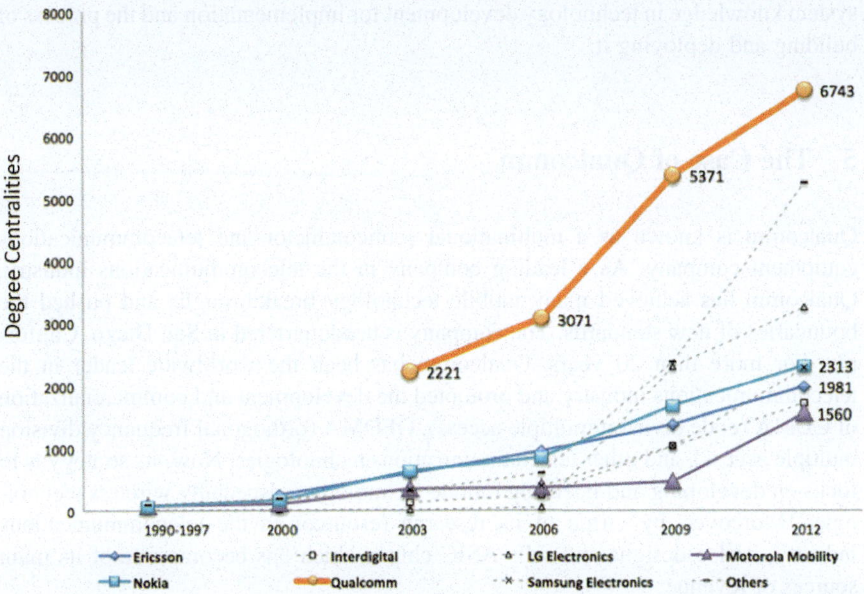

Fig. 4 Number of patent citations from firms' SEPs

telecommunications system similar to those of Nokia. This confirms that Qualcomm has shifted to vertical disintegration but still preserves the knowledge needed for system integration. We might conclude that system integration may become more strategic for vertically disintegrated firms when they are willing to become systematic technology providers, instead of merely selling product systems as traditional vertically integrated firms do.

Patent licensing may represent the main business for vertically disintegrated firms to provide their customers with technologies relevant to system knowledge. Undoubtedly, patent licensing is known as a means of knowledge transaction especially among firms, as they can quickly create demand by licensing their technologies to those actors who do not have sufficient system knowledge and relevant technologies for developing products.

Patent citations can be conceived as a measure of the usefulness of inventions (Yayavaram and Ahuja 2008), and Qualcomm's SEPs can be regarded as useful technologies that will most likely be accepted and assimilated by other firms. As discussed in Shiu and Yasumoto (2017b) and shown in Fig. 4, Qualcomm' SEPs have been cited by other firms approximately three times more than Nokia's SEPs (the degree centralities are calculated using UCInet to represent the numbers of firm SEPs cited by other firms). Even after Qualcomm left the infrastructure and mobile phone business, its technologies remained quite valuable to other firms in order to implement relevant technologies into products/services and thus contribute to demand creation. In the next section, we will examine the case of Qualcomm, one of the leading firms in the telecommunications industry, to investigate the role of system knowledge in technology development for implementation and the process of building and deploying it.

5 The Case of Qualcomm

Qualcomm is known as a multinational semiconductor and telecommunications equipment company. As a leading company in the telecommunications industry, Qualcomm has achieved many mobile technology breakthroughs and pushed the boundaries of new standards. The company is headquartered in San Diego, California. For more than 20 years, Qualcomm has been the worldwide leader in the telecommunications industry and promoted the development and commercialization of CDMA (code division multiple access), OFDMA (orthogonal frequency division multiple access), and other telecommunications technologies. Now, its strategy is to focus on developing and licensing mobile technologies, especially wireless technologies. Moreover, by virtue of its research resources in the telecommunications industry, it also designs and sells ASIC chips, which has become one of its main sources of revenue.

Qualcomm's corporate structure is based around two operating segments, QCT (Qualcomm CDMA Technologies) and QTL (Qualcomm Technology Licensing). QCT is a leading developer and supplier of integrated circuits and system software based on CDMA, OFDMA, and other telecommunications technologies. By using the fabless production model, QCT has concentrated on the development and design of integrated circuit products, providing advanced system solutions to the industry. A wide portfolio of products is manufactured and assembled by independent third-party suppliers. Nowadays, QCT is the industry's leading semiconductor and software developer and provider. The QCT segment shipped approximately 842 million mobile station modem (MSM) integrated circuits for CDMA- and OFDMA-based wireless devices in 2016. Compared to fiscal year 2015, even though its revenues decreased by 10%, QCT undoubtedly still retained a competitive edge on the hardware market.

As a wholly owned subsidiary of Qualcomm Incorporated, Qualcomm Technology Licensing (QTL) owns the vast majority of Qualcomm's patent portfolio. QTL licensing revenues include license fees and royalties, which are based on sales by licensees of products incorporating or using Qualcomm's intellectual property. Significant intellectual property applicable to products that implement CDMA and OFDMA, which are two of the main technologies currently used in wireless networks, is managed by the QTL segment. QTL's revenues made up 33% of total consolidated revenues in fiscal year 2016, showing the importance of Qualcomm's intellectual property portfolio. Through technology licensing, Qualcomm could gain benefits from the worldwide sales by licensees of CDMA phones, chipsets, and infrastructure.

Figure 5 confirms that the QCT and QTL segments have both achieved rapid growth since 2000. With the evolution of wireless technologies, the

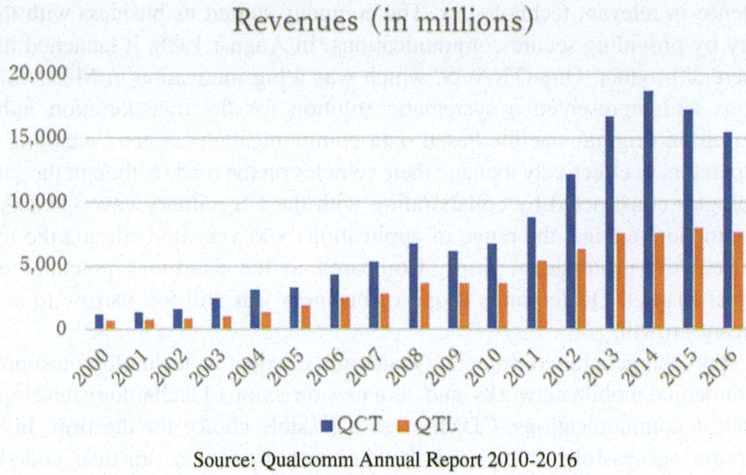

Fig. 5 Revenues of QCT and QTL (2000–2016). (Source: Qualcomm annual report 2010–2016)

telecommunications industries have had a significant economic impact on both developed and emerging markets. Moreover, benefiting from continuous investment in R&D and CDMA competences, Qualcomm has successfully led the standardization process of 3G and 4G, which has further increased its business in integrated circuit products, license fees, and royalties.

Qualcomm is known as the largest provider of wireless chipsets in the world. However, its main products are system solutions. In fact, Qualcomm is a standard system company. Thanks to its competences for systems integration, it has made significant contributions to standardization, driven the development of the wireless industry, and kept its core position in the evolution of telecommunications technologies.

Differently from its business competitors (Ericsson and Nokia), Qualcomm stopped possessing devices and infrastructures in the late 1990s. That is to say, it has successfully built and maintained its architectural control over telecommunications systems without any direct manufacturing. Inconsistency between the boundary of knowledge and business domain is evident in Qualcomm. To illustrate this aspect as well as capacity building for system integration, let us now explore in greater detail the development of Qualcomm and the features that play an important role in system integration.

5.1 *Qualcomm's Development*

Qualcomm was founded in 1985 by Irwin M Jacobs, Andrew Viterbi, Harvey White, Adelia Coffman, Andrew Cohen, Klein Gilhousen, and Franklin Antonio. The key founders of Qualcomm all had a solid academic background and extensive experience in relevant technologies. The company started its business with the US military by providing secure communications. In August 1988, it launched its first commercial product, OmniTRACS, which was a big innovation in M2M communications and represented a systematic solution for the transportation industry, relying on an original satellite-based data communications system, enabling truck fleet operators to effectively manage their vehicles on the road. Although the satellite technologies constructed by collaborating with the US military gave Qualcomm a huge competitive edge, the range of applications was very limited, and the market comprised few professional firms. Compared to the enormous potential of the consumer market, Qualcomm's scope of business was still too narrow to achieve significant growth.

To approach general consumers, Qualcomm devoted its technology resources to the commercial mobile networks, and, as a new direction of technology development in wireless communications, CDMA was a feasible choice for the firm. In 1989, Qualcomm successfully completed the demonstration of a practical code-based wireless system, proving the viability of CDMA. This ensured adequate resources for further development but also attracted the attention of the wireless industry's most influential leaders at that time. Moreover, in March 1993, Qualcomm launched the industry's first dual-mode CDMA-AMPS mobile phone, the CD-7000, which was the first CDMA-based cellular telephone designed by Qualcomm. The successful development of the CD-7000 meant that CDMA was no longer just a conceptual technology and opened up prospects for the growth of related industries. Also, in the same year, the United States Telecom Association (USTelecom) adopted CDMA as a cellular standard, thus further validating Qualcomm's business model.

In 1995, Qualcomm greatly expanded its scope of business and began to supply infrastructures and chipsets based on CDMA technologies, trying to construct a complete CDMA ecosystem, which evolved into the QCT segment in 1999. This expansion of business domain accelerated the process of adoption by major carriers and device manufacturers (Table 1).

To further accelerate the formation of peripheral markets and facilitate the diffusion of the CDMA technology, in October 1995 Qualcomm launched its technology licensing business unit, which evolves into QTL. Possessing competence regarding most of the core CDMA IP was certainly an advantage for Qualcomm, but, at the same time, it made it more difficult for other actors to enter this market. Therefore, diffusion of technology without losing its CDMA competitive advantage was essential. At first, the company had to develop a wide range of products by itself to create the market in the first place. Moreover, the core technology had not been modularized, and adoption costs were very high for other device makers.

In December 1999, Qualcomm announced an agreement with Kyocera to sell its terrestrial-based wireless CDMA consumer phone business, including its phone inventory and manufacturing equipment. As an important part of the agreement, Kyocera would purchase most of its chipsets and system software from Qualcomm for a period of 5 years and continue its existing royalty-bearing CDMA license agreement with Qualcomm.

Table 1 Key events in Qualcomm's evolution

1985	Foundation of Qualcomm
1988	Launch of OmniTRACS, a satellite-based data communications system for the transportation industry
1991	Qualcomm's IPO: Qualcomm becomes a public company
1993	Introduction of the industry's first dual-mode CDMA-AMPS mobile phone, the CD-7000
1995	Establishment of the CDMA ASIC Products Unit and launch of licensing business. Qualcomm begins supplying infrastructures and starts chipset and licensing divisions
1999	Acceptance of CDMA2000 1X and WCDMA as 3G wireless standards by ITU
	Selling of terrestrial-based wireless CDMA consumer phone business to Kyocera
	Agreement with Ericsson
2000	Release of first CDMA chipset to integrate GPS
2001	Introduction of BREW, the first open application platform for CDMA-based wireless devices
2007	Announcement of Snapdragon platform, the first chipset solution to break the gigahertz barrier

Following this step, Qualcomm was no longer the leading supplier of CDMA handsets and thus moved to the upper layers. Nonetheless, it continued to have strong relationships with suppliers in the sector. In addition, the strategic alliance with Kyocera provided Qualcomm with detailed feedback that is made up from its decreased manufacturing knowledge, while Kyocera could take advantage of Qualcomm's research capabilities on infrastructures for cellular and PC devices and wireless local loop areas.

As a new entrant, Qualcomm faced great challenges during the late 1990s, when leading firms were competing fiercely for global market shares. Especially in the process of standardization, its litigation with Ericsson over the intellectual property rights of CDMA versions had a significant influence on the wireless industry. Years of litigation led Ericsson and Qualcomm to rethink and reposition their capabilities. In March 1999, the two firms announced that they had reached an agreement. The critical part of the agreement was that Ericsson and Qualcomm could cross-license their respective Intellectual Property Rights (IPRs) for all CDMA technologies, which largely accelerated both global diffusion and standardization of CDMA. Under the agreement, the two firms would jointly support a single worldwide CDMA standard with three optional modes for 3G wireless communications. Moreover, Ericsson would purchase Qualcomm's terrestrial CDMA wireless infrastructure business, including its R&D resources. As one of the world's leading telecommunications equipment manufacturers, through this acquisition Ericsson could further expand its CDMA capabilities and build a more systemic telecommunications equipment solution.

Rather than competing with its own customers, Qualcomm strategically chose to shift its focus toward providing technologies to customers as an engineering company. Even though it built the ecosystem by developing systemic solutions in hardware and software, the company still had to solve the issues of fragmentation and limited compatibility between devices and systems. With the diversification of wireless products, manufacturers and application developers found it difficult to develop content across multiple devices. Moreover, test costs on multiple devices were still high, which limited opportunities for new entrants. To further integrate the ecosystem, Qualcomm announced BREW (Binary Runtime Environment for Wireless) in January 2001. BREW was a set of application programming interfaces (API), originally for CDMA devices. As it was a middle-tier development environment, developers and manufacturers could create various applications based on BREW's standardized environment, similar to the OS of PCs. Those applications could be implemented consistently across multiple devices.

The launch of BREW promoted the emergence of a growing community bringing together developers, operators, and device manufacturers. Moreover, its widespread adoption significantly reduced the cost of tests, lowering the development barrier to entry. Qualcomm's BREW solution greatly diversified the application layer of the ecosystem, and various applications enhanced the users' mobile data experience. Indeed, BREW also offered users the opportunity to personalize their wireless devices to suit their needs through application stores.

Back then, the adoption of CDMA-based chipsets was a prerequisite for product development by device manufacturers, and Qualcomm intended to provide an integrated platform for hardware development. In November 2007, Qualcomm introduced the Snapdragon platform and two chipset solutions, which were the first 1 GHz processors for mobile phones. Snapdragon is a series of system-on-chip (SoC) semiconductor products, which is Qualcomm's flagship suite in the realm of hardware. Based on ARM architecture, Snapdragon typically integrates CPU, graphics processing unit, wireless communication module, and other functional requirements that support the normal operation of mobile devices. Furthermore, Snapdragon semiconductors can be embedded in the devices of various systems, including mobile phones, tablet computers, cars, and other mobile devices.

Since Qualcomm released its first chipset solutions—i.e., the QSD8250 and QSD8650 processors in 2007—the Snapdragon product lineup has evolved from the 200 series to the 800 series, enhancing the performance of mobile devices and making multiple functions available while consuming less power. Snapdragon now has a wide portfolio of products ranging from low-end to high-end, and, primarily due to their excellent performance, more and more OEMs have begun to embed Snapdragon semiconductors in their consumer electronics products.

As for the OEMs, implementing the Snapdragon system-on-chip products significantly reduces design complexity and costs, in turn lowering the difficulty of independent development and promoting innovation. Consequently, OEMs can build new features into their products, thus responding to the needs of local customers, without diverting resources to the development of core parts. With the growth of shipment from emerging regions, Qualcomm has continued to maintain its global leadership in the market of mobile device chipsets.

By taking the above steps, Qualcomm has developed the CDMA ecosystem to a great extent, including hardware, software, and the main interface of the implementation layer. The company has also been providing third-party vendors with a complete range of CDMA-based solutions, with full technological support.

As previously explained, Qualcomm is now a system company that offers system solutions. However, its main focus is on the technology layer, which means that there is indeed inconsistency between its knowledge and business domain. Since the firm's knowledge stretches across the implementation layer and the technology layer, enabling it to develop system integration capabilities, it might be worth exploring how Qualcomm has managed to accumulate implementation layer knowledge without actually operating in that area and which features play an important role in system integration within the company.

5.2 *Bidirectional Learning Process*

Since Qualcomm strategically withdrew from the business domain linked to the implementation layer, it has been concentrating on its core competence, i.e., technology layer R&D. By underscoring its commitment to the upper layer, Qualcomm has signaled that it will no longer be a powerful competitor in the implementation layer sector, in which a growing number of vendors have flooded the market, thus diversifying CDMA peripheral products.

The implementation layer and technology layer are connected through standard specifications drafted by standardization bodies. The specifications mainly describe technology requirements, such as communications protocols and performance, and present concepts at an abstract level, which means that several manufacturers—especially new entrants, increasing in developing countries—cannot develop products directly according to the standard specifications. In fact, explanations about how to apply the specifications to the lower layer of products are extremely scarce, and, although standards have been adopted, they cannot be immediately implemented without a long-term commercialization process. Therefore, manufacturers are inclined to choose implementation layer solutions with full technological support and invest fewer resources to fully comprehend those complex specifications. Benefiting from accumulated CDMA knowledge, Qualcomm can integrate its solutions from the conceptual level and illustrate them in an approachable way, which other firms are unable to do. Thus, developing product by using Qualcomm's SEPs and proprietary patents is an attractive and feasible choice for them.

System integration on the interface is a crucial part of architectural control. Moreover, practical implementation knowledge outside the company is indispensable to build capabilities for system integration. Qualcomm has always attached great importance to fostering strategic relationships with its implementation layer partners, and, thanks to the precious contribution given in the early stages of CDMA, its links with both partners and third-party manufacturers have always been strong. Continual direct feedback from the implementation layer is a precious source of knowledge. For instance, the Qualcomm Engineering Services Group (ESG) has been providing technology consulting services to network operators since 2001, which is one of the most important sources of knowledge about the implementation layer. ESG is mainly engaged in technology training and engineering support to operators, including network optimization and technical upgrades of wireless network systems. In addition, direct access to the implementation layer gives Qualcomm the chance to better understand the needs of general customers.

Due to the complexity of specifications, manufacturers need to learn from Qualcomm how to integrate its hardware into their product development projects, and Qualcomm has to train them in doing so as effectively as possible. However, rather than one-way, the learning process is actually bidirectional, in that Qualcomm also has to learn from its consumers, which contributes enormously to its capability building for system integration. Other firms that have a manufacturing division are limited by the uncertainties and risks of their business and do not have sufficient resources to invest in technology layer R&D. Therefore, their sources of knowledge are not as various as those of Qualcomm.

With the expansion of the wireless telecommunications market, more and more firms are entering the industry and investing in the manufacturing of peripheral products. Furthermore, as they pay license fees to Qualcomm, they are also its customers. As said above, by maintaining close relationships with its implementation layer partners, Qualcomm has coordinated system design and manufacturing activities on the interface and built superior capabilities concerning system integration. Thus, the firm and its partners have been coevolving through bidirectional learning, which has expanded the relevant market and enabled Qualcomm to strengthen its capabilities without losing its implementation layer knowledge. As a result, even though it left the implementation layer business, it still retains all relevant knowledge about it. Hence, it can be concluded that, in the case of Qualcomm, knowledge boundaries are wider than organization boundaries.

5.3 The Role of System Engineers

Due to successful design and commercialization of CDMA, Qualcomm has accumulated systematic knowledge and a superior IP portfolio in the technology layer and has invested significant resources and engineering talent to maintain that advantage in the later generations (*Gs*) of wireless communications systems. Particularly, its IP at the early stages of CDMA influenced the process of standardization, on which additional technologies continue to be built. In the mid-1990s—while the CDMA technology was in the process of being fully adopted by the industry and had not yet been adopted for 3G—a team of engineers at Qualcomm had already started looking ahead and working on the first systems designed for data optimization. These early system designs were fundamental to the later versions of 3G and the 4G data-optimized systems.

The team working on data-optimized systems realized that CDMA would not scale well for data and started developing a new technology called orthogonal frequency division multiple access ("OFDMA"). In 2005, Qualcomm joined hands with Flarion Technologies, a start-up working on designing and prototyping of OFDMA. All the major standards that worked on the development of the 4G technology, including the widely deployed 3GPP Long-Term Evolution ("LTE") standards, selected OFDMA as their choice of technology for physical layer transmission.

Today, amidst talk of a smartly connected future, Qualcomm is building the technologies to make it a reality, by leading the system design for 5G, collaborating with industry leaders, and spearheading the research efforts that will create the next global wireless standard.

With the evolution of wireless generations, Qualcomm has continually updated and effectively exploited its IP portfolio, and system engineers have been playing a prominent role in system integration capability building. The company relies on software engineers, hardware engineers, and system engineers. Software engineers mainly focus on the development of embedded software and firmware for Qualcomm's range of wireless chips, also including the development of device drivers based on multiple telecommunications standards. Hardware engineers design and develop wireless devices and hardware solutions integrated with Qualcomm's competence for wireless connectivity, such as system-on-chip solutions. The role of system engineers is to design system architectures and interfaces linking to multiple peripherals. Taking advantage of the firm's strong technical background and cooperation with strategic allies, system engineers are committed to integrating the software, hardware, and wireless technologies related to standards into complete wireless connectivity systems across different areas of wireless technology. Moreover, to provide valuable solutions to customers, system engineers collaborate with third-party device manufacturers and developers, absorbing large amounts of implementation layer knowledge, especially in the process of standardization. In Qualcomm, system engineering is not an independent division but rather spreads across its various R&D divisions, including software, hardware, and system solutions.

System engineers at Qualcomm need to have a master's degree or PhD in communications, while solid implementation experience in platforms is a major plus. Additionally, they are required to possess capabilities for designing systems and integrating interfaces to ensure interoperability. System engineering is pivotal to effectively implementing architectural control across the whole industry, as systematic solutions for the mobile wireless industry are extremely complex, especially in the standard-setting process, and the unclear definition of tasks and objectives makes it difficult to assign engineers to each project. To overcome potential obstacles, Qualcomm does not have special leaders or coordinators for system engineering. Rather, software engineers, hardware engineers, and system engineers come together to undertake collaborative R&D, thus accelerating the R&D process itself. In general, related patents have to integrate multiple layers, from concept to application, which means that specialized and interdivisional knowledge is essential to properly link these different layers. It is notable that the engineers working in the firm all have extensive implementation layer experience and, as for system engineers, skills and experience in interface and product implementation are as important as system development knowledge.

As one of the most import features in terms of system integration capability, Qualcomm has built a superior technology layer IP portfolio. Critically, it has retained control of the key parts of the interfaces, which are essential to architectural control. The IP corporate culture of Qualcomm is quite different from that of other leading firms in the telecommunications industry, as it emphasizes spontaneous patent applications by employees. In order to deliver highly integrated system solutions to vendors, its software, hardware, and system engineers cooperate on each IP activity, generating content that covers both the technology layer and the implementation layer. Therefore, the formation of its IP portfolio is quite literally the result of cross-functional interactions.

Lastly, to prove how it has chased innovation since the beginning and what it has achieved so far, Qualcomm has created a *Patent Wall* in the front lobby of its San Diego headquarters, showcasing a small part of its IP portfolio. The Patent Wall displays patents according to the history of wireless connectivity and confirms the value of IP and the importance of IP protection in Qualcomm.

6 Discussion

Our data analysis of technology specifications indicates that system knowledge is important for technology diffusion to create demand. Furthermore, firms can effectively deal with technological complexity by enhancing their knowledge for system integration, i.e., system knowledge. The technologies of firms that possess such knowledge are most frequently cited, which means that they have a central role in technology diffusion to create demand.

The case of Qualcomm shows how a company can build knowledge for system integration. In the early stages of CDMA, Qualcomm had to develop a wide range of products by itself. With the growth of relevant markets, it strategically withdrew from the implementation layer business, setting itself apart from its competitors. Since then, it has mostly concentrated on technology layer R&D. Thanks to bidirectional learning from its partners regarding the implementation layer and intense interactions among system engineers, Qualcomm has acquired system knowledge related to both the implementation and technology layers and developed technologies for implementation based on system engineering.

These technologies shaped in accordance with system engineering reflect system knowledge for implementation, and, since they are developed from a systemic perspective and disclosed as SEPs, they support diffusion and creation of demand. Qualcomm's centrality in patent citation networks and paths of technology diffusion can be ascribed to its technology development in line with system engineering. By developing and diffusing technologies for implementation reflecting system knowledge, a leading firm can not only lay the technological foundations for demand creation but also preserve its knowledge and rights on system implementation.

Standardization has expanded the market for ICT products since the 1980s. The locus of competition has shifted from the overall package to the specific costs and performance characteristics of individual components (Matutes and Regibeau 1988; Economides 1989; Katz and Shapiro 1986, 1994). On the other hand, as explained by Shiu and Yasumoto (2017a, b, c), competition remains strong at the systematic level of product system too. Firms' SEPs, as inevitable technologies, are to be systematically implemented over open technology specifications. But, although technology information becomes publicly available through standards and patents, firms possess different degrees of property rights from a system perspective.

Furthermore, our evidence confirms that the scope of systematic implementation rights does not need to correspond to that of a firm's business. Despite possessing degrees of systematic implementation rights similar to those of Nokia, Qualcomm concentrates on providing technologies to its customers through patent licensing. Rather than total solution product systems, Qualcomm's technologies, which reflect system knowledge for implementation, are complementary to those of new entrants that want to seize the business opportunities offered by growing markets.

Brusoni et al. (2001) discuss the ways in which firms can maintain technological capabilities in a number of fields wider than those in which they decide to produce and examine the implications of this in terms of firm boundaries and vertical integration decisions. The case of Qualcomm shows that a firm's system knowledge and relevant technologies to drive demand creation need to cover areas that are wider than the firm's specific scope of business.

The architecture of CoPSs with technology specifications discloses know-what knowledge, i.e., typical standard specifications related to the requirements of product systems. At the same time, firms try to retain know-how knowledge for implementation (i.e., system integration capability) from the systematic perspective of property rights. Due to the open construct of know-what knowledge, with interoperability and connectivity as well-defined areas disclosed under standardization, firms may start to grow not only by absorbing know-what knowledge but also thanks to network effects.

On the other hand, know-how knowledge, which we defined as system knowledge for implementation, is more tacit and sticky than know-what knowledge. This chapter shows that, in the case of open CoPS, (1) product decomposability is not equal to knowledge decomposability (Dosi et al. 2003), (2) the price mechanism is not suitable for coordinating innovative activities (Brusoni et al. 2001; Chesbrough 2005), and (3) the more dispersed the production of knowledge and the more complex the products, the greater the requirement that firms embody explicit integrative capabilities (Prencipe 2005; Chesbrough 2005).

In this sense, firms providing systematic technologies for open CoPS should know how to build capabilities for system integration. As described in the case of Qualcomm, we see system integration of open CoPS not as limited to implementation capability on technology specifications but as more related to the strategic decision of whether to provide technologies reflecting system knowledge or total solution product systems. Close interactions between the IP division and system engineers in Qualcomm lead us to infer that system integration also needs to take property rights into account (i.e., systematic implementation rights).

7 Conclusions

This chapter shows that demand creation is encouraged by the technological foundations (i.e., sets of SEPs) residing in leading firms' system integration for implementation. When a product system consists of a variety of technologies that are open, retaining systematic implementation rights can help firms extend the scope of demand creation. Why are firms such as Qualcomm leading demand creation in the ICT industry? The reason is that they develop technologies while also accumulating system knowledge and retaining property rights. In other words, system integration is still required for providing technologies reflecting system knowledge in pursuit of demand creation. In addition, firms with system knowledge should deal with property rights by means of close interactions between patent managers and system engineers.

Vertical integration is not the only source of system integration capability. Indeed, even though Qualcomm left the implementation layer business, it still retains relevant knowledge through bidirectional learning. Frequent interactions across divisions and teams have significantly contributed to the accumulation of system knowledge.

Our case study has shed some light on how demand creation occurs, but several aspects remain to be examined for a thorough exploration of the topic. Future research should look at the influence of non-SEPs on firms' capabilities to bolster demand creation. Moreover, the different strategies followed by firms for systematic implementation rights should be explored further. Lastly, extensive comparisons across firms, e.g., between new entrants and incumbents, might prove useful to determine the generalizability of the implications illustrated here.

References

Baldwin CY (2010) When open architecture beats closed: the entrepreneurial use of architectural knowledge. Harvard Business School Finance Working Paper 10-063

Baldwin CY, Woodard CJ (2009) The architecture of platforms: a unified view. In: Gawer A (ed) Platforms markets and innovation. Edward Elgar, London, pp 19–44

Bekkers R (2001) Mobile telecommunications standards: GSM, UMTS, TETRA, and ERMES. Artech House, London

Bekkers R et al (2002a) Intellectual property rights, strategic technology agreements and market structure: the case of GSM. Res Policy 31(7):1141–1161

Bekkers R et al (2002b) Intellectual property rights and standardization: the case of GSM. Telecommun Policy 26:171–188

Bezerra J et al (2015) The mobile revolution: how mobile technologies drive a trillion-dollar impact. BCG perspectives. https://www.bcg.com/publications/2015/telecommunications-technology-industries-themobile-revolution.aspx. Accessed 7 July 2018

Blind K, Mangelsdorf A (2016) Motives to standardize: empirical evidence from Germany. Technovation 48-49:13–24

Blind K, Thumm N (2004) Interrelation between patenting and standardization strategies: empirical evidence and policy implications. Res Policy 33:1583–1598

Boudreau K (2010) Open platform strategies and innovation: granting access vs. devolving control. Manag Sci 56(10):1849–1872
Brusoni S et al (2001) Knowledge specialization, organizational coupling, and the boundaries of the firm: why do firms know more than they make? Adm Sci Q 46(4):597–621
Cargill CF (1997) Open systems standardization: a business approach. Prentice Hall, Upper Saddle River
Chesbrough HW (2005). Open innovation: the new imperative for creating and profiting from technology: Harvard Business Press First Trade Paper Edition
David PA, Greenstein S (1990) The economics of compatibility standards: an introduction to recent research. Econ Innov New Technol 1(1–2):3–41
Davies A (1996) Innovation in large technical systems: the case of telecommunications. Ind Corp Chang 5(4):1143–1180
Davies A, Brady T (2000) Organizational capabilities and learning in complex product systems: towards repeatable solutions. Res Policy 29(7–8):931–953
Dosi G et al (2003) The economics of systems integration: towards an evolutionary interpretation. Bus Sys Integr:95–113
Economides N (1989) Desirability of compatibility in the absence of network externalities. Am Econ Rev:1165–1181
European Commission (2014) Patents and standards: a modern framework for IPR-based standardization. European Union, Brussels
Farrell J, Saloner G (1988) Coordination through committees and markets. RAND J Econ 19 (2):235–252
Funk JL (2002) Global competition between and within standards. Palgrave, Basingstoke
Garud R, Kumaraswamy A (1993) Changing competitive dynamics in network industries: an exploration of sun Microsystems' open systems strategy. Strateg Manag J 14:351–369
Garud R, Kumaraswamy A (1995) Technological and organizational designs for realizing economies of substitution. Strateg Manag J 16(1):93–109
Garud R et al (2002) Institutional entrepreneurship in the sponsorship of common technological standards: the case of sun Microsystems and java. Acad Manag J 45(1):196–214
Gawer A, Cusumano MA (2002) Platform leadership: how Intel, Microsoft, and Cisco Drive Industry Innovation. Harvard Business School Press, Boston
Granstrand O (1999) The economics and management of intellectual property. Edward Elgar Publishing, Northampton
He ZL et al (2006) Entry and competitive dynamics in the mobile telecommunications market. Res Policy 35(8):1147–1165
Helpman E (1998) General purpose technologies and economic growth. MIT Press, Cambridge, MA
Henderson RM, Clark KB (1990) Architectural innovation: the reconfiguration of existing product technologies and the failure of established firms. Adm Sci Q 35:9–30
Henkel J, Baldwin CY (2009) Modularity for value appropriation: drawing the boundaries of intellectual property. Harvard Business School Working Paper 90-097
Hobday M (1998) Product complexity, innovation and industrial organization. Res Policy 26:689–710
Iansiti M, Levien R (2004) Strategy as ecology. Harv Bus Rev 82(3):68–81
Jaffe AB, Trajtenberg M (2002) Patents, citations, and innovations: a window on the knowledge economy. MIT Press, Cambridge, MA
Kang B, Motohashi K (2015) Essential intellectual property rights and inventors' involvement in standardization. Res Policy 44(2):483–492
Katz ML, Shapiro C (1986) Technology adoption in the presence of network externalities. J Polit Econ 94(4):822–841
Katz ML, Shapiro C (1994) Systems competition and network effects. J Econ Perspect 8(2):93–115
Kristiansen E, Thum M (1997) R&D incentives in compatible networks. J Econ 65:55–78

Leiponen A (2008) Competing through cooperation: the organization of standard setting in wireless telecommunications. Manag Sci 54(11):1904–1919

Lipsey RG et al (2005) Economic transformations: general purpose technologies and long-term economic growth. Oxford University Press, New York

Matutes C, Regibeau P (1988) "Mix and match": product compatibility without network externalities. Rand J Econ 19:221–234

Moore JF (1993) Predators and prey: a new ecology of competition. Harv Bus Rev 71(3):75–83

Office of Science and Technology (1990) Technology foresight progress through partnership: defence and aerospace. Office of Science and Technology, London

Pavitt K (2005) Specialization and systems integration: where manufacture and services still meet. In: Prencipe A et al (eds) The business of systems integration. Oxford University Press, New York, pp 78–91

Pon B et al (2014) Android and the demise of operating system-based power: firm strategy and platform control in the post-PC world. Telecommun Policy 38(11):979–991

Prencipe A (2005) Corporate strategy and systems integration capabilities: managing networks in complex systems industries. In: Prencipe A et al (eds) The business of systems integration. Oxford University Press, New York, pp 114–132

Rysman M, Simcoe T (2008) Patents and the performance of voluntary standard-setting organizations. Manag Sci 54(11):1920–1934

Schilling M (2009) Protecting or diffusing a technology platform: tradeoffs in appropriability, network externalities, and architectural control. In: Gawer A (ed) Platforms, markets and innovation. Edward Elgar Publishing, London, pp 192–218

Shiu JM, Yasumoto M (2015) Investigating firms' knowledge management in the standardization: The analysis of technology specification-declared essential patent networks on telecommunication industry. MMRC Discussion Paper Series, University of Tokyo, p 465

Shiu JM, Yasumoto M (2017a) Exploring the architectural control over opened system-goods: analysis of technology specifications and standard-essential patents in the telecommunication industry. Presented at the AOM (academy of management) annual meeting, Atlanta

Shiu JM, Yasumoto M (2017b) Investigating knowledge spillovers under standardization: the examination of the patent-citation networks in the telecommunication industry. J Manag Policy Pract 18(2):81–104

Shiu JM, Yasumoto M (2017c) Exploring the source of architectural control of complex system-goods under standardization. Presented at the SMS (strategic management society) annual meeting, Huston

Simcoe TS (2006) Open standards and intellectual property rights. In: Chesbrough H et al (eds) Open innovation: researching a new paradigm. Oxford University Press, Oxford, pp 161–183

Simcoe TS (2012) Standard setting committees: consensus governance for shared technology platforms. Am Econ Rev 102(1):305–336

Simcoe TS et al (2009) Competing on standards? Entrepreneurship, intellectual property, and platform technologies. J Econ Manag Strateg 18(3):775–816

Steinmuller WE (2003) The role of technical standards in coordinating the division of labour in complex system industries. In: Prencipe A et al (eds) The business of systems integration. Oxford University Press, New York, pp 133–151

Teece DJ (2007) Explicating dynamic capabilities: the nature and microfoundations of (sustainable) enterprise performance. Strateg Manag J 28(13):1319–1350

Van de Kaa G, de Bruijn H (2015) Platforms and incentives for consensus building on complex ICT systems: the development of WiFi. Telecommun Policy 39(7):580–589

Weiss M, Cargill C (1992) Consortia in the standards development process. J Am Soc Inf Sci 43 (8):559–565

West J (2003) How open is open enough? Melding proprietary and open source platform strategies. Res Policy 32:1259–1285

West J (2006) The economic realities of open standards: black, white and many shades of gray. In: Greenstein S, Stango V (eds) Standards and public policy. Cambridge University Press, Cambridge, pp 87–122

West J, Dedrick J (2001) Open source standardization: the rise of Linux in the network era. Knowl Technol Policy 14(2):88–112

Xia M et al (2012) Enhancing value via cooperation: firms' process benefits from participation in a standard consortium. Ind Corp Chang 21(3):699–729

Yayavaram S, Ahuja G (2008) Decomposability in knowledge structures and its impact on the usefulness of inventions and knowledge-base malleability. Adm Sci Q 53(2):333–362

The Impact of Platform Providers' Knowledge on Interfirm Division of Labor: The Case of the Mobile Phone Industry

Jing-Ming Shiu and Masanori Yasumoto

Abstract To achieve higher performance in the interfirm division of labor, it is crucial to understand how to assign decision-making rights properly. Past research indicates that decision-making rights between a firm and its partner are determined by how much knowledge the firm and partner have. Indeed, the more knowledge a firm has compared with its partner, the more decision-making rights should be assigned to it. On the contrary, the more knowledge the firm's partner has, the more decision-making rights should be assigned to the firm's partner. This study extends the above scope of analysis and examines the case of a firm and its partner relying on third-party knowledge. From the point of view of interfirm division of labor, how should they assign decision-making rights to pursue higher performance? We examined 50 smartphone product development projects carried out by brand firms and Taiwanese contract manufacturers from 2012 to 2014 and found greater performance enhancement when both aimed to acquire different platform provider knowledge for their decision-making rights, based on their respective business-oriented activities. At the same time, decision-making rights induce opportunism and a disincentive effect that should be controlled by means of close interaction and communication.

1 Introduction

Normally, firms are expected to outsource certain activities—and, consequently, grant decision-making rights—to business partners depending on the capabilities and amount of knowledge the partners have (Barzel 1997; Foss and Foss 1998; Hart and Moore 1990; Windsperger 1996). If a firm possesses more knowledge than its

J.-M. Shiu (✉)
National Cheng Kung University, Tainan, Taiwan
e-mail: jingmingshiu@mail.ncku.edu.tw

M. Yasumoto
Yokohama National University, Yokohama, Japan

T. Fujimoto, F. Ikuine (eds.), *Industrial Competitiveness and Design Evolution*, Evolutionary Economics and Social Complexity Science 12,
https://doi.org/10.1007/978-4-431-55145-4_14

partner, then the firm should be assigned the decision-making rights to carry out the work. On the contrary, if the firm's partner is more knowledgeable, then it should be granted more decision-making rights (Windsperger 1996, 2001, 2002a, 2004, 2009; Windsperger and Dant 2006). According to this principal agency perspective, in the interfirm division of labor, firms are expected to pursue efficiency and performance by paying attention to the *location of knowledge* and delegating decision-making rights accordingly.

However, the dyadic interfirm relationship mentioned in past research should be extended to cases of triadic interfirm division of labor. This is because in the ICT (Information and Communications Technology) industry (including personal computers, mobile phones, liquid crystal displays, etc.), *knowledge* is not a proprietary asset owned by firms or their partners, but it is possessed by platform providers as third parties. In the traditional dyadic model, either the firms or their partners can assign decision-making rights depending on their comparative degrees of knowledge. Nowadays, however, besides marketing the core components of ICT products (e.g., CPUs, chipsets, etc.), platform providers possess proprietary knowledge able to shape the business relationships between firms and their partners.[1] In this sense, in the interfirm division of labor, decision-making rights are likely to be influenced by how firms and their partners assimilate knowledge from platform providers. Yet, academic studies on the interfirm division of labor have not yet shed sufficient light on the above aspects.

Comprehensive quantitative research on these topics is needed in the ICT industry. Indeed, previous studies have discussed many examples of integral product architecture (e.g., automobile) and its interfirm division of labor.[2] On the other hand, as for the ICT industry, scholars have focused on the relationship between leading brand firms and contract manufacturers,[3] the influence of platform providers on product architecture change (i.e., from integral to modular), and platform providers' strategies. Yet, discussion is lacking about how the brand firms and contract manufacturers carry out interfirm product development in order to achieve higher performance. In fact, it is very difficult to develop core components without using

[1]For example, when platform provider Intel releases a new CPU, it invites hardware and software engineers from PC brand firms and contract manufacturers to come together and solve technical issues.

[2]For example, Toyota has entrusted not only automobile production but also car body parts development to contract manufacturers, such as Kanto Auto Body. According to Seike (1995) and Shioji (1996), Toyota set up the same development organization in these body manufacturers so as to immediately check the status of development and production.

[3]For example, Doig et al. (2001) state that in the ICT industry brand firms have actively outsourced since the second half of the 1980s. In the United States, PC and network equipment firms, such as IBM, Nortel, Apple, 3Com, Hewlett Packard, Maxtor, Lucent, etc., have outsourced electronic board and product assembly to contract manufacturers. Also, since the 1990s, European firms, including Nokia, Ericsson, Alcatel, etc., have outsourced the production of telecommunication equipment and mobile phone devices to contract manufacturers, such as Solectron, Flextronics, and SCI. In 2000, it was reported that approximately 75% of brand firms' products in the ICT industry were produced and shipped by their contract manufacturers.

platform providers' technologies. Brand firms have to rely on platform providers' core components and technologies to outsource new product development to contract manufacturers.

This research investigates the aspects of knowledge and decision-making rights among brand firms, contract manufacturers, and platform providers in the mobile phone industry, focusing specifically on how platform providers' knowledge may impact on brand firms and contract manufacturers. In a situation in which platform providers' core components and technologies circulate on the market, how can brand firms increase new product development performance when dividing labor between themselves and their contract manufacturers? In the next section, we will review how past research has discussed the interfirm division of labor in the ICT industry, especially for what concerns knowledge and decision-making rights. We will then point out any research gaps and illustrate our analysis framework and hypotheses for further examination. Section 3 will concentrate on Taiwanese contract manufacturers and brand firms in the mobile phone industry and describe the variables used to examine the triadic interfirm division of labor. Section 4 will present the results of our analysis, while Sects. 5 and 6 will discuss our findings and key implications for practitioners and future research.

2 Theoretical Background and Hypotheses

2.1 The Debate About the Interfirm Division of Labor in the ICT Industry

In order to examine the triadic interfirm division of labor in ICT, we looked at the concept of knowledge as presented in previous studies. *Knowledge* is an intangible asset and a corporate capability (Conner and Prahalad 1996; Grant 1996; Kogut 2000; Kogut and Zander 1996; Leonard-Barton 1992; Nonaka and Takeuchi 1995). For an appropriate division of labor and the attainment of higher performance, both leading brand firms and their partners need to possess and constantly create knowledge related to their projects. Moreover, this will help establish better collaboration among them (Jacobides 2005, 2008; Jacobides and Hitt 2005; Jacobides et al. 2006; Jacobides and Billinger 2006).

Product development revolves around two key concepts, i.e., *architectural knowledge* and *core component knowledge*. Architectural knowledge is knowledge about how components are connected with one another to fulfill the design requirements of products (Henderson and Clark 1990). Thus, architectural knowledge is more context-specific and related to implementation in new product development (i.e., system engineering; Vincenti 1990, Takeishi 2001). On the other hand, core component knowledge is knowledge about the main concepts of a product, which is constructed and combined based on market needs and technologies (Clark 1985). That is to say, core component knowledge is shaped in the stages of concept

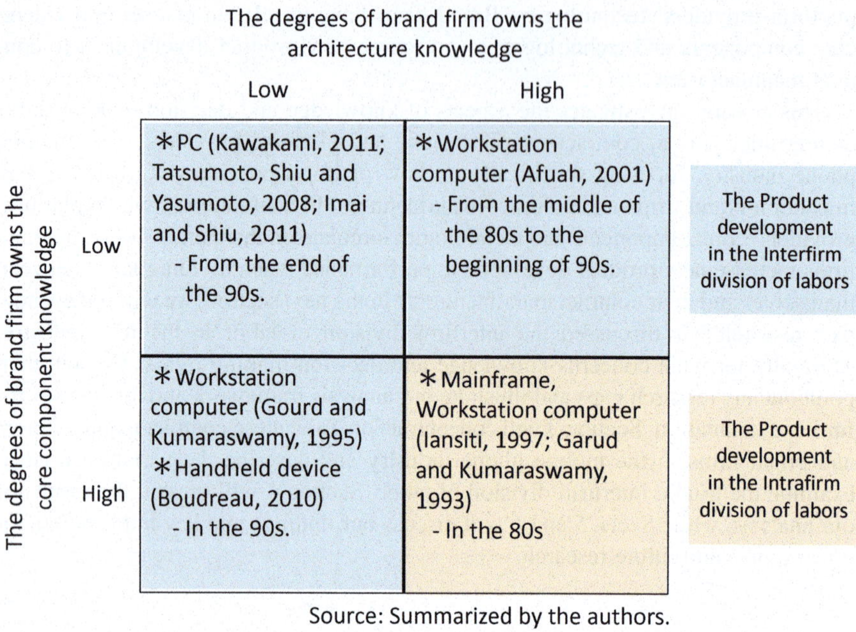

Fig. 1 The knowledge ownership of firms

development and technology development (Anderson and Tushman 1990; Tushman and Anderson 1986; Tushman and Murmann 1998).

In terms of the knowledge types (i.e., core component knowledge and architectural knowledge) and its ownership by firms, existing research also considers decision-making rights as firm characteristics in the ICT industry's interfirm division of labor. First, the lower right quadrant of Fig. 1 shows the intrafirm division of labor for mainframes and workstation computers in the 1980s, when the technology development divisions and product development divisions of firms owned both core component knowledge and architectural knowledge (Iansiti 1997, 1998, 1999; Garud and Kumaraswamy 1993). This allowed them to control the technological progress of CPUs to acquire more benefits than their competitors through economies of substitution (Garud and Kumaraswamy 1993, 1995). Likewise, it is said that IBM's competitive advantage in workstations (i.e., System Z/Architecture) is based on core component knowledge of the CISC (Complex Instruction Set Computing) CPU and on architectural knowledge of manufacturing processes, packaging, ceramics, heat dissipation technologies, and so on.

On the other hand, in the lower left quadrant of Fig. 1, brand firms own the core component knowledge, while product development projects are licensed to and implemented by other firms. For example, in the workstation computer industry of the 1990s, Sun Microsystems controlled the workstation product architecture by embedding electrical signals and algorithms between the workstation's CPU and memory in its own chipset and operation system. Then, it licensed product development to other firms (Baldwin and Clark 1997). In this way, Sun Microsystems devoted itself to the development of core components, while transferring the decision-making rights of product development projects to other firms. As a consequence, workstation computers were put on the market under Sun Microsystems's proprietary control over core components (Garud and Kumaraswamy 1993, 1995). Similarly, as Boudreau (2010) explains, in the handheld devices sector, Palm retained decision-making rights over some core components (e.g., graphical user interface, operating system, kernel, software drivers, analog device control utility, etc.) while delegating to other firms' decision-making rights over noncore components, including board-level circuit design, industrial design, and performance tests. As a result, Palm was able to quickly and successfully launch its handheld devices.

The upper right quadrant of Fig. 1 describes the situation in which brand firms own the architectural knowledge for the development of workstation computers, while component suppliers possess core component knowledge, which enables them to influence the allocation of decision-making rights in product development. By examining a total of 81 projects for 25 workstation computers from 1986 to 1993, Afuah (2001) found that, when discontinuous technology changes occurred in core components (e.g., from CISC CPU to RISC CPU), in-house CPU development capabilities had a positive effect on product development performance. Eisenhardt and Tabrizi (1995) investigated a total of 72 product development projects by European, Asian, and US PC manufacturers and showed that the firms that involved their suppliers in the early stages of product development were able to shorten their product development lead time. Other studies also point to higher product development performance when suppliers actively participate in product development from the start (Eisenhardt 1989; D'Aveni 1994; Eisenhardt and Tabrizi 1995; Galunic and Eisenhardt 2001). It can thus be inferred that firms' decision-making rights in product development are influenced by external core component knowledge. In other words, by involving core component suppliers in the early stages of new product development, firms may acquire core component knowledge and strengthen their decision-making rights.

The upper left quadrant of Fig. 1 refers to a situation, often observed in the modern ICT industry, in which firms own very little or no architectural and core component knowledge. For example, in the PC industry (Kawakami 2011; Tatsumoto et al. 2008) and mobile phone industry (Shiu 2010), the core component suppliers provide firms with not only core component knowledge but also architectural knowledge. In such circumstances, when core component suppliers possess the comprehensive knowledge needed for the product development projects of brand firms, they do not act as simple suppliers but as *platform leaders* (Gawer and Cusumano 2002), with absolute decision-making rights on both concept and implementation. In addition, the modularization of products induced by such platform providers eventually changes interfirm relationships in product development and, as a result, every firm might be able to develop components independently (Henderson and Clark 1990; Langlois and Robertson 2002; Baldwin and Clark 2000).

In sum, research on the ICT industry has focused on the dyadic interfirm division of labor between brand firms and platform providers or brand firms and their license partners (e.g., contract manufacturers). Implications concerning the aspects of knowledge and decision-making rights have occasionally been explored, but open issues remain as to how brand firms and contract manufacturers deal with interfirm division of labor to enhance the performance of their product development projects. In particular, our purpose here is to investigate what happens when brand firms and contract manufacturers rely on the knowledge possessed by platform providers and how they can pursue higher product development performance by suitably allocating decision-making rights.

2.2 Hypotheses

Our interviews with several engineers from platform providers revealed that platform providers try to design their core components so as to be accepted and integrated into their customers' product development projects. However, as they have limited resources, they will prioritize the technical support offered according to their customers' sales, technological capabilities, etc. For example, when a brand firm develops a product with an innovative concept, the platform providers will offer stronger technical support to that brand firm. On other hand, when a contract manufacturer adopts a new core component before a brand firm uses it, the platform providers will adjust their technical support priorities in favor of the contract manufacturer. That is to say, the decision-making rights of brand firms or contract manufacturers may be affected by the knowledge of platform providers through the level of technical support given. Consequently, decision-making rights in the interfirm division of labor between brand firms and contract manufacturers will be influenced by how much knowledge they are able to acquire from platform providers.

Decision-making rights in product development can be divided into two types. The first set of decision-making rights concerns the planning stage of product development, which determines product specifications, such as core concept, price, and functions included. The second set of decision-making rights has to do with product specifications related to the implementation stage of product development. Furthermore, we assume that acquiring core component knowledge and architectural knowledge from platform providers is inevitable. The core concept of a product is a top-level design decision that determines its main function and structure, but it also reflects consumer needs coming from the market. Therefore, when either brand firms or contract manufacturers acquire a lot of core component knowledge from a platform provider, they can acquire greater decision-making rights regarding concept planning in product development. On the other hand, when either brand firms or contract manufacturers acquire a large amount of architectural knowledge from a platform provider, they can acquire greater decision-making rights regarding the

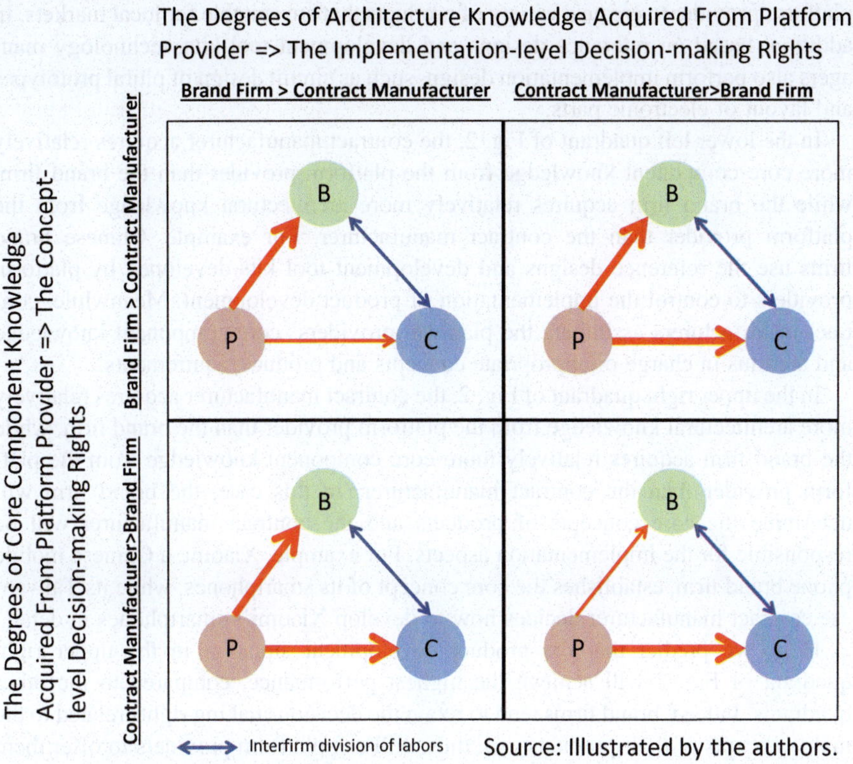

Fig. 2 Types of triadic interfirm division of labor

implementation level of product development. Hence, four patterns can be identified, as illustrated in Fig. 2.

In the upper left quadrant of Fig. 2, the brand firm acquires relatively more core component knowledge and architectural knowledge from the platform provider than the contract manufacturer. So, in the interfirm division of labor, the brand firm will have relatively stronger decision-making rights for what concerns both planning and implementation. For example, in the case of the iPhone, Apple has more core component knowledge and architectural knowledge compared to contract manufacturers, so that its decision-making rights will be greater in the planning stage as well as in the implementation stage of iPhone development.

In the lower right quadrant of Fig. 2, the contract manufacturer acquires relatively more core component knowledge and architectural knowledge from the platform provider than the brand firm. In this case, compared to the brand firm, the contract manufacturer will have stronger decision-making rights at both the planning and implementation levels of product development. For example, Taiwanese and Chinese contract manufacturers develop a variety of smartphone prototypes and present them to Southeast Asian brand firms. In particular, the technology managers and salesmen of contract manufacturers follow the technology road maps elaborated by platform provider to create a variety of products that are suitable for local markets. In addition, based on reference designs and development tool kits, technology managers also perform implementation design, such as circuit design of plural prototypes and layout of electronic parts.

In the lower left quadrant of Fig. 2, the contract manufacturer acquires relatively more core component knowledge from the platform provider than the brand firm, while the brand firm acquires relatively more architectural knowledge from the platform provider than the contract manufacturer. For example, Chinese brand firms use the reference designs and development tool kits developed by platform providers to control the implementation of product development. Meanwhile, contract manufacturers assimilate the platform providers' core component knowledge and are thus in charge of appropriate concepts and product requirements.

In the upper right quadrant of Fig. 2, the contract manufacturer acquires relatively more architectural knowledge from the platform provider than the brand firm, while the brand firm acquires relatively more core component knowledge from the platform provider than the contract manufacturer. In this case, the brand firm will determine the core concepts of products and the contract manufacturer will be responsible for the implementation aspects. For example, Xiaomi, a Chinese mobile phone brand firm, establishes the core concept of its smartphones, while its Taiwanese contract manufacturer decides how to develop Xiaomi's smartphones in detail.

Here, we predict that the product development situation in the upper right quadrant of Fig. 2 will achieve the highest performance, compared to the other quadrants. Indeed, brand firms tend to retain the decision-making rights related to the main concepts of their products and, thus, will ask platform providers to offer them more core component knowledge. The reason for this is that the quality of product planning greatly affects product development, and core component knowledge is the most crucial part of product hierarchy design as well as the main concept that reflects market needs (Clark 1985). Therefore, brand firms differentiate themselves from their competitors by concentrating on core concepts, functional requirements, pricing, etc.

The importance to brand firms of possessing core component knowledge has frequently been discussed in technology innovation studies. According to Tushman and Murmann (1998), a product system consists of a core subsystem and noncore subsystems (peripheral subsystems). The core subsystem is a critical bottleneck of technology (Baldwin 2015), owned by firms as property rights (Baldwin 2015; Boudreau and Hagiu 2009; Strahilevitz 2006). The core subsystem has the greatest impact on product specifications and performance (Constant 1980; Hughes 1983; Ulrich and Eppinger 1995) and contributes greatly to product innovation (Tushman and Murmann 1998; Tushman and Rosenkopf 1992). Furthermore, considering the upper and lower layers of a product system, core component knowledge is the upper-layer knowledge, which influences other lower-layer knowledge. For instance, in a nested hierarchy of product development (Alexander 1964; Clark 1985; Marple 1961), problems found in the lower-layer peripheral parts will sometimes need to be solved at the upper layer (Dosi 1982; Dosi et al. 2003). Thus, core component knowledge will help brand firms not only decide how to design suitable product concepts but also examine whether components are well developed together under the core concept of the product (Henderson and Clark 1990).

On the other hand, innovation in how to connect components is regarded as architectural innovation. Possessing and constantly reviewing architectural knowledge can help firms make the most of architectural innovation—and this is why architectural knowledge can also be thought of as a dynamic capability (Baldwin 2015; Henderson and Clark 1990; Henderson and Cockburn 1994). According to Baldwin (2010), firms can use architectural knowledge to (1) implement the product functions, (2) develop the methods to link components, and (3) monitor product responsiveness when customers use it both properly and improperly. Moreover, when product complexity increases and product development becomes ever more difficult, brand firms need to possess architectural knowledge to ensure product quality standards (e.g., to determine the most efficient testing methods and decide whether developers should use existing assets or develop new ones).

Contract manufacturers typically pursue profit by trying to secure production orders from brand firms. When manufacturing gross profit declines, contract manufacturers attempt to extend their business scope by providing brand firms with more value-added activities. In addition, since they have to compete with other contract manufacturers, they propose better contract conditions to brand firms. Product development services are a type of value-added activities. Furthermore, in order to enhance production quality, contract manufacturers are usually involved in the stages of product design and development. Hence, they tend to focus on decision-making rights at the implementation level of product development with brand firms. In other words, in the interfirm division of labor with brand firms, contract manufacturers are likely to acquire more architectural knowledge from platform providers.

To sum up, in the interfirm division of labor, brand firms and contract manufacturers try to consolidate the decision-making rights related to the concept level and implementation level of product development, respectively. Under these circumstances, it is reasonable for platform providers to offer core component knowledge (i.e., road maps of core components) and architectural knowledge (i.e., reference designs and development tool kits) to brand firms and contract manufacturers to speed up product development. Thus, we propose the following hypothesis, Hypothesis 1.

Hypothesis 1 In the interfirm division of labor, when brand firms and contract manufacturers, respectively, acquire and hold relatively higher degrees of core component knowledge (i.e., concept-level decision-making rights) and architectural knowledge (i.e., implementation-level decision-making rights) from platform providers, they can achieve higher product development performance.

Normally, assignment of appropriate decision-making rights is deemed to contribute to product development performance in the interfirm division of labor. Yet, brand firms may use their concept-related decision-making rights to ask contract manufacturers for extra design changes, while contract manufacturers may use their implementation-related decision-making rights to cut development costs by adopting only generic and already existing design assets. This is due to uncertainties in the market environment within the ICT industry, which leads to *incomplete contract* problems between brand firms and contract manufacturers. In addition, if the tasks required are very risky but not completely specified in the contract, the agent will tend to adopt a "you get what you pay for" attitude (Gibbons 2005; Gibbons and Roberts 2013; Holmstrom and Milgrom 1991; Prendergast 1999). The "you get what you pay for" behavior is opportunism, has a disincentive effect (Windsperger 2003), and makes it difficult for brand firms to extract residual income from contract manufacturers. This leads to Hypothesis 2.

Hypothesis 2 The decision-making rights of brand firms and contract manufacturers may cause disincentive effects, which reduce product development performance in the interfirm division of labor.

Due to the common situation of incomplete contracts, brand firms and contract manufacturers need to establish a mechanism for close interaction and coordination to prevent opportunism and the abovementioned disincentive effects. According to Asanuma (1988) and Clark and Fujimoto (1991), automobile manufacturers can improve their product development performance by involving suppliers in the early stages of new product development. Research also indicates that, in the Japanese copier industry, camera industry, PC industry (Imai et al. 1985), and mainframe and minicomputer industry (Eisenhardt and Tabrizi 1995), early problem-solving with suppliers is a crucial success factor in product development. Indeed, establishing close communication with suppliers has become indispensable for product development in the interfirm division of labor (Ancona and Caldwell 1992; Brown and Eisenhardt 1998; Clark and Fujimoto 1991; Takeishi 2001). Thanks to early problem-solving and close interaction and communication, brand firms will be able to reduce opportunism and disincentives in the interfirm division of labor. This translates into our final two hypotheses, Hypotheses 3 and 4.

Hypothesis 3 By establishing early problem-solving, brand firms can suppress opportunism and disincentive effects to increase product development performance in the interfirm division of labor.

Hypothesis 4 By establishing close interaction and communication, brand firms can suppress opportunism and disincentive effects to increase product development performance in the interfirm division of labor.

3 Research Method

3.1 Sample

In order to verify the above hypotheses, we conducted a questionnaire survey targeting contract manufacturers. We adopted individual smartphone product development projects between contract manufacturers and brand firms as our basic unit of analysis. Minor rework projects on the exterior of existing smartphones were excluded. Although it is clearly important to consider the management of multiple projects (Aoshima 2002; Cusumano and Nobeoka 1998; Nobeoka 1996), we decided to focus on how to manage single projects.

We selected Taiwanese contract manufacturers as the target of our analysis to examine their interfirm division of labor with leading brand firms. Taiwanese contract manufacturers are indispensable and engaged as brand firms' business partners in the global value chain in many ICT sectors, including PCs, mobile phones, digital cameras, semiconductors, and so on.[4]

Prior to conducting the questionnaire survey, we interviewed 27 engineers and project managers from 8 Taiwanese contract manufacturers. We discussed their projects over the previous 3 years, identified key projects for each firm, and determined the approach of the questionnaire survey. In May 2015, we presented our online questionnaire, using Google Form, to 8 contract manufacturers and received 93 answers. Duplicate replies about the same project were eliminated, and our final data sample consisted of 50 projects in 8 firms over the previous 3 years. The period distribution of the 50 projects was 15 in 2012, 16 in 2013, and 19 in 2014.

As Table 1 shows, approximately two projects per contract manufacturer were carried out each year. Currently, it takes between 6 and 10 months to develop a new smartphone, so it can be inferred that these 8 contract manufacturers simultaneously developed several major models. Obviously, our data do not offer a comprehensive picture, but the 50 projects investigated here are representative of the work carried out by the 8 Taiwanese contract manufacturers in question between 2012 and 2014.

Despite the absence of some leading brands (e.g., ZTE, LG, Huawei, Motorola, TCL, Coolpad, Xiaomi, and Samsung), our sample does cover the outsourcing projects of major smartphone producers, like Nokia, Apple, Lenovo, and Sony. Moreover, Samsung does not outsource its projects, and ZTE, Huawei, TCL, Coolpad, and Xiaomi mainly outsource to Chinese contract manufacturers, which are not included in this research. In any case, we believe these 50 projects to well represent outsourcing in smartphone development by brand firms.

[4]These Taiwanese contract manufacturers, also called ODM (original design manufacturers), grew by approximately 26% every year in the 1990s by doing business with US, European, and Japanese brand firms (Kawakami 2011). Taiwanese contract manufacturers are important collaboration partners for brand firms, which are keen on outsourcing portions of their business. Besides, the phenomenon of contract manufacturers being engaged in both production and new product development is found not only in the ICT industry. For example, a contract manufacturer named Magna Steyr, a Canadian-based Austrian firm, has been in collaboration with BMW not only to develop components/system modules but whole vehicles. Thus, contract manufacturers' business activity scopes seem to overlap brand firms' value chains, which is a non-negligible factor in the worldwide interfirm division of labor.

Table 1 Data sample

		Brand firms													
		Acer	Apple	Asus	Blackberry	Dell	HP	HTC	Lenovo	NEC	Nokia	Sharp	Sony	Channel Vendor (8 firms)	No. of projects of contract manufacturers
Contract manufacturers	Compal			1(2012)				1(2013) 1(2014)	1(2012) 2(2013)		1(2012) 1(2013) 1(2014)		2(2014)		11
	FIH	1(2014)			1(2014)						1(2012)	2(2012) 1(2013)	1(2012) 1(2014)	1(2013) 1(2013)	10
	Pegatron		1(2012) 1(2013) 1(2014)	1(2014)		1(2012)	2(2012)							1(2012) 1(2013) 1(2013)	10
	Arima									1 (2013)			1(2012) 1(2013) 5(2014)		8
	Quanta			1(2013) 2(2014)										1(2012) 1(2013)	5
	Wistron				1(2013) 1 (2014)									1(2014)	3
	Inventec						1(2012)						1(2014)		2
	Qisda								1(2013)						1
	No. of outsourcing projects of Brand Firms	1	3	5	3	1	3	2	4	1	4	3	12	8	50

Table 2 Correlations among variables

	Descriptive statistics				Correlations				
	Mean	S.D.	Min	Max	IDR	CDR	EPS	CIC	PDP
IDR	0.00	0.98	−2.37	2.02					
CDR	0.00	1.18	−2.54	1.87	−0.169				
EPS	0.00	0.47	−1.41	0.99	−0.148	0.548***			
CIC	0.00	0.78	−1.86	1.34	−0.112	0.346**	0.289**		
PDP	0.00	0.75	−1.87	1.51	0.08	0.524***	0.488***	0.527***	
TN	0.10	0.30	0.00	1.00	0.172	0.173	0.12	−0.095	0.182

$N = 50$, * p-value < 0.1, **p-value < 0.05, ***p-value < 0.01

3.2 Variables for Examination[5]

Our dependent variable, product development performance (PDP), is based on the product design quality variables of Takeishi (2001, 2002, 2003). First, we asked product managers to evaluate a certain smartphone development project in terms of product performance, cost, quality, function, structure, and so on, according to 11 questionnaire items (Cronbach $\alpha = 0.928$). Then, in order to better reflect the competitiveness of the market, we used the same 11 questionnaire items to ask product managers to evaluate the projects by comparing them with other smartphone development projects (Cronbach $\alpha = 0.945$). After that, we calculated the PDP by using the mean values of the two evaluations by product managers from contract manufacturers.

In order to test the four hypotheses put forward in the previous section, five independent variables were chosen. A description of each variable is provided below. We referred to the index of technical novelty of Takeishi (2001, 2002, 2003) and used the Technological Newness (TN) of core components as a control variable. Descriptive statistics and correlations among the variables are shown in Table 2 below.

First, to more objectively measure the allocation of decision-making rights between brand firms and contract manufacturers at the planning level and at the implementation level of product development, we asked the product managers of contract manufacturers to assess the degrees of decision-making rights that they had in their projects compared to brand firms. By referring to Yasumoto and Shiu (2007), we created questionnaire items on contract manufacturers' decision-making rights related to core component knowledge and architectural knowledge acquired from platform providers.

A total of 12 questionnaire items explored the decision-making rights linked to (1) product development schedule (Cronbach $\alpha = 0.861$), (2) quality testing methods (Cronbach $\alpha = 0.95$), (3) approaches to problem-solving (Cronbach $\alpha = 0.95$), (4) modular design (Cronbach $\alpha = 0.773$), (5) components integration (Cronbach

[5]Due to page limitation, all questionnaire items are listed in Shiu (2016).

$\alpha = 0.768$), and (6) reuse of existing design assets (Cronbach $\alpha = 0.856$). By averaging the scores of the items (1)–(6), we calculated the implementation-level decision-making rights (IDR) of contract manufacturers.

Next, we developed 15 questionnaire items and asked the product managers of contract manufacturers to assess the concept-level decision-making rights of brand firms, which are based on core component knowledge from platform providers, and state whether they were stronger than their own. These questionnaire items concerned (1) product release schedule (Cronbach $\alpha = 0.947$), (2) quality standards (Cronbach $\alpha = 0.951$), (3) problem-solving schedule (Cronbach $\alpha = 0.958$), (4) definition of product functionalities (Cronbach $\alpha = 0.940$), and (5) selection of components (Cronbach $\alpha = 0.954$). After that, by averaging the scores of the items (1)–(5), we calculated the concept-level decision-making rights (CDR) of brand firms.

As mentioned in the previous section, due to incomplete contract problems, opportunism and disincentive effects may arise when contract manufacturers and brand firms use their decision-making rights improperly. Moreover, it is believed that opportunism and disincentive effects may become stronger when both brand firms and contract manufacturers possess higher degrees of decision-making rights. In other words, we presume that when one side uses its decision-making rights to pursue self-interest, then the other side will use its decision-making rights to fight back. Therefore, we calculated the interaction between IDR and CDR as a new variable, i.e., IDR $\times$ CDR.

Last, some questionnaire items addressed early problem-solving (EPS) and close interaction and communication (CIC), to see whether these two variables can be appropriate means to reduce opportunism and disincentive effects and contribute to better performance in product development between brand firms and contract manufacturers. First, as for EPS, 15 questionnaire items asked the product managers of contract manufacturers how prepared and able brand firms were in setting out the principles of product development, product specifications, budget, DFM (design for manufacturing), problem-solving approaches, and so on, in collaboration with their contract manufacturers at the beginning of product development projects (Cronbach $\alpha = 0.649$). The mean values of the 15 questionnaire items were used to calculate EPS. Second, with regard to CIC, five questionnaire items dealt with the frequency with which brand firms communicated with their contract manufacturers by means of videoconferences, teleconferences, conference calls, sync-up meetings, bug review meetings, and so on (Cronbach $\alpha = 0.826$). We averaged the values of the five questionnaire items to obtain CIC.

4 Analysis of the Results

This section illustrates our results in relation to Hypotheses 1, 2, 3, and 4. In Fig. 3, the horizontal axis represents the contract manufacturers' relative degrees of implementation-level decision-making rights (IDR), which are based on the architectural knowledge acquired from platform providers, compared to those of brand

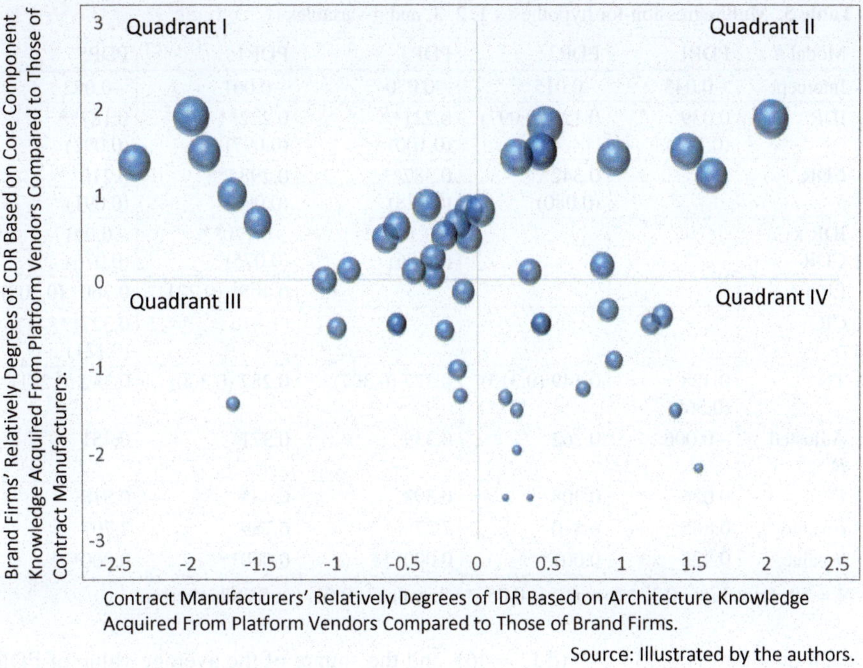

Fig. 3 Relation among IDR, CDR, and PDP

firms. The vertical axis represents the brand firms' relative degrees of concept-level decision-making rights (CDR), which are based on the core component knowledge acquired from platform providers, compared to those of contract manufacturers. We recalculated the IDR and CDR by using mean centering to divide the PDPs of each project into four quadrants. Since IDR, CDR, and PDP are confidential information, the names of the brand firms and contract manufacturers do not appear.

We did not use mean centering for the visual representation of each project's product development performance (PDP). Thus, the larger the size of the blue spheres in Fig. 3, the higher the product development performance (PDP). This approach allowed us to observe that the blue spheres (i.e., PDP) in quadrant I and quadrant II are bigger than those in quadrant III and quadrant IV. Next, we tried to identify any statistical differences among the PDP values for each project in quadrant I, quadrant II, quadrant III, and quadrant IV using ANOVA (analysis of variance). The PDP values were calculated by mean centering for ANOVA, and their averages for quadrant I, quadrant II, quadrant III, and quadrant IV are 0.31, 0.38, −0.62, and −0.26, respectively.[6] Also, the square of the average value of PDP in

[6]Quadrant I has 16 projects, in which maximum PDP is 1.51, minimum PDP is −1.87, standard deviation is 0.82, and averaged value is 0.31. Quadrant II has 11 projects, in which maximum PDP is 1.11, minimum PDP is −0.76, standard deviation is 0.59, and averaged value is 0.38. Quadrant III has 9 projects, in which maximum PDP is 0.28, minimum PDP is −1.41, standard deviation is 0.53,

Table 3 Multiregression for hypotheses 1, 2, 3, and 4 variables

Model #	PDP1	PDP2	PDP3	PDP4	PDP5
Intercept	−0.043	−0.015	−0.070	−0.061	−0.052
IDR	0.039 (0.111)	0.123 (0.097)	0.221** (0.100)	0.222** (0.097)	0.187** (0.092)
CDR		0.342** (0.080)	0.387*** (0.078)	0.298*** (0.091)	0.210** (0.091)
IDR x CDR			−0.191** (0.076)	−0.170** (0.075)	−0.091 (0.076)
EPS				0.405* (0.221)	0.360* (0.208)
CIC					0.327*** (0.121)
TN	0.428 (0.360)	0.149 (0.315)	0.327 (0.307)	0.282 (0.300)	0.342 (0.281)
Adjusted R^2	−0.006	0.262	0.339	0.371	0.451
R^2	0.036	0.308	0.393	0.435	0.518
F value	0.865	6.810	7.27	6.789	7.707
p-value	0.428	0.001***	0.000***	0.000***	0.000***

$N = 50$, **p*-value < 0.1, ***p*-value < 0.05, ****p*-value < 0.01, ( ): Standard error

each quadrant is 0.433 *** (d.f. = 46), and the square of the average value of PDP between each quadrant is 2.549 *** (d.f. = 3). Furthermore, since Levene's test was not significant at 0.416, a difference in the average value of PDP between quadrants was found (F ratio = 5.893). These results indicate that, when brand firms have relatively more concept-level decision-making rights (CDR) and contract manufacturers have relatively more implementation-level decision-making rights (IDR) (i.e., quadrant II), the best product development performance can be achieved.

We used multiple regression analysis to examine the relation among CDR, IDR, and PDP for Hypothesis 1. The results of Hypotheses 2, 3, and 4 are also shown together in Table 3. In model PDP1, it is not statistically significant that the implementation-level decision-making rights (IDR) of contract manufacturers contribute to product development performance (PDP). In model PDP2, although the above is still not statistically significant, the concept-level decision-making rights (CDR) of brand firms do contribute to product development performance (PDP), as confirmed by the statistically significant coefficient (i.e., 0.342 ***). However, when we consider the opportunism and disincentive effect caused by the interaction between IDR and CDR, all the variables become statistically significant (the coefficients of IDR, CDR, and IDR x CDR are 0.221**, 0.387***, and −0.191**) in model PDP3. Moreover, the adjusted R^2 of model PDP3 is 0.339, and P value is 0.000 ***, which is statistically significant over the adjusted R^2 of PDP2 (0.262).

and averaged value is −0.62. Quadrant IV has 14 projects, in which maximum PDP is 0.47, minimum PDP is −1.22, standard deviation is 0.57, and averaged value is −0.26.

We know from model PDP3 that the implementation-level decision-making rights (IDR) of contract manufacturers contribute to product development performance (PDP), as do the concept-level decision-making rights (CDR) of brand firms. That is to say, the greater IDR and CDR are, the higher the PDP between brand firms and contract manufacturers, which means that Hypothesis 1 is supported. In addition, we know that the pros and cons of decision-making rights should be taken into account in the interfirm division of labor. Differently from the positive influence of IDR and CDR on PDP, the interaction of IDR and CDR has a negative impact on PDP linked to opportunism and disincentive effects, which become more significant when both IDR and CDR increase. Therefore, Hypothesis 2 is also supported.

Next, we analyzed the two variables early problem-solving (EPS) and close interaction and communication (CIC) between brand firms and contract manufactures for model PDP4 and model PDP5. In these two models, both IDR and CDR are still statistically significant. Adjusted R^2 of model PDP5 is 0.451, and P value is 0.000***, which means that PDP5 is more statistically significant than models PDP1 to PDP4. Therefore, by interpreting the results of models PDP3 to PDP5, we found that both brand firms' and contract manufacturers' decision-making rights (IDR and CDR) still have a positive and statistically significant influence on PDP, which means again that Hypothesis 1 is supported.

More specifically, the adjusted R^2 of model PDP4 is 0.371 (P value is 0.000***), which is higher than 0.339 (P value is 0.000***) of model PDP3. The coefficients of IDR and CDR are 0.222** and 0.298***, respectively. Also, it emerges that early problem-solving (EPS) has a positive influence on PDP (EPS coefficient = 0.405 *). The coefficient of IDR × CDR is −0.170**, which is lower than the coefficient of IDR x CDR in model PDP3 (−0.191**). Next, when included CIC in model PDP5, we found that the adjusted R^2 of model PDP5 is 0.451 (P value is 0.000***), higher than 0.371 (P value = 0.000***) of model PDP4. IDR, CDR, EPS, and CIC are 0.187**, 0.210***, 0.360*, and 0.327***, respectively, all statistically significant and all having a positive influence on PDP. Lastly, the coefficient of IDR × CDR is −0.091, which is not statistically significant. According to these results, Hypotheses 3 and 4 are well supported.

5 Discussion

The traditional model known as dyadic interfirm division of labor to examine new product development performance between brand firms and contract manufacturers is extended here to develop a triadic approach. In fact, in our setting, both brand firms and contract manufacturers rely on platform providers' knowledge to develop new products. First, differently from products like automobiles, characterized by integral architecture, the role of platform providers may make ICT product architecture become more modular, which decreases communication efforts in the interfirm division of labor. However, as our results reveal, although product architecture is changing toward modular, firms and their partners still need to enhance their knowledge, so as to retain their comparative advantages and properly allocate decision-making rights for successful new product development projects.

As confirmed by the ANOVA results in Fig. 3, instead of surrendering all decision-making rights to their contract manufacturers (i.e., quadrant IV), when brand firms retain concept-level decision-making rights and delegate implementation-level decision-making rights to their contract manufacturers (i.e., quadrant II), the highest performance in product development is achieved. In addition, in such a decentralized interfirm division of labor, by establishing early problem-solving and close interaction and communication, brand firms can reduce opportunism and disincentive effects and successfully develop new products. To do this, it is important for brand firms to acquire knowledge from platform providers, which, within the framework of this book, can be regarded as *capability building*.

It is assumed by many that firms should possess internally any knowledge related to outsourcing work (Brusoni and Prencipe 2001; Brusoni et al. 2001; Lincoln et al. 1998; Takeishi 2001, 2002, 2003). However, we believe that knowledge may be possessed by different firms according to their business specialization, which turns out to be the driving force behind the proper shaping of interfirm division of labor. In other words, firms should consider what sort of knowledge is more important and suitable for their core business activities, compared to their partners. Indeed, by adopting an approach based on comparative advantages for capability building, firms can successfully collaborate with their partners and achieve higher performance in the interfirm division of labor.

The comparative advantage perspective is useful to understand how interfirm division of labor may work when knowledge becomes an exogenous variable. The principal-agent theory assumes that knowledge is allocated between the principal and the agent and this influences their decision-making rights. However, we assert that, when the knowledge in question is possessed neither by the principal nor by the agent but by a third party, then the allocation of decision-making rights may be reconsidered from the perspective of comparative advantages, in pursuit of higher performance in the interfirm division of labor. Furthermore, when firms attempt to build their capabilities for the interfirm division of labor from the perspective of comparative advantages, they may also reduce opportunism and disincentive effects by establishing closer and more interactive communication with their partners.

In other words, brand firms can enjoy the efficiency of the interfirm division of labor by delegating their decision-making rights to their contract manufacturers, but not without some risks. This is because it is difficult for brand firms in the ICT industry to predict the number of products that will be sold in advance and, when the environment and competition conditions change, their relationship with contract manufacturers becomes unstable. This relationship is also influenced by whether the decision-making rights possessed by brand firms and contract manufacturers are suitably balanced.

For example, there are different reasons behind the opportunism and disincentive effects in quadrant I and quadrant IV of Fig. 3. In quadrant I, contract manufacturers have more limited decision-making rights than brand firms at the concept and implementation levels of product development. Ideally, contract manufacturers should be able to concentrate their efforts on finding out the root causes of problems in product development. However, when they have limited decision-making rights compared to brand firms, they may be less motivated to carry out additional experiments. They may use the excuse that they do not have a good relationship with platform provider and adopt a *not my responsibility* attitude in problem-solving activities. On the other hand, in quadrant IV, contract manufacturers have stronger decision-making rights at the concept and implementation levels of product development, compared to brand firms. In this case, contract manufacturers may use their decision-making rights for their self-interest (e.g., using existing design assets instead of doing customized design for brand firms).

6 Conclusions

As shown in past research (e.g., Alcacer and Oxley 2014[7]), contract manufacturers need to choose good brand firms to learn from in order to secure further growth. However, when contract manufacturers can also strengthen their capabilities by learning from nonbrand firms, i.e., platform providers, the interplay of elements like knowledge, decision-making rights, and so on must be reconsidered to understand performance dynamics in the interfirm division of labor. This chapter examined 50 smartphone product development projects carried out by brand firms and Taiwanese contract manufacturers from 2012 to 2014. We found that the performance of these modular product development projects was more significantly enhanced when brand firms and contract manufacturers aimed to build their capabilities, which are different from each other. Ideally and practically, opportunism and disincentive effects should be kept under control by means of close and interactive communication.

However, some challenges still remain. First, it is not clear how brand firms can build their capabilities by using the knowledge that platform providers have, and this should be clarified in future studies. Second, Samsung and Huawei started to develop Android smartphones using their own core components (i.e., CPUs) from November 2009 and October 2012, respectively. Such in-house core components should be included in the discussion of interfirm division of labor and also be the subject of further research. Third, software core components should also be taken into account. As Boudreau (2010) and Shiu and Yasumoto (2016) explain, brand firms can gain knowledge about software core components by becoming involved in the Android open-source community. A detailed analysis of these cases may shed light on how brand firms can acquire the necessary core component knowledge.

[7] Alcacer and Oxley (2014) investigate 114 contract manufacturers that provided product development services to brand firms in the mobile phone industry from 1994 to 2010. It comes to the conclusion that the choice of which brand firm to learn from does indeed affect the growth of contract manufacturers.

References

Abernathy WJ (1978) The productivity dilemma: roadblock to innovation in the automobile industry. Johns Hopkins University Press, Baltimore

Afuah A (2001) Dynamic boundaries of the firm: are firms better off being vertically integrated in the face of a technological change? Acad Manag J 44:1211–1228

Alcacer J, Oxley J (2014) Learning by supplying. Strateg Manag J 35:204–223

Alexander C (1964) Notes on the synthesis of form. Harvard University Press, Cambridge, MA

Ancona DG, Caldwell DF (1992) Bridging the boundary: external process and performance in organizational teams. Adm Sci Q 37:634–665

Anderson P, Tushman M (1990) Technological discontinuities and dominant designs: a cyclical model of technological change. Adm Sci Q 35:604–633

Aoshima Y (2002) Transfer of system knowledge across generation development: empirical observations from Japanese automobile development. Ind Relat 41(4):605–628

Asanuma B (1988) Japanese manufacturer-supplier relationships in international perspective: the automobile case. Working Paper 8, Faculty of Economics, Kyoto University

Baldwin CY (2010) When open architecture beats closed: the entrepreneurial use of architectural knowledge. Harvard Business School Finance Working Paper 10-063

Baldwin CY (2015) Bottlenecks, modules and dynamic architectural capabilities. Harvard Business School Finance Working Paper 15-028

Baldwin CY, Clark KB (1997) Managing in the ages of modularity. Harv Bus Rev 75(5):84–93

Baldwin CY, Clark KB (2000) Design rules: the power of modularity. MIT Press, Cambridge, MA

Barzel Y (1997) Economic analysis of property rights. Cambridge University Press, New York

Boudreau K (2010) Open platform strategies and innovation: granting access vs. devolving control. Manag Sci 56(10):1849–1872

Boudreau K, Hagiu A (2009) Platform rules: multi-sided platforms as regulators. In: Gawer A (ed) Platforms, markets and innovation. Edward Elgar, London, pp 163–191

Brown SL, Eisenhardt KM (1998) Competing on the edge: strategy as structured chaos. Harvard Business School Press, Boston

Brusoni S, Prencipe A (2001) Managing knowledge in loosely coupled networks: exploring the links between product and knowledge dynamics. J Manag Stud 38(7):1019–1035

Brusoni S et al (2001) Knowledge specialization, organizational coupling, and the boundaries of the firm: why do firms know more than they make? Adm Sci Q 46(4):597–621

Clark KB (1985) The interactions of design hierarchies and market concepts in technology evolution. Res Policy 14:235–251

Clark KB, Fujimoto T (1991) Product development performance: strategy, organization and management in the world auto industry. Harvard Business School Press, Boston

Conner KR, Prahalad CK (1996) A resource-based theory of the firm: knowledge versus opportunism. Organ Sci 7(5):477–501

Constant EW (1980) The origins of the turbojet revolution. John Hopkins University Press, Baltimore

Cusumano M, Nobeoka K (1998) Thinking beyond lean. Free Press, New York

Cusumano M et al (1992) Strategic maneuvering and mass-market dynamics: the triumph of VHS over Beta. Bus Hist Rev 66:51–94

D'Aveni RA (1994) Hypercompetition. Free Press, New York

Dedrick J, Kraemer KL (1998) Asia's computer challenge: threat or opportunity for the United States and the world? Oxford University Press, New York

Doig S et al (2001) Has outsourcing gone too far? McKinsey Q 4:25–37

Dosi G (1982) Technological paradigms and technological trajectories. Res Policy 11(3):147–162

Dosi G et al (2003) The economics of systems integration: toward an evolutionary interpretation. In: Prencipe A et al (eds) The business of systems integration. Oxford University Press, New York

Dyer JH (1996) Does governance matter? Keiretsu alliances and asset specificity as sources of Japanese competitive advantage. Organ Sci 7(6):649–666

Eisenhardt KM (1989) Making fast strategic decisions in high velocity environments. Acad Manag J 32:543–576

Eisenhardt KM, Tabrizi BN (1995) Accelerating adaptive processes – product innovation in the global computer industry. Adm Sci Q 40(1):84–110

Foss K, Foss NJ (1998) The market process and the firm toward a dynamic property rights perspective. DRUID Working Paper 98-14, DRUID, Copenhagen Business School, Department of Industrial Economics and Strategy/Aalborg University, Department of Business Studies

Galunic DC, Eisenhardt KM (2001) Architectural innovation and modular corporate forms. Acad Manag J 44:1229–1249

Garud R, Kumaraswamy A (1993) Changing competitive dynamics in network industries: an exploration of Sun Microsystems' open systems strategy. Strateg Manag J 14:351–369

Garud R, Kumaraswamy A (1995) Technological and organizational designs for realizing economies of substitution. Strateg Manag J 16(1):93–109

Gawer A, Cusumano MA (2002) Platform leadership: how Intel, Microsoft, and Cisco drive industry innovation. Harvard Business School Press, Boston

Gereffi G et al (2005) The governance of global value chains. Rev Int Polit Econ 12(1):78–104

Gibbons R (2005) Incentives between firms (and within). Manag Sci 51(1):2–17

Gibbons R, Roberts J (2013) Economic theories of incentives in organizations. In: Gibbons R, Roberts J (eds) The handbook of organizational economics. Princeton University Press, Princeton, pp 56–99

Grant RM (1996) Toward a knowledge based theory of the firm. Strateg Manag J 17(2):109–122

Hart O, Moore J (1990) Property rights and the nature of the firm. J Polit Econ 98(6):1119–1158

Henderson RM, Clark KB (1990) Architectural innovation: the reconfiguration of existing product technologies and the failure of established firms. Adm Sci Q 35:9–30

Henderson R, Cockburn I (1994) Measuring competence? exploring firm effects in pharmaceutical research. Strateg Manag J 15(1):63–84

Holmstrom B, Milgrom P (1991) Multitask principal-agent analyses: incentive contracts, asset ownership, and job design. J Law Econ Org 7:24–52

Hughes TP (1983) Networks of power: electrification in western society, 1880–1930. John Hopkins University Press, Baltimore

Iansiti M (1997) Technology integration: making critical choices in a western electric show and convention. Harvard Business Review Press, Boston

Iansiti M (1998) Technology integration: making critical choice in a dynamic world. Harvard Business School Press, Boston

Iansiti M (1999) How the incumbent can win: managing technological transitions in the semiconductor industry. Manag Sci 46(2):169–185

Iansiti M, Clark KB (1994) Integration and dynamic capability: evidence from product development in automobiles and mainframe computers. Ind Corp Chang 3(3):557–605

Iansiti M, Khanna T (1995) Technological evolution, system architecture and the obsolescence of firm capabilities. Ind Corp Chang 4:331–336

Imai K et al (1985) Managing the new product development process: how Japanese firms learn and unlearn. In: Clark KB et al (eds) The uneasy alliance: managing the productivity-technology dilemma. Harvard Business School Press, Boston, pp 337–375

Jacobides MG (2005) Industry change through vertical disintegration: how and why markets emerged in mortgage banking. Acad Manag J 48(3):465–498

Jacobides MG (2008) How capability differences, transaction costs, and learning curves interact to shape vertical scope. Organ Sci 19(2):306–326

Jacobides MG, Billinger S (2006) Designing the boundaries of the firm: from "make, buy, or ally" to the dynamic benefits of vertical architecture. Organ Sci 17(2):249–261

Jacobides MG, Hitt LM (2005) Losing sight of the forest for the trees? productive capability and gains from trade as drivers of vertical scope. Strateg Manag J 27(1):1209–1227

Jacobides MG, Winter SG (2005) The co-evolution of capabilities and transaction costs: explaining the institutional structure of production. Strateg Manag J 26(5):395–413

Jacobides MG et al (2006) Benefiting from innovation: value creation, value appropriation and the role of industry architecture. Res Policy 35(8):1200–1221

Kawakami M (2011) Interfirm dynamics in notebook PC value chains and the rise of Taiwanese original design manufacturing firms. In: Kawakami M, Sturgeon TJ (eds) The dynamics of local learning in global value chains: experiences from East Asia. Palgrave Macmillan, New York

Kogut B (2000) The network as knowledge: generative rules and the emergence of structure. Strateg Manag J 21:405–425

Kogut B, Zander U (1996) What firms do? coordination, identity, and learning. Organ Sci 7 (5):502–518

Langlois RN, Robertson PL (2002) Firms, markets and economic change. Routledge, London

Leonard-Barton D (1992) Core capabilities and core rigidities: a paradox in managing new product development. Strateg Manag J 13:111–125

Lincoln JR et al (1998) Organizational learning and purchase-supply relations in Japan: Hitachi, Matsushita, and Toyota compared. Calif Manag Rev 40:241–264

Marple D (1961) The decisions of engineering design. IEEE Trans Eng Manag 2:55–71

Nobeoka K (1996) Multi project senryaku-post lean no seihinkaihatsu management. Yuhikaku, Tokyo

Nonaka I, Takeuchi H (1995) The knowledge-creating firm. Oxford University Press, New York

Prendergast C (1999) The provision of incentives in firms. J Econ Lit 37(1):7–63

Seike A (1995) Nihongata soshikikan kankei no management. Hakuto Shobo, Tokyo

Shioji H (1986) Toyota jiko niokeru itakuseisan no tenkai: 1960nendai Toyota no tameigara shiyou ryousan kikou (2). Keizai Ronsou 138(5,6):52–77

Shioji H (1996) "Itaku" automotive production: an aspect of the development of full-line and wide-selection production by Toyota in the 1960s. Kyoto Univ Econ Rev 65(1):19–42

Shiu JM (2010) Chugoku keitaidenwa sangyo niokeru suichoku hitougou no keisei hattenyouin no kousatsu. Innov Manag 7:51–71

Shiu JM (2016) Keitaidenwa sangyou niokeru seihinkaihatsu nitsuite genjyouhoukoku. Akamon Bus Rev 16(3):179–208

Shiu JM, Yasumoto M (2016) Benefitting from contributions to the Android open source community. Ann Bus Admin Sci 15:239–250

Sturgeon TJ, Lee JR (2001) Industry co-evolution and the rise of a shared supply-base for electronics manufacturing. Paper presented at the Nelson and Winter Conference, Aalborg Denmark, June 2001

Sturgeon TJ, Lester RK (2004) The new global supply-base: new challenges for local suppliers in East Asia. In: Shahid Y et al (eds) Global production networking and technological change in East Asia. The World Bank and Oxford University Press, Washington DC

Sturgeon TJ, Memedovic O (2010) Looking for global value chains in trade statistics: mapping structural change and compressed development in the world economy. UNIDO Working Paper, Vienna

Strahilevitz LJ (2006) Information asymmetries and the rights to exclude. Mich Law Rev 104:1835–1898

Tatsumoto H et al (2008) Chishiki to kigyo no kyoukai no chousei to modularity no kouchiku: pasokon sangyou niokeru gijyutsu platform kaihatsu no jirei. Soshikikagaku 42(2):19–32

Takeishi A (2001) Bridging inter- and intra-firm boundaries: management of supplier involvement in automobile product development. Strateg Manag J 22(5):403–433

Takeishi A (2002) Knowledge partitioning in the interfirm division of labor: the case of automotive product development. Organ Sci 13(3):321–338

Takeishi A (2003) Bungyou to kyousou:kyousouyuui no outsourcing management. Yuhikaku, Tokyo

Tushman M, Anderson P (1986) Technological discontinuities and organizational environments. Adm Sci Q 31(3):439–465

Tushman M, Murmann JP (1998) Dominant designs, technology cycles, and organizational outcomes. Res Organ Behav 20:231–266

Tushman M, Rosenkopf L (1992) Organizational determinants of technological change: toward a sociology of technological evolution. In: Cummings L, Staw B (eds) Research in organizational behavior 14:311–347. JAI Press, Greenwich

Ulrich K, Eppinger S (1995) Product design and development. McGraw-Hill, New York

Vincenti W (1990) What engineers know and how they know it. Analytical studies from aeronautical history. Johns Hopkins University Press, Baltimore

Windsperger J (1996) The nature of franchising: a property rights approach. Rev Econ 2:130–143

Windsperger J (2001) Allocation of decision and ownership rights in franchising. Proceedings of the International Society of Franchising Conference, Las Vegas

Windsperger J (2002a) The structure of ownership rights in franchising networks: an incomplete contracting view. Eur J Law Econ 13:129–142

Windsperger J (2002b) Allocation of decision and ownership rights in franchise relationships: empirical results from the Austrian franchise sector. J Mark Channels 10:19–38

Windsperger J (2003) Complementarities and substitutabilities in franchise contracting: some results from the German franchise sector. J Manag Gov 7:291–313

Windsperger J (2004) Centralization of franchising networks: evidence from the Austrian franchise sector. J Bus Res 57:1361–1369

Windsperger J (2009) Allocation of decision rights in joint ventures. Manag Decis Econ 30 (8):491–501

Windsperger J, Dant R (2006) Contractibility and ownership redirection in franchising: a property rights view. J Retail 82:259–272

Yasumoto M, Shiu JM (2007) An investigation into collaborative novel technology adoption in vertical disintegration: interfirm development processes for system integration in the Japanese, Taiwanese, and Chinese mobile phone handset industries. Ann Bus Admin Sci 6:35–69

Conclusion

Takahiro Fujimoto and Fumihiko Ikuine

Abstract In this concluding chapter of the present book, we first summarize the research results of the previous 14 chapters, including theoretical frameworks (Chapters "A Design-Information-Flow View of Industries, Firms, and Sites," "The Nature of International Competition Among Firms," "Product Variety for Effective Demand Creation," "Capability Building and Demand Creation in 'Genba-Oriented Firms'," and "Evolution of Business Ecosystems"), cases of capability building of firms and sites in global competition (Chapters "Evolution of Organizational Capabilities in Manufacturing: The Case of the Toyota Motor Corporation," "The Nature of Surviving Japanese Factories in the Global Competition: An Empirical Analysis of Electrical and Electronics Factories," "The Effectiveness of Group Leaders in the Lean Production System: Time Study and Agent-Based Model of Leaders' Behavior," "The Diversity and Reality of Kaizen in Toyota," "Balancing Standardization and Integration in ICT Systems: An Architectural Perspective"), and cases of architectural strategies and demand creation by firms in digital industries (Chapters "Creating New Demand: The Impact of Competition, the Formation of Submarkets," "Decline in Demand Creation: The Development Productivity Dilemma and Its Consequences," "Investigating the Creation and Diffusion of Knowledge for Demand Creation: The Case of the Telecommunications Industry," and "The Impact of Platform Providers' Knowledge on Interfirm Division of Labor: The Case of the Mobile Phone Industry").

The original version of this chapter was revised: This chapter was previously published non-open access and has now been changed to open access. The correction to this chapter is available at https://doi.org/10.1007/978-4-431-55145-4_16.

T. Fujimoto (✉)
Graduate School of Economics, The University of Tokyo, Bunkyo-Ku, Tokyo, Japan
e-mail: fujimoto@e.u-tokyo.ac.jp

F. Ikuine
Faculty of Engineering, Information and Systems, University of Tsukuba, Tsukuba, Ibaraki, Japan
e-mail: fikuine@sk.tsukuba.ac.jp

T. Fujimoto, F. Ikuine (eds.), *Industrial Competitiveness and Design Evolution*,
Evolutionary Economics and Social Complexity Science 12,
https://doi.org/10.1007/978-4-431-55145-4_15

After that, we present an augmented version of the capability-architecture-performance framework, which was initially proposed in chapter "A Design-Information-Flow View of Industries, Firms, and Sites." That is, based on our research results, which demonstrate the multilayered nature of capability building and demand creation, we reconstruct our capability-architecture framework as follows. On the supply side, the evolution of manufacturing capabilities and production technologies leads to capability building and productivity improvements at the level of sites, firms, and industries, which makes it possible for production and development sites in high-wage nations to survive intense global cost competition. On the demand side, the evolution of technologies and architectures at the level of products, components, and platforms results in great product variety and quality, which drives up the products' demand curves and expands their effective demand quantities. Capability-building capability for the former and architecture-building capability for the latter are the key dynamic capabilities allowing both firms and sites to survive and grow in the long run.

Thus, our conclusion is that the design-flow view of manufacturing and the capability-architecture-performance framework for analyzing the evolution of industries and firms can explain rather consistently the main industrial phenomena of the late twentieth century and early twenty-first century, i.e., intense global competition and rapid digitalization.

1 Summary of the Book

1.1 Explaining Evolution in the Age of Globalization and Digitalization

The present book illustrates an evolutionary framework based on the social sciences for analyzing the dynamics of firms, industries, and industrial sites. In its empirical and historical application, we focus on globalization and digitalization of firms and industries between the 1990s and 2010s, since we believe these to be the two trends that most significantly changed our society during that period. In other words, this book tries to identify the basic logic that can consistently explain what happened to national economies, industries, firms, and their sites (*genba*) looking mostly at Japan, one of the countries most heavily affected by this coincidence of globalization and digitalization.

More specifically, we explore three main trends of this period: (i) *global cost competition between advanced nations and emerging nations*, whose average wage rates are extremely low (e.g., roughly speaking $100 per month in China versus $2000 per month in Japan); (ii) *minute interindustrial trade*, which is significantly affected by the firms' choice of locations for both design sites and production sites; and (iii) *new forms of competition and complementation in digital (ICT) industries*, in which rapid demand creation is achieved by effective platform formation, in addition to effective product development. This book aims to identify a theoretical framework that may thoroughly interpret the industrial phenomena mentioned above.

For the purpose of explaining the phenomena occurring in high-wage industrialized national economies during the era of global competition and digitalization, the authors present empirical, historical, and theoretical evidence, including field studies and statistical analyses. We investigate both conventional physical goods sectors and information/computer/software sectors, capability building and demand creation, product competition and platform competition, as well as advanced nations and emerging economies.

1.2 *Summary of the Chapters*

Before proposing an overall framework for analyzing the aforementioned industrial phenomena of the late twentieth and early twenty-first century, let us summarize the previous 14 chapters of the present book.

1.2.1 Theoretical Foundations

The first part of this book, chapters "A Design-Information-Flow View of Industries, Firms, and Sites," "The Nature of International Competition Among Firms," "Prod uct Variety for Effective Demand Creation," "Capability Building and Demand Creation in 'Genba-Oriented Firms'," and "Evolution of Business Ecosystems," explored theoretical concepts and frameworks that may explain the evolution of economic entities, such as industries, firms, and manufacturing sites (genba), as well as that of economic artifacts, such as products, components, and platforms.

Although different chapters focused on subtly different topics, they shared certain common concepts, among which industries as flows of economic value, organizational routines and capabilities that govern such flows, productivity and overall productive performance as effectiveness of such flows. cost competition as a handicapped productivity race with international wage gaps, the manufacturing site as the foundation of a firm/industry/region, demand creation for stable employment, design as a source of value added, technology and architecture as the two sides of design, industries as networks of competing/complementing/transacting products, and firms' strategies that take into account the capabilities of manufacturing sites and/or architectures of platforms/products/components.

Chapter "A Design-Information-Flow View of Industries, Firms, and Sites" proposed an overall evolutionary framework for analyzing firms, industries, and manufacturing sites (genba), as well as products and platforms. Manufacturing sites were broadly defined as the places where value-carrying design information flows to the customers. Within this manufacturing site-based view, a firm was regarded as a collection of various sites under the control of single capital, whereas an industry was defined as a collection of similar sites in terms of their product design information.

By dynamically reinterpreting the classical concept of comparative advantage and applying it not only to production costs but also to design costs, this introductory chapter tried to explain what happened to the trade goods industrial sectors of postwar Japan in terms of economic growth, labor shortage, yen appreciation, capability building, international competition during the Cold War, global competition after the Cold War, and relative wage/productivity divergence and convergence vis-à-vis other advanced/emerging nations.

Chapter "The Nature of International Competition Among Firms" presented a modern and dynamic reinterpretation of the Ricardian-Sraffian theories of international values and trade (Ricardo 1951, Fujimoto and Shiozawa 2012) and applied it to the cases of *handicapped productivity competition* with large wage gaps between advanced and emerging nations. We started from Ricardo's original two-country, two-commodity cases and moved on to the determination of international values in M-country, N-commodity cases (mainly $M < N$), defined by a spanning tree (a term from graph theory) of competitive production techniques, which we may call a "new theory of international values." This theory of international values was also applied to some actual industrial and trade phenomena during the period of global competition and digitalization, mainly between the 1990s and 2010s.

Chapter "Product Variety for Effective Demand Creation" introduced a conceptual framework explaining *effective demand creation by product variety*. Using a coverage function linking product variety with the coverage ratios of effective demand achieved as a percentage of potential demand, we identified the optimal product variety that can maximize cumulative profit, as a firm increases the number of products introduced to the market. This tool makes it possible to predict how optimal product variety is affected by such strategic factors as fixed costs, variable costs, mark-up ratios, product cycle positions, and potential sales volumes.

Chapter "Capability Building and Demand Creation in 'Genba-Oriented Firms'" argued that building capabilities for productivity improvements (mostly chapters "A Design-Information-Flow View of Industries, Firms, and Sites" and "The Nature of International Competition Among Firms") and enhancing product design variety and quality for effective demand creation (Chapter "Product Variety for Effective Demand Creation") are not separate issues but interconnected activities for certain types of firms, i.e., *genba-oriented firms*. The analysis relied on a simple approach with Ricardian linear cost functions (flat supply curves) and product differentiation by design (downward-sloping demand curves). We illustrated a four-quadrant model with four main variables, supply price (P), demand quantity (X), number of employees (N), and wage rates (W), which we called *PXNW model*. By analyzing it, we argued that genba-oriented firms may reach a steady state that ensures the achievement of target profit (mark-up) rate and target number of employees at the same time. Moreover, they can move from one steady state to another, improved steady state with higher wage-to-price ratio by combining (i) capability building for enhancing physical productivities and (ii) product design improvements for generating additional demand at the firm level.

Chapter "Evolution of Business Ecosystems" turned to the evolution of *business ecosystems* under global digitalization, where the platforms and/or products in question are characterized by open-modular (open) architectures with many complementary goods. The key to this industrial transformation is modularization, or the decomposition of a complex artifact into functionally nearly complete (i.e., nearly decomposable) modules by creating open industry-standard interfaces among them. In such an open-modular system of products and components, or *platform*, the interrelations among products and/or components are not only competitive or transactive but also complementary. In this case, cumulative effects among mutually complementary goods (i.e., network externalities) make for extremely rapid demand creation through the cumulative expansion of successful platforms.

In business ecosystems, proactive strategic changes to product architectures made by platform-leading firms (e.g., establishing global standard interfaces among complementary products) can cause drastic and rather swift changes in the industry's structures. From this point of view, clear differences are detected between the patterns of industrial competition in traditional product-to-product sectors (discussed in chapters "The Nature of International Competition Among Firms," "Product Variety for Effective Demand Creation," and "Capability Building and Demand Creation in 'Genba-Oriented Firms'"), in which the relative quality and variety of a firm's products vis-à-vis those of the competitors are key to success, and in the emerging platform-based environments illustrated in chapter "Evolution of Business Ecosystems," in which creating large clusters of complementary products and firms against rival platforms is the crucial factor.

Our capability-architecture framework can be effectively applied to both product competition and platform competition. On the side of architecture, we think that the difference between the two can be attributed to the architectural transformations of the products and platforms in question; on the side of capability, we argue that capability-building capabilities are important in product competition, whereas architecture-building capabilities are critical in platform competition. In any case, it is our opinion that these two industrial phenomena must be analyzed together, partly because product competition continues inside the ecosystem and partly because platform competition sometimes emerges in the middle of intense product competition in today's world of globalization and digitalization.

1.2.2 Capability Building of Firms and Sites Facing Global Competition

Chapters "Evolution of Organizational Capabilities in Manufacturing: The Case of the Toyota Motor Corporation," "The Nature of Surviving Japanese Factories in the Global Competition: An Empirical Analysis of Electrical and Electronics Factories," "The Effectiveness of Group Leaders in the Lean Production System: Time Study and Agent-Based Model of Leaders' Behavior," and "The Diversity and Reality of Kaizen in Toyota" discussed manufacturing capability building to improve physical productivity at multiple levels, including firms, factories, and organizational units within them.

Chapter "Evolution of Organizational Capabilities in Manufacturing: The Case of the Toyota Motor Corporation" investigated the formation of a firm's complex operation-based competency, such as that of the Toyota Motor Corporation in manufacturing. We suggested combining the resource-capability view of the firm and the evolutionary framework, i.e., a dynamic perspective that separately explains an observed system's survival (functional logic) and its formation (genetic logic). To shed light on the evolution of complex manufacturing systems, two main concepts were proposed: *multipath system emergence*, for analyzing unpredictable variations in such manufacturing systems, and *evolutionary learning capability*, or simply evolutionary capability, for explaining why certain firms can develop complex manufacturing capabilities faster and enjoy sustainable competitive advantages longer than their rivals. The chapter tried to apply these concepts and frameworks to a historical analysis of organizational routines in the manufacturing system at Toyota.

Chapter "The Nature of Surviving Japanese Factories in the Global Competition: An Empirical Analysis of Electrical and Electronics Factories" discussed manufacturing capabilities and competitiveness mostly at the level of factories (manufacturing sites). To do so, the authors used questionnaire-based surveys of about 100 Japanese factories in the electrical and electronics industries, the Japanese sectors most affected by the combination of globalization and digitalization in the period between the 1990s and 2010s.

As for competitiveness, the surveys measured both productive performance (deep-level competitiveness) and market performance (surface-level competitiveness), introduced in chapter "A Design-Information-Flow View of Industries, Firms, and Sites." In relation to the former, the questionnaires focused on the perceived relative performance levels of industrial sites regarding quality, productivity, cost, delivery, flexibility, and development vis-à-vis their toughest competitors, including foreign factories within the same company. In terms of the latter indicator, price and service competitiveness was evaluated against that of main rivals.

The results of the surveys indicated that, even in the case of apparently declining industries in Japan, manufacturing capabilities and productive performance were still at world-class level, but this was not true for unit production costs. This evidence confirms the basic nature of global cost competition as a productivity race with international wage gaps as handicaps (Chapter "The Nature of Interna tional Competition Among Firms"). On the other hand, high responsiveness to customer demands was these factories' most effective capability to overcome unit cost disadvantages. Moreover, multifunctionality of production sites was a supporting factor to boost the performance of these struggling factories, and new product proposals/development improved shop-floor organizational openness and effectively activated shop-floor communication.

Chapter "The Effectiveness of Group Leaders in the Lean Production System: Time Study and Agent-Based Model of Leaders' Behavior" explored the level of work groups and group leaders inside individual factories. Although the literature on Japanese manufacturing, the Toyota System, and Lean Production has often emphasized the importance of group leaders in manufacturing capability building (e.g., *kaizen*), there has been little research on the role and behavior of real group leaders based on in-depth field studies.

Against this background, the chapter combined detailed time studies of actual group leaders inside a Toyota-style factory and simulation analyses of the cellular automation model that may reproduce the behavioral patterns found in the empirical results. The model focused on four main variables: task flow continuity in the production line, task load, problem occurrence, and help actions by group leaders. The results of the simulation suggested that help from a leader significantly improves task flow continuity when task workload is high and problem frequency is low, but a slight increase in problem occurrence disrupts such continuity. The findings of the field studies confirmed that high-productivity lines, supported by help actions from group leaders, tended to suffer from sudden drops in performance when additional inexperienced workers increased problem frequency and the leaders were forced to do heavy work.

Chapter "The Diversity and Reality of Kaizen in Toyota" focused on continuous capability building, or *kaizen*, by work organizations and their leaders, as well as production engineers at the factory or corporate level. It pointed out that the ranges or spans of coordination for a particular kaizen activity may differ depending upon its nature, despite the stereotyped image of kaizen activities as a collection of mutually independent incremental process innovations solely at the level of work groups on the shop floor. A long-term observation of kaizen activities indicated that they may also include product innovations and rather larger-scale innovations in budget, human inputs, and outcomes not only at the level of work groups but also in the higher layers of production organizations. Furthermore, individual kaizen activities may be interconnected, since solutions from a certain kaizen activity may trigger another problem-solving cycle.

The results of the field studies in this chapter also underlined the important role of shop-floor production engineers as *staff-in-line organization*, i.e., the interface for vertical and horizontal coordination between shop-floor work organizations and corporate production engineers, as well as between smaller and larger kaizen activities. Thus, kaizen activities in real factory contexts are more diversified, multifaceted, and multilayered than their stereotyped notion suggests.

Chapter "Balancing Standardization and Integration in ICT Systems: An Archi tectural Perspective" discussed the role of IT (information technology) in supporting both manufacturing and capability building of firms and sites. Based on field observations of better-performing cases in the manufacturing industries, this chapter concentrated on the challenges of globalization and digitalization. It argued that firms should aim to achieve a good balance between a Global Standard IT System (GSIS), provided as a standard package by global IT firms, and an Integrated Manufacturing IT System (IMIS), which is more site-specific or firm-specific, and merge them into a Global Integrated Manufacturing IT System (GIMIS) as an effective hybrid of both.

The overall IT system supporting a firm's manufacturing, or flows of design information to customers, is likely to adopt GSIS or IMIS or both as its components. Thus, we predict that an IT system adopting GSIS as its standard component will tend toward an open-modular architecture, while an IT system adopting IMIS will be architecturally more closed-integral. It follows from this that firms adopting GIMIS as a balanced solution are likely to go for hybrid architectures with open and integral subsystems. For instance, the Japanese firm Komatsu, known as a leader in competitive manufacturing IT, actually uses GSIS (e.g., SAP) for its global accounting system but has chosen an internally developed (i.e., firm-specific or site-specific) IT system, or IMIS, for its engineering and manufacturing bills of materials (E-BOM and M-BOM). We can regard this combination as a typical case of GIMIS with both open and integral components.

1.2.3 Architectural Evolution and Platform Formation in Digitalized Industries

Whereas Part II of this book dealt mostly with capability building for higher productive performance (i.e., better flows of value-carrying design information) by firms and sites manufacturing physical goods and facing intense global competition, Part III (Chapters "Creating New Demand: The Impact of Competition, the Forma tion of Submarkets," "Decline in Demand Creation: The Development Productivity Dilemma and Its Consequences," "Investigating the Creation and Diffusion of Knowledge for Demand Creation: The Case of the Telecommunications Industry," and "The Impact of Platform Providers' Knowledge on Interfirm Division of Labor: The Case of the Mobile Phone Industry") turned to another major trend of this period, i.e., the global emergence of the digital economy. We discussed design and architectural strategies for cumulative demand creation at the level of economic artifacts, including products and their key components, as well as platforms and their ecosystems, which involve many complementary products (Chapter "Evolution of Business Ecosystems").

We first looked at the prototypical case of platforms and ecosystems in the Japanese home video game industry in chapters "Creating New Demand: The Impact of Competition, the Formation of Submarkets" and "Decline in Demand Creation: The Development Productivity Dilemma and Its Consequences" and then analyzed more advanced and global cases, i.e., cellular phones and smartphones, in chapters "Investigating the Creation and Diffusion of Knowledge for Demand Creation: The Case of the Telecommunications Industry" and "The Impact of Platform Providers' Knowledge on Interfirm Division of Labor: The Case of the Mobile Phone Industry." Today's platform leaders are mostly US-based companies, such as Intel, Microsoft, Google, Apple, and Facebook. Yet, some Japanese companies may be seen as precursor platform leaders, e.g., Nintendo, Sega, and Sony Computer Entertainment. They set up a basic business model relying on the simplest form of platform and created huge demand worldwide. To start our analysis of the digital economy, we examined this old but simple type of platform business and revealed the mechanisms behind it.

Chapter "Creating New Demand: The Impact of Competition, the Formation of Submarkets" focused on the early 1980s, when an entirely new market, that of home video game software (console game software), was established in Japan. The so-called Family Computer (or NES, Nintendo Entertainment System) was a platform with (quasi) open-modular architecture that induced more than 100 firms to enter this market. Such firms released home video game software titles in rapid succession, and some managed to develop new submarkets within the home video game industry. They competed with one another by differentiating their products, and product diversity grew considerably thanks to this competition, attracting many consumers and generating huge demand.

Chapter "Decline in Demand Creation: The Development Productivity Dilemma and Its Consequences" also examined the case of home video game software to investigate the consequences of demand creation by firms. Firms that succeeded in creating demand invested in development activities for further growth. To meet the strong demand of the new market, they accumulated knowledge on new product development (development know-how). As time passed, they tended to create ever more similar products, relying on their accumulated development know-how to achieve higher development productivity. At the same time, firms with rich development know-how started to have an implicit bias against working on completely new products, because it was difficult for them to move away from the old ideas inherent in their development know-how.

The above resulted in a stalemate in product diversification, and customers began to spend their time and money on other forms of entertainment, because the repetitiveness of titles failed to provide the expected enjoyment. This is what we regard as a decline in the power to create demand. In particular, when an open-modular architecture is adopted, customers may move to other areas of consumption through the platform. As a result, the boundaries of the industry as a place for demand creation become blurred, and the sector eventually dissolves by merging with other industries.

Chapter "Investigating the Creation and Diffusion of Knowledge for Demand Creation: The Case of the Telecommunications Industry" looked at "modern" open architecture, i.e., a more complex case of open architecture. As theoretically shown in chapter "Evolution of Business Ecosystems," entry barriers preventing access to an industry are lowered by the appearance of a platform with open architecture. Many new entrants aggressively work by trial and error to achieve rapid product evolution. This, in turn, allows consumers to easily find desirable products, and demand booms over a short period of time. The mobile phone industry is such a case, and this chapter examined it to understand the factors behind rapid industry evolution and demand creation.

Our analysis of SEPs (standard-essential patents) revealed the importance of a particular firm, Qualcomm, which supported the evolution of its industry and products by disclosing or licensing its technologies. Qualcomm changed its business model at the end of the 1990s and specialized in technological consulting, licensing of technology, and the provision of system knowledge. It continued to learn from the product development activities of other companies while providing them with the technological knowledge needed for such product development. Through this mutual learning, Qualcomm accumulated unique system knowledge and secured a strong competitive advantage. At the same time, it promoted the diffusion of technological foundations by providing SEPs and related information to other companies. That is, Qualcomm simultaneously attained capability building, diffusion of technological foundations, and increased demand creation within its industry.

Chapter "The Impact of Platform Providers' Knowledge on Interfirm Division of Labor: The Case of the Mobile Phone Industry" focused on how firms boost capability building under the open architecture framework and how the capability building of each company affects collaboration among companies, i.e., the interfirm division of labor. To shed light on these issues, the results of questionnaire surveys of 50 smartphone product development projects were presented and analyzed.

In the smartphone industry, there are three types of firms. The *brand firms* launch products under their own brand, the *contract manufactures* undertake development and manufacturing of products on behalf of the brand firms, and the *platform providers* provide both technological information and core components. As confirmed by the questionnaire surveys, both the brand firms and the contract manufactures derive the necessary information from the platform provider and utilize it to build their organizational capabilities. Our evidence also showed that the performance of development projects increases by allocating decision-making rights to match organizational (development) capabilities. In addition, the organizational capabilities of brand firms and contract manufactures not only enhance the performance of development activities but also prevent opportunistic behavior by their partners. The implication of this chapter is that, even under an open architecture framework, the capability building of firms is still fundamental.

2 Theoretical Implications

2.1 *Revisiting Our Evolutionary Framework of Industries and Firms*

2.1.1 Augmented Capability-Architecture Framework

After summarizing the chapters of our book, let us discuss the overall framework that may explain our research findings in a consistent way. In Part I, we proposed a design-based evolutionary framework for analyzing the dynamics of industries and firms that consists of three major components: (1) *capability* of manufacturing sites, (2) *architecture* of products and platforms, and (3) *competitiveness* of industries and firms (Fig. 3 of chapter "A Design-Information-Flow View of Industries, Firms, and Sites"). This framework aims to answer our research questions regarding some seemingly important industrial phenomena occurred during the post-Cold-War period (mostly the 1990s, 2000s, and 2010s), including intense global cost competition between advanced and emerging nations, minute interindustrial trade, and new forms of competition and complementation in digital industries.

We also proposed additional theoretical models to explore the relations among manufacturing capabilities, physical productivities, and comparative production costs (Chapter "The Nature of International Competition Among Firms"), product design variety for additional demand creation (Chapter "Product Variety for Effec tive Demand Creation"), interactions between capability building and demand

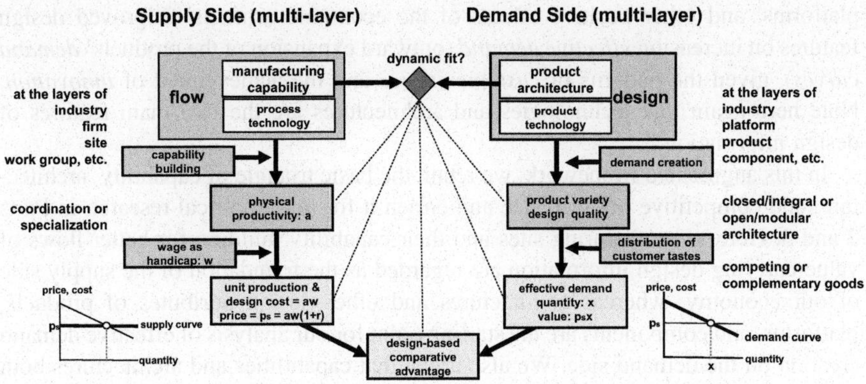

Fig. 1 Multilayered framework of capability building and demand creation

creation in firms that aim to achieve both profit and employment (Chapter "Capability Building and Demand Creation in 'Genba-Oriented Firms'"), and ecosystem formation and cumulative demand creation in open-modular architecture industries (Chapter "Evolution of Business Ecosystems").

The present book then tried to apply these theoretical tools to the actual evolution of industries and firms during the period following the end of the Cold War. Specifically, we chose two industrial phenomena to be explained in detail: (i) *capability building* for productivity improvements at manufacturing sites in advanced nations facing intense global cost competition vis-à-vis their lower-wage rivals in emerging nations and (ii) cumulative *demand creation* through the multiplying effect of network externalities in digital industries, characterized by products and platforms with open-modular architecture. Part II of this book dealt with the first topic (i.e., capability building) on the supply side, whereas Part III tackled the second topic (i.e., demand creation) on the demand side.

In order to apply our analytical framework to these two topics in relation to global competition and digitalization over the past few decades, the framework itself had to be enriched, so as to better explain the abovementioned industrial phenomena in a consistent way.

Figure 1 displays the augmented capability-architecture framework. The left side shows how *manufacturing capability building* that aims at better *flows of value-carrying design information*, even when process technologies are unchanged, can enhance physical productivities (i.e., input coefficients) and other productive performance features by which industrial sites in advanced nations, with the *handicap of higher wage rates*, try to prevail in the intense *global competition* against sites in lower-wage countries by lowering the *supply curves*, given the target profit (mark-up) ratios.

On the other hand, the right side (demand side) of Fig. 1 shows how the firms' *design-based efforts for demand creation*, such as *architectural strategies* and development of new *product technologies*, bring about improvements in *design quality*, expansion of *product variety*, formation of open-modular architecture platforms, and other positive effects of the economic artifacts' improved design features on increasing *effective demand* (outward expansion of the products' *demand curves*), given the patterns of *customer tastes* and the achievement of *innovation*. Note here again that technologies and architectures are the two main features of design information.

In this augmented framework, we retain the basic triangle of capability, architecture, and competitive performance and enrich it for our empirical research in Parts 2 and 3. Here, manufacturing sites and their capability building for better flows of value-carrying design information are regarded as the foundation of the supply side of our economy, whereas architectures and other design attributes of products, platforms, and components are the starting point for our analysis of effective demand creation on the demand side. We also argue that capabilities and architectures both evolve over time and that the comparative advantage of a certain country's products is affected by dynamic interactions between the capabilities of sites and the architectures of products (Chapter "A Design-Information-Flow View of Industries, Firms, and Sites").

Let us now explain the supply (left) side and demand (right) side of Fig. 1 in greater detail.

2.1.2 Evolution of Manufacturing Capabilities

The left side of Fig. 1, or the supply side, comprises such key concepts as manufacturing capability building, productive resources, flow of value-carrying design information, physical productivity, unit production, design cost, as well as other indicators of productive performance.

On the supply side, our basic notion of industries and firms is both site-based and product-based. The *manufacturing sites* (genba) can be analyzed as systems of value-carrying design information that flows between productive resources and eventually to the customers in the market. It is worth underlining that, since today's firms are essentially multiproduct, multi-industry, and multinational, a national industry is not a collection of firms but rather a collection of manufacturing sites (genba). In other words, sites can be seen as *flows* of value-carrying design information and *stocks* called productive resources. Using key concepts by E. Penrose (1959), we regard a firm or its manufacturing site as a collection of productive resources as stocks of potential values, which generate productive services as *flows of value-carrying design information to the customers*. In this context, it is the *manufacturing capability* that connects the underutilized resources (stocks) and realized productive services (flows) of firms and sites.

We also argue that the effectiveness of productive services can be measured by certain indicators of productive performance, including physical productivity, production lead times, and manufacturing quality (Fujimoto 1999). In the case of physical goods, or hardware products, structural design information is embedded in tangible (material-based) media, whereas in software products, it is embedded in intangible, electronic media. In the case of services, functional design information is embedded in intangible (energy-based) media (Chapter "A Design-Information-Flow View of Industries, Firms, and Sites"). Thus, our broad concept of manufacturing covers both hardware and software products and both manufacturing and service industries.

In our evolutionary framework, *capability* is a collection of organizational routines that govern the flow of value-carrying design information, whereas *competitive performance* is conceived as the ability of some objects or agents to be selected by other agents under the condition of free choice by the latter. Capabilities may be not only static but also dynamic or evolutionary. First, *static capability*, or ordinary manufacturing capability, is an organization's ability to maintain certain levels of productive performance, or speed, efficiency, accuracy, and flexibility of the flows of value-carrying design information within a given set of productive resources. Second, *dynamic capability* is an organization's ability to improve its performance by changing its flows, as well as its productive resources and organizational capabilities themselves. Thus, our definition of dynamic capability is broad, in that it includes organizational capability for continuous and repetitive improvements (kaizen, chapter "The Diversity and Reality of Kaizen in Toyota"), enhancement of human resources (Chapters "The Nature of Surviving Japanese Factories in the Global Competition: An Empirical Analysis of Electrical and Electronics Factories" and "The Effectiveness of Group Leaders in the Lean Production System: Time Study and Agent-Based Model of Leaders' Behavior"), redeployment of productive resources, and evolution of manufacturing capabilities and routines themselves (Teece and Pisano 1994; Fujimoto 1999, Chap. 6).

In this context, *evolutionary capability* refers to a firm's ability to improve performance in the long run, even when it does not know if it needs deliberate actions/decisions based on ex ante rational plans or emergent processes that may capture unintended results in the next phases of its competition and evolution.

2.1.3 Evolution of Architectures

The right side of Fig. 1, or the demand side, is related to demand creation, design attributes such as product architecture and product technology, product variety, design quality, and effective demand quantity. In other words, it aims to describe the evolution of the design aspects of products, platforms, and other economic artifacts.

In this context, again, a *product* is an artifact with economic value that can be exchanged among economic agents. As such, a product tends to be structurally coherent and functionally complete. A *component* is a structural part of the product, which may also be exchangeable. A *platform* means a collection of products or components that are functionally and structurally linked to one another by certain common design information (e.g., industry-standard interfaces, application programming interfaces (APIs), common component/module designs) shared among firms or across the whole industry. As such, a platform includes products that are complementary to one another, or complementary goods, among which cumulative network externality effects are often observed.

A platform is shared by a collection of firms and sites, which may constitute a *business ecosystem* (Chapter "Evolution of Business Ecosystems"). More than one platform with similar functions may coexist and compete within an industry, and the architecture-building capability of competing platform-leading firms (platform leaders) is critical for generating cumulative network effects among complementary goods.

The design information/knowledge of the abovementioned artifacts, such as products, processes, parts, and platforms, has two aspects: *technology*, which is related to concrete causal knowledge or information between the structures and functions of certain artifacts, and *architecture*, which refers to abstract graphical relations or correspondences between the structures and functions of various artifacts (Chapter "A Design-Information-Flow View of Industries, Firms, and Sites"). Both technologies and architectures that are related to an industry or firm tend to evolve over time, and, in order to analyze the evolution of industries and firms, we have to look at the dynamics of both technological and architectural aspects of the artifacts related to them.

The concept of architecture consists of two basic types: integral and modular architecture. An artifact's overall (macro) architecture is *integral* when its functional and structural elements are connected by means of many-to-many correspondences. It is *modular* when the functional-structural connections are linked through one-to-one correspondences. Actual artifacts are usually placed somewhere in between these two extremes of the architectural spectrum.

Another important distinction regarding product architecture is between closed and open (open-modular). When the basic design information is shared only within a single firm (e.g., firm-specific, common component designs shared across its products), this is a case of *closed architecture*. When the design information is open to the public (e.g., industry-standard interfaces linking components and products shared across firms), we call it open architecture. Thus, by combining the integral/modular and closed/open dimensions of architectures, we can identify three basic types: closed-integral, closed-modular, and open (open-modular) architecture.

An additional aspect is that of internal and external architecture. A component's *internal architecture* refers to the architecture of its design, while its *external architecture* is that of its client's product architecture. If the external architecture of a product or component is open-modular, the item is likely to be an industry-standard product/component. If it is externally closed-integral, the item is likely to be custom-designed and optimized for the upper system. To the extent that design activities involve coordination between an artifact's functions and structures, we may regard a product with integral architecture as relatively coordination-intensive, whereas a product with open-modular architecture is relatively coordination-saving (Chapter "The Impact of Platform Providers' Knowledge on Interfirm Division of Labor: The Case of the Mobile Phone Industry").

As for the *platform* (i.e., a collection of interconnected products), when it is open-modular, core products (e.g., smartphones) and complementary products (e.g., software products) enhance one another through the cumulative effect of network externalities. In other words, increases in demand for the core product will boost demand for complementary periphery products in the case of a platform with open-modular architecture. Thus, in the traditional *inter-product competition* with product differentiation (i.e., a downward-sloping demand curve for each product model), increases in demand for a product tend to cause decreases in demand for competing products, whereas *within a business ecosystem characterized by an open-modular architecture platform*, expansion of effective demand for a product is likely to result in greater demand for complementary products within the same platform (Chapters "Evolution of Business Ecosystems" and "Creating New Demand: The Impact of Competition, the Formation of Submarkets").

Firms will choose different competitive strategies as the above types of architectural positioning vary. For instance, a platform-leading firm is likely to choose an open architecture platform strategy to set up a rapidly growing platform/ecosystem and profit from it (Chapter "Investigating the Creation and Diffusion of Knowledge for Demand Creation: The Case of the Telecommunications Industry").

2.2 *Multilayered Evolution of Capabilities and Architectures*

2.2.1 The Hierarchical Nature of Complex Economic Artifacts

One of the main features of the augmented evolutionary framework explained above (Fig. 1) is that efforts in terms of both capability building and demand creation are essentially multilayered. This is a natural consequence of the fact that today's products and processes, seen as economic artifacts, are becoming increasingly complex, as well as of the logic that complex artifacts tend to become hierarchical (Simon 1969).

To the extent that a complex artifact is conceived as a hierarchical system (Simon 1969; Langlois and Robertson 1992), we can analyze its internal and external architectures (e.g., products, components). Moreover, an industry can be seen as interconnected hierarchies of artifacts on the stock side (Clark and Fujimoto 1991; Fujimoto 1999). We can lay out such hierarchies of artifacts moving along the design information flow from upstream to downstream: hierarchies of product functional design (perceived needs), product structural design, process structural design, actual production process, actual product structure, and actual process structure. Firms and sites interact with each other within this system of multiple hierarchies. Hence, in this book, we see an industry as a system of interconnected hierarchies of value-carrying artifacts, as well as flows and interactions among them.

Within the aforementioned industry as interconnected multiple hierarchies, firms, sites, and products interact with each other in three basic ways: transaction, competition, and complementation (Brandenburger and Nalebuff 1996). Specifically, we may define the following interactions in an industry as multiple hierarchies: *physical transaction*, mostly along the vertical axis of the hierarchy; d*esign transaction*, between the upstream and downstream hierarchies; *complementation*, mostly along the horizontal axis within a hierarchy; and *competition*, mostly among alternative hierarchies. When there are alternative platforms that the customers can use for similar purposes, *inter-platform competition* will involve developing additional complementary goods to enhance the attractiveness of the platform as a whole—a very different pattern of industrial competition compared with traditional inter-product competition (Chapters "A Design-Information-Flow View of Industries, Firms, and Sites" and "Evolution of Business Ecosystems").

Even in these cases, however, ordinary inter-product or inter-component competition does exist at the lower levels of the hierarchies. Product or component firms competing within the platform may need to deal with both traditional competition and the challenge of making other firms accept their industry-standard interfaces. In other words, they may need to carry out traditional capability-building competition internally and implement newer types of architectural strategies externally at the same time. In any case, depending upon the architecture of products, processes, components, and platforms, the behavior of the firms in an industry (or a network of connected industries) will have to be analyzed in terms of capability building at their sites and architectural choices for their products.

2.2.2 Multilayered Capability Building to Increase Productivity

As shown in the top-left area of Fig. 1 (supply side), *manufacturing capability* affects the competitive performance of manufacturing sites, or the efficiency and accuracy of the flows of value-carrying design information to the customers, including physical productivity (inverse of input coefficient), unit production cost, and unit design cost. When global cost competition is intense, with market prices decreasing rapidly, physical productivity increases through capability building are crucial for the survival of the domestic manufacturing sites in high-wage countries, as such efforts drive down the supply curve of the products in question (bottom-left part of Fig. 1).

Our definitions of production, as the transfer of value-carrying design information from the process to the product, and of physical productivity, as multiplication of design-information-transfer density and speed, provide some insight into the events happening at many manufacturing sites in Japan, one of the high-wage countries facing intense global cost competition.

Physical labor productivity (unit per person-hour) equals the *speed* (unit per value-adding time) times the *density* (value-adding time per person-hour) of the design information transfer from the workers to the product. Continuous improvements that increase the density of the design information transfer by n times can increase the physical productivity by n times, other things being equal. That is, dramatic increases in physical productivities (e.g., 5 times in 5 years), through capability building for reducing non-value-adding time (*muda* in Japanese) and increasing density of design information transfer, played a critical role in the survival of many manufacturing sites in Japan in the post-Cold-War global competition with large international wage gaps as handicaps (e.g., salaries of $100 versus $2000 per month), as suggested by the surveys and case studies in this book (Chapters "The Effectiveness of Group Leaders in the Lean Production System: Time Study and Agent-Based Model of Leaders' Behavior" and "The Diversity and Reality of Kaizen in Toyota"). In addition, productivity improvements occurred even when the introduction of new production/process technologies to boost the speed of design information transfer was restricted, which was partly due to limited financial resources during the recession.

Lastly, such capability building for survival was observed at multiple levels of the supply side, including the firm as a whole (e.g., Toyota, chapter "Evolution of Organizational Capabilities in Manufacturing: The Case of the Toyota Motor Cor poration"), factories as manufacturing sites (Chapter "The Nature of Surviving Japanese Factories in the Global Competition: An Empirical Analysis of Electrical and Electronics Factories"), and work organizations inside each factory (Chapters "The Effectiveness of Group Leaders in the Lean Production System: Time Study and Agent-Based Model of Leaders' Behavior" and "The Diversity and Reality of Kaizen in Toyota"). These research results suggest that *multilayered capability building* is necessary for national economies, industries, firms, factories, and other genba in high-wage advanced countries facing global cost competition as *handicapped productivity competition*, due to large wage differences vis-à-vis low-wage emerging nations (Chapter "The Nature of International Competition Among Firms").

2.2.3 Multilayered Demand Creation in Mature National Economies

On the demand side, design information that carries value added plays a pivotal role in *demand creation*. As shown in the right-hand part of Fig. 1, an increase in the quality or variety of product designs, given the distribution of the customers' tastes, will generate additional demand quantity through the outward expansion of the demand curve in question. That is, a new and functionally improved product design with new technological innovation, or a new combination of the product's functions and structures, will shift its demand curve outward. The introduction of an additional product with a new design may reduce the actual demand for competing (functionally similar) products, under the circumstances of product differentiation. However, the net effect of the expansion of product variety on the total demand of the whole industry in question (i.e., a collection of functionally similar products) is likely to be positive (Chapter "Product Variety for Effective Demand Creation").

The research results in this book also suggest the possibility of *multilayered demand creation* or multilevel efforts for effectively creating demand by nations, regional industries, firms, platforms, products, and components. This is particularly true in advanced nations, including Japan in the 1990s and 2000s, where total demand growth was rather slow.

During the Great Depression of the 1930s, J.M. Keynes emphasized the importance of demand creation by national governments through fiscal and other policies

(Keynes 1936). This book argues that, to the extent that regional industries, firms, and sites aim at survival and stable employment, these economic entities may all pursue demand creation through product/business diversification, product/process innovation, platform formation, and other marketing efforts.

J. Schumpeter identified discontinuous innovations by entrepreneurs as the driver of national economic development (Schumpeter 1934). The present book pays special attention to *grassroots innovations* for stable employment at the level of regions, industries, firms, and manufacturing sites (Lecler et al. 2012, Chap. 4) and points out that multilayered demand creation is an essential tool for mature national economies that need to secure employment by generating additional demand.

2.2.4 Multilayered Demand Creation in Open-Modular Architecture Platforms

The previous section discussed *multilayered demand creation* by local manufacturing sites, site-oriented firms, regional industries, and national governments within traditional product competition in mature national economies. Let us now turn to the newer industrial phenomenon of *inter-platform competition* with *platform-leading firms* (platform leaders) and its hierarchical nature. As explained repeatedly in other sections, when the platform in question is architecturally open-modular, with industry-standard interfaces linking complementary products, the addition of a new product will drive up the demand for complementary products through the effect of network externalities, which may result in rapid demand expansion across the whole industry as a collection of competing platforms (Chapter "Evolution of Business Ecosystems").

In this volume, we discussed the case of a prototypical open-modular platform that emerged in the 1980s, i.e., console game hardware and its software (Chapters "Creating New Demand: The Impact of Competition, the Formation of Submarkets" and "Decline in Demand Creation: The Development Productivity Dilemma and Its Consequences"). Although the sector was first developed in the USA, fully fledged platform-leading firms were established mainly in Japan (e.g., Nintendo, Sega, and Sony). The video game industry of those days was characterized by a relatively simple *two-layer platform* with complementary demand creation between the core products (i.e., console game hardware) and their complementary peripheral products, (i.e., game software; see Fig. 2). It ought to be underlined that the core products, the console game hardware, were architecturally closed-inside and were developed/produced/sold by the platform-leading firms, a situation similar to more recent cases of core products tending toward the closed-inside architecture (e.g., Apple's iPhone).

As suggested here, however, there exist more complex cases, such as *three-layer open platform* with open-inside core products and closed-inside core components (Fig. 3). Empirically, we can apply this model to such cases as personal computer systems (e.g., IBM PCs), conventional cellular phone systems (e.g., GSM), current smartphone systems (e.g., Android phones), and so on.

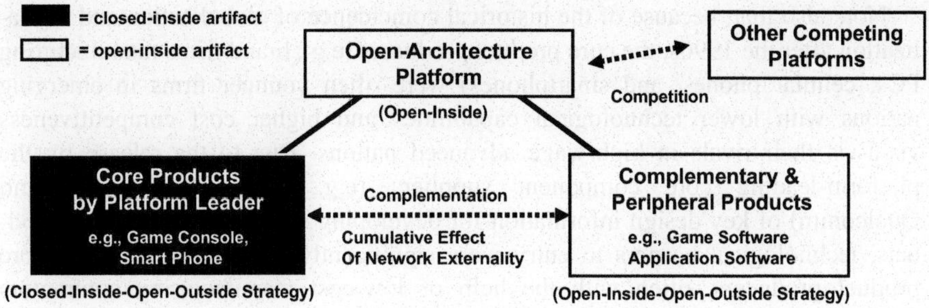

Fig. 2 Two-layer (simple) platform

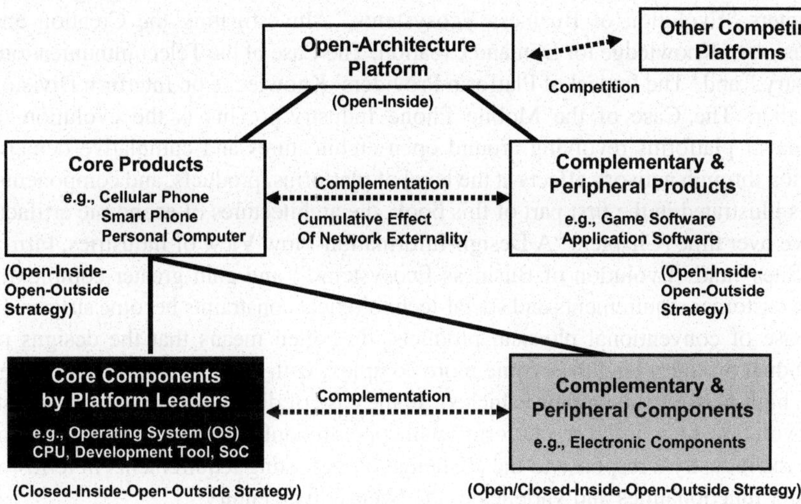

Fig. 3 Three-layer (complex) platform

It is important to emphasize once more that the architectural positioning of *core products*, such as personal computers, cellular phones, and Android phones, is fundamentally open-inside in the second layer. These products heavily rely on industry-standard hardware/software components, APIs, and development tools that are provided by the core component suppliers as platform-leading firms (e.g., Intel, Microsoft, Media-Tech, Google, and Qualcomm). On the other hand, the core component suppliers adopt a closed-inside-open-outside architectural strategy in the third layer (see the lower part of Fig. 3), by disclosing a portion of the design information about the core components to other firms while keeping the rest secret. In other words, there is a clear contrast between the *classic* platform strategy (Fig. 2) and the *modern* platform strategy (Fig. 3). Whereas the platform-leading firms are positioned in the second layer as core product suppliers in the simple, two-layer platform (Fig. 2), their position is in the third layer as core component suppliers in the case of a three-layer platform (Fig. 3).

Note also that, because of the historical coincidence of globalization and digitalization after the 1990s, the core product producers (e.g., brand firms manufacturing PCs, cellular phones, and smartphones) were often younger firms in emerging nations with lower technological capabilities and higher cost competitiveness vis-à-vis their rivals in high-wage advanced nations. Due to the release by the platform-leading core component suppliers (e.g., Intel, Media-Tech, and Qualcomm) of key design information for developing such open-inside core products, technological barriers to entry were significantly lowered, so that the core product producers, often with the help of low-cost manufacturing contractors based in emerging countries, enjoyed strong cost competitiveness. The prices of the core products dropped accordingly, which further accelerated cumulative demand creation by network externality in both advanced and emerging nations (Chapters "Evolution of Business Ecosystems," "Investigating the Creation and Diffusion of Knowledge for Demand Creation: The Case of the Telecommunications Industry," and "The Impact of Platform Providers' Knowledge on Interfirm Division of Labor: The Case of the Mobile Phone Industry"). This is the evolution of multilayer platforms revolving around open architectures and cumulative demand creation through network effects at the level of platforms, products, and components.

As illustrated in the first part of this book, the architectures of economic artifacts evolve over time (Chapters "A Design-Information-Flow View of Industries, Firms, and Sites" and "Evolution of Business Ecosystems") and gain greater complexity when customer requirements and social-technological constraints become stricter. In the case of conventional physical products, this often means that the designs of individual products tend to become more complex, with more integral architectures (e.g., high-performance motor vehicles). In the case of digital products, by contrast, the evolution of multilayer platforms with open-modular architectures has so far been the industry's response to the challenge of increasing requirements in terms of product functionalities and varieties in the twenty-first century.

2.2.5 Capability-Building Capability and Architecture-Building Capability

Within the context of the capability-architecture-performance framework summarized in Fig. 1, it is worth focusing also on the dual nature of the so-called dynamic capabilities: *capability-building capability* and *architecture-building capability* (Penrose 1959; Teece and Pisano 1994; Fujimoto 1999; Teece 2007). In our book, we predicted that the dynamic fit between the organizational capabilities of industries, firms, and sites (left-hand side of Fig. 1) and the architectures of platforms, products, and components (right-hand side of Fig. 1) will positively affect their competitive performance, including productive performance of the manufacturing sites, market performance of the products, and profit performance of the firms (Chapter "A Design-Information-Flow View of Industries, Firms, and Sites"). We also explained that competitive performance refers to the ability of sites, products, firms, and industries to be selected by other economic entities (e.g., firms, customers, investors). As such, competitive performance is itself multilayered.

Regarding this capability-architecture fit, we argue that, to the extent that design and production are coordinative activities of organizations, we should pay special attention to the *coordination richness* of the sites and firms, on the one hand, and *coordination intensity* of the products' design and production, on the other hand. In other words, it is important to consider the allocation of coordination in relation to artifacts and organizations. If one allocates the coordination embedded in the artifacts (products, components, and processes) by designing technology and/or architecture, the coordination requirements of the organization may become lighter. A typical case is that of digital products (e.g., Android phones), where the platform-leading core component suppliers, core product producers, and low-cost manufacturing contractors from emerging countries join forces and attain strong competitiveness. Conversely, if one allocates the coordination embedded in the organizations (well-organized routines, strong dynamic capabilities, and superior evolutionary capability) by management, both firms and sites can handle complex artifacts efficiently and flexibly. We can see such a case in highly complex and integral products (e.g., high-performance motor vehicles). The features mentioned above, i.e., well-organized routines, strong dynamic capabilities, and superior evolutionary capability, are called here *capability-building capabilities*.

Using this concept of coordination in organizations, we proposed the framework of *design-based comparative advantage*. A country endowed with coordination-rich manufacturing sites (e.g., Japan) will tend to enjoy Ricardian comparative advantages in terms of unit design cost in coordination-intensive products, that is, tradable goods with relatively integral architectures. On the other hand, a country endowed with human resources possessing high levels of specialized expertise (e.g., the USA) will probably have design-based comparative advantages in relatively coordination-saving or modular products (Chapter "The Nature of International Competition Among Firms").

Besides, when the architecture of the system of complementary goods is open-modular, this product system will tend to evolve into a platform, in which the platform-leading firms, with profit models based on the size of the platform, will enjoy exceptional profit performance, mostly thanks to the high-speed growth of the platform through the cumulative effect of network externalities (Chapter "Evolution of Business Ecosystems"). Thus, the architectures of products, components, and platforms will significantly affect the ways in which firms and sites compete, as well as their performance. In addition, the platform itself evolves and is shaped through an emergent process, and this is why no firm can design and produce all the artifacts in the ecosystem. So, firms must determine the position of their sites within the ecosystem as such process occurs. In other words, while platforms are being shaped, firms need to evaluate their sites' capabilities and performance and make strategic choices as to where they should be positioned within the platform (i.e., platform positioning). We regard the capabilities linked to this strategic choice as *architecture-building capabilities*.

For what concerns the targets of the changes, let us explore the effects of the two types of dynamic capabilities. *Capability-building capabilities* enhance performance in competition among sites and products, including kaizen capability and evolutionary capability (Teece and Pisano 1994; Fujimoto 1999). A*rchitecture-building capabilities* help establish industry standards, generate network externalities, and be successful in inter-platform competition (Teece 2007). Indeed, in many of today's forms of competition in platform-driven industries, both types of capabilities are needed for survival and growth. In conventional market competition among products, capability-building capabilities are required for long-term success, while in the newer platform competition, architecture-building capabilities become key. Motor vehicles are a good example to illustrate conventional competition among physical products. Given that their architectures were relatively stable, capability-building capabilities were of crucial importance to the firms' long-term success (Fujimoto 1999). In the new competition among digital platforms for products such as smartphones, instead, the architectures are less constrained by physical and technological characteristics and more easily shaped by the potential platform-leading firms (e.g., setting industry standards), which increases the relevance of architecture-building capabilities. In other words, focusing only on product competition and capability building may not guarantee sustainable firm growth in this situation.

However, this does not mean that today's firms should disregard capability building and shift their efforts entirely toward architecture building. As mentioned later, even as digital transformation changes the nature of industrial competition, what both new and existing firms need for long-term survival and growth is a combination of tenacious capability building at their genba and smart architectural strategy at their headquarters.

In this sense, our capability-architecture-based evolutionary framework seems to be reasonably effective in providing broader and more balanced views. By using this framework, we can consider both the physical and digital domains of industries, closed- and open-modular architectures, capability building and architecture building, as well as product competition and platform competition.

3 Conclusion

3.1 *Empirical Questions and Our Tentative Answers*

The findings of this book seem to be generally consistent with our evolutionary capability-architecture-performance framework, as shown in Fig. 1, which involves multilayered efforts on both the supply and demand side. In other words, our empirical and historical research on the late twentieth and early twenty-first century reveals the presence of both capability building for productivity increases on the supply side and demand creation by developing new products, components, or platforms and enhancing variety and/or design quality on the demand side. Indeed, the period in question is characterized by the *evolution of capabilities* of manufacturing industries and firms facing intense global competition and by the *evolution of architectures* of products and platforms facing rapid technological digitalization.

As already mentioned, our volume tackles three main questions in order to understand and analyze the significant changes happening between the 1990s and 2010s:

1. *Intense global competition between high-wage and low-wage nations*
2. *Minute-level intraindustrial trade mainly between advanced nations*
3. *New forms of competition and complementation in digitalized industries*

By applying our design-flow-based evolutionary framework of industries and firms, we reinterpret these issues as follows.

1. The *global cost competition* of this period is essentially *physical productivity competition with the handicap of international wage gaps*, as suggested by our new Ricardian model of comparative advantages and international values (Chapter "The Nature of International Competition Among Firms"). Organizational capability building for improved flows of value-carrying design information to the customers (i.e., manufacturing in a broad sense) is key for the survival of manufacturing firms (Chapters "A Design-Information-Flow View of Indus tries, Firms, and Sites" and "Evolution of Organizational Capabilities in Manufacturing: The Case of the Toyota Motor Corporation").

   Besides, our genba-based view of industries also suggests that local firms and sites facing global competition are, in many cases, motivated not only by profitability and survival, achieved through cost reductions, but also by stable employment, pursued through demand creation. After all, many of the Japanese manufacturing firms struggling to cope with global competition, particularly small and medium ones, are genba-oriented and community-oriented, aiming at target profit rates and stable employment at the same time (Chapter "Capability Building and Demand Creation in 'Genba-Oriented Firms'").
2. *Minute-level intraindustrial trade* occurs when there are internationally competing products that are functionally similar but mutually differentiated in their designs (Chapters "The Nature of International Competition Among Firms" and "Product Variety for Effective Demand Creation"), i.e., when product design-development sites for certain functionally similar products are located in multiple countries. In this situation, the question is not only "where to produce" but also "where to design" the products in question, so as to predict the international trade structure. Thus, the *design-based comparative advantage* framework is proposed here to analyze international trade phenomena of this type (Chapter "A Design-Information-Flow View of Industries, Firms, and Sites").
3. As for *global digitalization*, a thorough explanation of *cumulative demand creation in the digital industries* (e.g., PCs, video game consoles, cellular phones, smartphones, Internet services) includes the following concepts and logics (Chapter "Evolution of Business Ecosystems"): platforms comprising many complementary goods, industry-standard interfaces among them, open (open-modular)

architectural strategies by platform-leading firms, cumulative demand creation through network externalities among complementary goods, and extremely rapid growth of platform-leading firms and business ecosystems that consist of many competing/complementing/transacting firms (Brandenburger and Nalebuff 1996).

Two relevant aspects of most digital business ecosystems play a central role: the architecture of the platforms themselves is essentially open-modular with industry-standard interfaces, while the architecture of the subsystems inside a given platform, such as products, key components, and production processes, may still be closed-inside and/or integral-inside. In other words, the overall structure of a platform may be hierarchical and complex, with mixed architectures in its various layers. Therefore, when practitioners and researchers focus on an individual layer, the degree of openness of the platform observed by them might vary, the corresponding business ecosystem may be different, and it may seem maneuverable in different ways. Thus, the business ecosystem may still be complex, with an intertwined network of competition, complementation, and transaction among firms and products.

By tackling the three main questions above through both theoretical and empirical studies, we reach the tentative conclusion that, even when digital technology rapidly shifts the nature of the industrial game toward inter-platform competition, by setting industry standards and pursuing cumulative demand creation thanks to mutually complementary players, traditional inter-product competition and continuous capability building keep being vital for many of the manufacturing and service firms involved in the platform in question. What has happened in the age of globalization and digitalization is not the simple substitution of conventional product competition by completely new platform competition, but rather an evolution toward more complex industry and firm structures that involve both product and platform competition among competing, complementary, and transacting firms.

3.2 Toward a Framework for Exploring Our Complex World

The main observation put forward in this volume is that today's industries and firms are growing in complexity. Product designs as well as manufacturing processes and capabilities have become complex, but also the process of demand creation (i.e., how value-carrying design information flows to the customers in the market) and the nature of industrial competition clearly appear more intricate. In order to capture such complex and rapidly evolving phenomena, our analytical framework has to be dynamic, multilayered, and multifaceted. In addition to our analytical framework, other key concepts that have guided our analysis are as follows.

Emergent View Our evolutionary framework needs to capture both the intended and unintended behavior of firms and sites, since complex systems cannot be created through deliberate (i.e., ex ante rational) decision-making alone (Chapters "A Design-Information-Flow View of Industries, Firms, and Sites" and "Evolution of Organizational Capabilities in Manufacturing: The Case of the Toyota Motor

Corporation"). The process of generating complex systems involves both deliberate and emergent courses of actions by firms and their sites (Mintzberg and Waters 1985; Fujimoto 1999).

Multilayered Perspective To the extent that complex systems are designed and built hierarchically (Simon 1969), our framework needs to be multilayered on both the supply and demand side. On the supply side, our framework includes not only national/global economies, industries, and firms but also manufacturing site or genba, in which value-carrying design information flows to the customers in the market (Chapter "A Design-Information-Flow View of Industries, Firms, and Sites"). Our multilayered framework featuring genba has led to the concept of genba-oriented firms. On the demand side, our design-based view that product design information is the source of economic value added has led to our notion of effective demand creation at multiple levels (economies, industries, firms, and sites). Another important concept in the present book is that product design quality, variety, and architecture can effectively generate additional demand at multiple levels (Chapters "Product Variety for Effective Demand Creation," "Capability Building and Demand Creation in 'Genba-Oriented Firms'," and "Evolution of Business Ecosystems").

Conventional product competition entails rather simple causality, in that greater design quality and product variety expand effective demand, whereas in the current platform and/or product competition, we must assume rather complex causality between design quality and effective demand. In the latter case, positive interactions among complementary goods trigger a cumulative process of demand creation—the design quality of the core product (e.g., smartphones) attracts many complementary products (e.g., application software) and increases variety, which in turn enhances the attractiveness of the core product to customers. Considering the difference between conventional competition and the current form of competition, our multilayered framework of platform, product, and component designs seems to capture the evolution of both the physical and digital domains of today's industrial economy reasonably well.

As already noted, products and components that are part of an open-modular architecture platform may have some closed and/or integral features. Thus, we need a multilayered approach to the firms' architectural strategies regarding which parts of the platforms and/or products are open (or closed). In fact, many rather successful manufacturing firms actually adopt a mixed architecture strategy (i.e., closed-inside and open-outside) by adopting for different architectures at different layers.

Multifaceted Perspective Our evolutionary framework is multifaceted so as to cover a wide range of today's industrial phenomena in a consistent way, including both physical and digital industries, open-modular and closed-integral architectures, product competition and platform competition, technology and architecture as the concrete and abstract sides of design information, flows and forms of value-carrying design information to the customers, both manufacturing and nonmanufacturing industries. capability building and demand creation, profit and employment, and industries and communities. In other words, our design-based and site-based framework of capability-architecture-performance for analyzing the evolution of industries, firms, and sites seems to be a reasonably effective tool to understand today's complex industrial phenomena.

Redefinition of an Industry We have chosen to adopt a multifaceted design-flow-based view of an industry, rather than describing it as an abstract intersection of supply and demand curves, That is, we regard an industry as aggregate *flows of similar value-carrying design information* among productive resources, such as product concepts, product designs, process designs, actual processes, direct materials, work-in-process, and actual products. Since each productive resource can be described as a hierarchy of a system and its components, we may also see it as interconnected multiple hierarchies of design information.

On the supply side, an industry is a *collection of manufacturing sites* (*genba*) as a physical place, each of which has flows of value-carrying design information to the customers. A genba is also an organizational unit that belongs to a firm, an industry, and a local community at the same time. In other words, a genba is the linking pin among firms, industries, and communities. On the demand side, an industry is conventionally a *collection of products* with similar designs. The products in an industry (e.g., cars) are functionally similar (e.g., mobility), and they may be decomposed into functionally complementary components (e.g., engines, bodies, suspensions, etc.).

According to the above conventional definition, an industry is a collection of functionally similar products, while an open-modular architecture platform is a collection of functionary complementary products. Hence, a *platform* can be regarded as a concept akin to that of industry, in that it is *a collection of functionary complementary products* that customers can buy in the market (e.g., smartphones and application software). Thus, we may redefine an industry broadly as a collection of either functionally similar or functionally complementary products, including the concept of platform.

To sum up, our concept of industry is indeed multifaceted, but it is consistently based on the concepts of *design* as a source of economic value and of its *flow* to customers.

Interdisciplinary Approach Finally and theoretically, our analytical framework is rather multidisciplinary, in that it combines evolutionary economics, Ricardian classical theories of production and international trade, theories of product differentiation and monopolistic competition, theories of network externality, axiomatic design theories in engineering sciences, flow-oriented theories of operations management (e.g., Toyota System), and resource-capability view of strategic management. We have tried to integrate these theories into the capability-architecture-performance framework (Fig. 1) in order to explain in an internally consistent manner the variety of phenomena that happened during the period under investigation.

We have not adopted the core part of standard neoclassical economics, including the general equilibrium theory, not necessarily because we disagree with it theoretically but simply because the theories used here, such as the Ricardian trade theory, seem to be able to explain both the industrial phenomena and economic aspects of production observed more effectively. In short, the concept of equilibrium is difficult to apply to the reality of this period, characterized by rapid and continuous changes in both capabilities and architectures.

After all, the main purpose of this book is empirical, i.e., explaining the diversity and dynamics of the industrial phenomena happening in this period in a consistent way, and our theoretical framework is not a set of assumptions but rather the result of our empirical research. By reflecting on what we have observed in the countries most affected by globalization and digitalization, we have developed a capability-architecture framework that has turned out to be interdisciplinary, evolutionary, design-based, flow-based, and genba-based.

3.3 For Future Research

The present book explores the evolution of industries and firms during the late twentieth century and the early twenty-first century from the perspective of both capability and architecture. We have paid special attention to two major trends characterizing this period: global competition and digitalization.

Accordingly, after presenting our theoretical concepts and frameworks in Part I, Part II (Chapters "Evolution of Organizational Capabilities in Manufacturing: The Case of the Toyota Motor Corporation," "The Nature of Surviving Japanese Facto ries in the Global Competition: An Empirical Analysis of Electrical and Electronics Factories," "The Effectiveness of Group Leaders in the Lean Production System: Time Study and Agent-Based Model of Leaders' Behavior," "The Diversity and Reality of Kaizen in Toyota," and "Balancing Standardization and Integration in ICT Systems: An Architectural Perspective") focused on multilayered capability building efforts by firms, factories, and work groups facing global competition. Then, Part III (Chapters "Creating New Demand: The Impact of Competition, the Formation of Submarkets," "Decline in Demand Creation: The Development Productivity Dilemma and Its Consequences," "Investigating the Creation and Diffusion of Knowledge for Demand Creation: The Case of the Telecommunications Industry," and "The Impact of Platform Providers' Knowledge on Interfirm Division of Labor: The Case of the Mobile Phone Industry") mainly discussed cumulative demand creation in the digitalized industries with open-modular architecture. Part II dealt with product competition in physical goods industries, while Part III tackled platform competition. As for dynamic capabilities, the former emphasized capability-building capability (e.g., evolutionary capability) under the condition of architectural stability, while the latter highlighted architecture-building capability in the context of rapid architectural evolution. We aimed to analyze these two apparently different research themes as consistently as possible through the framework of capability and architecture or, more fundamentally, the design-flow view of industries. By contrast, in the existing literature, capability building in physical products and architectural transformation in digital industries are mostly discussed separately.

However, the emerging trend in the 2010s is the connection between the digital layer (e.g., Internet, ICT) and the physical layer (e.g., motor vehicles, factory equipment) by means of an interface layer. We may describe this through an analogy by calling *high sky* the digital/cyber/ICT layer, *ground* the physical layer, and *low sky* the interfacing (cyber-physical) layer. Connected cars, automatic driving, connected factories, and the Internet of Things (IoT) are popular words that capture this emerging trend of *sky-ground* connection.

When the *high sky*, in which revolutionary architectural changes and explosive demand creation take place, and the *ground*, which is severely constrained by physical laws and environment/energy/safety regulations, come together, the resulting total system and its behavior will be extremely complex. In order to understand this complexity of physical-digital industrial systems, we need a comprehensive research framework that can cover in a consistent manner both physical and digital layers, closed- and open-modular architectures, product competition and platform competition, continuous improvement (kaizen) and discontinuous innovation, organizational capability and architectural strategies, as well as capability building and demand creation.

The present book mainly aims to develop an evolutionary economic framework that can be applied not only to the industrial changes of the recent past (e.g., globalization and digitalization) but also to the aforementioned emerging trend of *sky-ground* integration. This may also be seen as an attempt to rebuild what Sir John Hicks called plutology, or *production economy* in modern terms (Hicks 1976). About 100 years after the publication of *Industry and Trade* by Alfred Marshall (Marshall 1919), it may be a good time to put forward a new version of the production economy framework for analyzing industrial performance and evolution in the early twenty-first century.

References

Brandenburger AM, Nalebuff BJ (1996) Co-opetition. Doubleday, New York

Clark KB, Fujimoto T (1991) Product development performance. Harvard Business School Press, Boston

Fujimoto T (1999) The evolution of a manufacturing system at Toyota. Oxford University Press, New York

Fujimoto T, Shiozawa Y (2012) Inter and intra company competition in the age of global competition: a micro and macro interpretation of Ricardian trade theory. Evol Inst Econ Rev 8(2):193–231

Hicks JR (1976) 'Revolutions' in economics. In: Latsis S (ed) Method and appraisal in economics. Cambridge University Press, Cambridge, pp 207–218. https://doi.org/10.1017/CBO9780511572203.009

Keynes JM (1936) The general theory of employment, interest and money. Macmillan, London

Langlois RN, Robertson PL (1992) Networks and innovation in a modular system: lessons from the microcomputer and stereo component industries. Res Policy 21:297–313
Lecler Y et al (eds) (2012) The dynamics of regional innovation: policy challenges in Europe and Japan. World Scientific, Singapore
Marshall A (1919) Industry and trade: a study of industrial technique and business organization. Macmillan, London
Mintzberg H, Waters JA (1985) Of strategies, deliberate and emergent. Strateg Manag J 6 (3):257–272
Penrose ET (1959) The theory of the growth of the firm. Basil Blackwell, Oxford
Ricardo D (1951[1817–21]) On the principles of political economy and taxation. In: Sraffa P (ed) The works and correspondence of David Ricardo, vol 1. Cambridge University Press, Cambridge. https://books.google.co.jp/books/about/On_the_Principles_of_Political_Economy_a.html?id=cUBKAAAAYAAJ&redir_esc=y
Schumpeter JA (1934) The theory of economic development: an inquiry into profits, capital, credit, interest, and the business cycle. Transaction Publishers, London
Simon HA (1969) The science of the artificial. MIT Press, Cambridge, MA
Teece DJ (2007) Explicating dynamic capabilities: the nature and microfoundations of (sustainable) enterprise performance. Strateg Manag J 28(13):1319–1350
Teece DJ, Pisano G (1994) The dynamic capabilities of firms: an introduction. Ind Corp Chang 3 (3):537–556

Correction to: Industrial Competitiveness and Design Evolution

Takahiro Fujimoto and Fumihiko Ikuine

Correction to:
T. Fujimoto, F. Ikuine (eds.), *Industrial Competitiveness and Design Evolution*, Evolutionary Economics and Social Complexity Science 12, https://doi.org/10.1007/978-4-431-55145-4

Chapter 01: A Design-Information-Flow View of Industries, Firms, and Sites
Chapter 15: Conclusion

The above mentioned chapters were originally published non-open access. They have now been made available open access © 'The Author(s)' under a CC BY-NC-ND 4.0 license. The book has also been updated with the change.

The updated version of these chapters can be found at
https://doi.org/10.1007/978-4-431-55145-4_1
https://doi.org/10.1007/978-4-431-55145-4_15

T. Fujimoto, F. Ikuine (eds.), *Industrial Competitiveness and Design Evolution*,
Evolutionary Economics and Social Complexity Science 12,
https://doi.org/10.1007/978-4-431-55145-4_16

The manufacturer's authorised representative in the EU is Springer Nature Customer Service Centre GmbH, Europaplatz 3, 69115 Heidelberg, Germany. If you have any concerns regarding our products, please contact ProductSafety@springernature.com

Printed and bound by CPI Group (UK) Ltd, Croydon, CR0 4YY
01/10/2026
02225871-0014